HISTORICAL FIGURES

AUDUBON THE NATURALIST

A HISTORY OF HIS LIFE AND TIME

VOLUME I

HISTORICAL FIGURES

Additional books and e-books in this series can be found on Nova's website under the Series tab.

HISTORICAL FIGURES

AUDUBON THE NATURALIST

A HISTORY OF HIS LIFE AND TIME

VOLUME I

FRANCIS HOBART HERRICK

snova
New York

Additional color graphics may be available in the e-book version of this book.

Library of Congress Cataloging-in-Publication Data

ISBN: 978-1-53615-587-7

Published by Nova Science Publishers, Inc. † New York

TO
ELIZABETH
MY
SISTER

CONTENTS

Figure 1. John J. Audubon. After the Rare Engraving by C. Turner, A.R.A., of the Miniature Painted by Frederick Cruikshank, About 1831; Published for the Engraver by Robert Havell, London, 1835.

Nothing extenuate,
Nor set down aught in malice....
Shakespeare, Othello to his biographers.

Time, whose tooth gnaws away everything else, is powerless against truth.
Huxley.

What a curious, interesting book, a biographer, well acquainted with my life, could write; it is still more wonderful and extraordinary than that of my father.
Audubon, in letter to his wife, March 12, 1828.

PREFACE*

The origin of the gifted ornithologist, animal painter, and writer, known to the world as John James Audubon, has remained a mystery up to the present time. In now lifting the veil which was cast over his early existence, I feel that I serve the cause of historical truth; at the same time it is possible to do fuller justice to all most intimately concerned with the story of his life and accomplishments.

The present work is in reality the outcome of what was first undertaken as a holiday recreation in the summer of 1903. While engaged upon a research of quite a different character, I reread, with greater care, Audubon's *Ornithological Biography*, and after turning the leaves of his extraordinary illustrations, it seemed to me most strange that but little should be known of the making of so original and masterful a character. As I was in England at the time some investigations were undertaken in London, but, as might have been expected, with rather barren results. After my return to America in the following year the search was continued, but as it proved equally fruitless here, the subject was set aside. Not until 1913, when this investigation was resumed in France, did I meet with success.

Every man, however poor or inconsequential he may appear or be, is supposed to possess an estate, and every man of affairs is almost certain to leave behind him domestic, professional, or commercial papers, which are, in some degree, a mark of his attainments and an indication of his character and tastes. In the summer of 1913 I went to France in search of the personal records of the naturalist's father, Lieutenant Jean Audubon, whose home had been at Nantes and in the little commune of Couëron, nine miles below that city, on the right bank of the Loire. The part which Lieutenant Audubon played in the French Revolution was fully revealed in his letters, his reports to the Central Committee, and numerous other documents which are preserved in the archives of the Préfecture at Nantes; while complete records of his naval career both in the merchant marine and governmental service (*service pour l'État*) were subsequently obtained at Paris; but at Nantes his name had all but vanished, and little could be learned of his immediate family, which had been nearly extinct in France for over thirty years.

Again the quest seemed likely to prove futile until a letter, which I received through the kindness of Mr. Louis Goldschmidt, then American Consul at Nantes, to M. Giraud Gangie,

---

* This is an augmented, edited and reformatted version of "Audubon the Naturalist: A History of His Life and Time. Volume I" by Francis Hobart Herrick, originally published by New York London, D. Appleton-Century Company, Inc., dated 1938. The views, opinions, and nomenclature expressed in this book are those of the authors and do not reflect the views of Nova Science Publishers, Inc.

conservateur of the public library in that city, brought a response, under date of December 29, 1913, informing me that two years before that time, he had met by chance in the streets of Couëron a retired notary who assured him that he held in possession numerous exact records of Jean Audubon and his family. The sage Henry Thoreau once remarked that you might search long and diligently for a rare bird, and then of a sudden surprise the whole family at dinner. So it happened in this case, and since these manuscript records, sought by many in vain on this side of the Atlantic, are so important for this history, the reader is entitled to an account of them.

Upon corresponding with the gentleman in question, M. L. Lavigne, I was informed that the documents in his possession were of the most varied description, comprising letters, wills, deeds, certificates of births, baptisms, adoptions, marriages and deaths, to the number, it is believed, of several hundred pieces. This unique and extraordinary collection of Audubonian records had been slumbering in a house in the commune of Couëron called "Les Tourterelles" ("The Turtle Doves") for nearly a hundred years, or since the death of the naturalist's stepmother in 1821.

Since I was unable to judge of the authenticity of the documents or to visit France at that time, my friend, Professor Gustav G. Laubscher, who happened to be in Paris, engaged in investigating Romance literary subjects, kindly consented to go to Couëron for the purpose of inspecting them. Monsieur Lavigne had already prepared for me, and still held, a number of photographs of the most important manuscripts, which are now for the first time reproduced, and, with the aid of a stenographer, in the course of two or three days they were able to transcribe the most essential and interesting parts of this voluminous material. But at that very moment sinister clouds were blackening the skies of Europe, and my friend was obliged to leave his task unfinished and hasten to Paris; when he arrived in that city, on the memorable Saturday of August 1, 1914, orders for the mobilization of troops had been posted; it was some time before copies of the manuscripts were received from Couëron, and he left the French capital to return to America.

These documents came into the hands of Monsieur Lavigne through his wife, who was a daughter and legatee of Gabriel Loyen du Puigaudeau, the second, son of Gabriel Loyen du Puigaudeau, the son-in-law of Lieutenant and Mme. Jean Audubon. Gabriel Loyen du Puigaudeau, the second, who died at Couëron in 1892, is thought to have destroyed all letters of the naturalist which had been in possession of the family and which were written previous to 1820, when his relations with the elder Du Puigaudeau were broken off; not a line in the handwriting of John James Audubon has been preserved at Couëron.

In June and July, 1914, Dr. Laubscher had repeatedly applied to the French Foreign Office, through the American Embassy at Paris, for permission to examine the *dossier* of Jean Audubon in the archives of the Department of the Marine, in order to verify certain dates in his naval career and to obtain the personal reports which he submitted upon his numerous battles at sea, but at that period of strain it was impossible to gain further access to the papers sought.

Having told the story of the way in which these unique and important records came into my possession, I wish to express my gratitude to Professor Laubscher for his able cooperation in securing transcriptions and photographs, and to Monsieur Lavigne for his kind permission to use them, as well as for his careful response to numerous questions which arose in the course of the investigation.

In dealing with letters and documents, of whatever kind, in manuscript, I have made it my invariable rule to reproduce the form and substance of the record as it exists as exactly as possible; in translations, however, no attempt has been made to preserve any minor idiosyncrasies of the writer. The source of all scientific, literary or historical material previously published is indicated in footnotes, and the reader will find copious references to hitherto unpublished documents, which in their complete and original form, with or without translations, together with an annotated Bibliography, have been gathered in Appendices at the end of Volume II. For convenience of reference each chapter has been treated as a unit so far as the footnotes are concerned, and the quoted author's name, with the title of his work in addition to the bibliographic number, has been given in nearly every instance.

Besides the many coadjutors whose friendly aid has been gladly acknowledged in the body of this work, I now wish to offer my sincere thanks, in particular, to the Misses Maria R. and Florence Audubon, granddaughters of the naturalist, who have shown me many courtesies, and to the Hon. Myron T. Herrick, late American Ambassador to France, for his kindly assistance in obtaining documentary transcripts from the Department of the Marine at Paris. I am under special obligations also to the librarians of the British Museum and Oxford University, the Linnæan and Zoölogical Societies of London, the Jardin des Plantes at Paris, the Public Libraries of Boston and New York, and the libraries of the Historical Societies of New York, Philadelphia, Boston and Louisiana, as well as to the Director of the Museum of Comparative Zoölogy of Harvard University, and to the American Museum of Natural History in New York City, for photographs of paintings and other objects, for permission to read or copy manuscripts, and for favors of various sorts. Furthermore, I am indebted to the good offices of Mr. Ferdinand Lathrop Mayer, Secretary of Legation, Port-au-Prince, and of M. Fontaine, American Consular Agent at Les Cayes, Haiti, for a series of photographs made expressly to represent Les Cayes as it appears today. I would also acknowledge the courtesy of the Corporation of Trinity Parish, New York, through Mr. Pendleton Dudley, for an excellent photograph of the Audubon Monument.

I cannot express too fully my appreciation of the hearty response which the publishers of these volumes have given to every question concerned with their presentation in an adequate and attractive form, and particularly to Mr. Francis G. Wickware, of D. Appleton and Company, to whose knowledge, skill, and unabated interest the reader, like myself, is indebted in manifold ways.

My friend, Mr. Ruthven Deane, well known for his investigations in Audubonia and American ornithological literature, has not only read the proofs of the text, but has generously placed at my disposal many valuable notes, references, pictures, letters and other documents, drawn from his own researches and valuable personal collections. I wish to express in the most particular manner also my appreciation of the generous spirit in which Mr. Joseph Y. Jeanes has opened the treasures in his possession, embracing not only large numbers of hitherto unpublished letters, but an unrivaled collection of early unpublished Audubonian drawings, for the enrichment and embellishment of these pages. For the loan or transcription of other original manuscript material, or for supplying much needed data of every description, I am further most indebted to Mr. Welton H. Rozier, of St. Louis; Mr. Tom J. Rozier, of Ste. Geneviève; Mr. C. A. Rozier, of St. Louis; the Secretary of the Linnæan Society of London, through my friend, Mr. George E. Bullen, of St. Albans; Mr. Henry R. Rowland of the Buffalo Society of Natural Sciences, of Buffalo; Mr. William Beer, of the Howard Memorial Library, of New Orleans; and Mr. W. H. Wetherill, of Philadelphia. For the use of new photographic and other illustrative material, I am further indebted to Mr. Stanley Clisby Arthur, of the Conservation Commission of Louisiana, and to *Cassinia*, the medium of publication of the Delaware Valley Ornithological Club.

Through the kindness of Messrs. Charles Scribner's Sons I have been permitted to draw rather freely from *Audubon and His Journals*, by Miss Maria R. Audubon and Elliott Coues, and to reproduce three portraits therefrom; original photographs of two of these have been kindly supplied by Dr. R. W. Shufeldt. I also owe to the courtesy of the Girard Trust Company, of Philadelphia, the privilege of quoting certain letters contained in William Healey Dall's *Spencer Fullerton Baird*.

To my esteemed colleague, Professor Benjamin P. Bourland, I am under particular obligations for his invaluable aid in revising translations from the French and in the transliteration of manuscripts, as well as for his kindly assistance in correspondence on related subjects. I have derived much benefit also from my sister, Miss Elizabeth A. Herrick, who has made many valuable suggestions. To all others who have aided me by will or deed in the course of this work I wish to express my cordial thanks.

Francis H. Herrick.
Western Reserve University, Cleveland.
July 2, 1917.

ILLUSTRATIONS IN VOLUME I

Chronology

1785

April 26.—Fougère, Jean Rabin, or Jean Jacques Fougère Audubon, born at Les Cayes, Santo Domingo, now Haiti.

1789

Fougère, at four years, and Muguet, his sister by adoption, at two, are taken by their father to the United States, and thence to France.

1794

March 7 (17 ventose, an 2).—Fougère, when nine years old, and Muguet at six, are legally adopted as the children of Jean Audubon and Anne Moynet, his wife.

1800

October 23 (1 brumaire, an 9).—Baptized, Jean Jacques Fougère, at Nantes, when in his sixteenth year.

1802-1803

Studies drawing for a brief period under Jacques Louis David, at Paris.

1803

First return to America, at eighteen, to learn English and enter trade: settles at "Mill Grove" farm, near Philadelphia, where he spends a year and begins his studies of American birds.

1804

December 15.—Half-interest in "Mill Grove" acquired by Francis Dacosta, who begins to exploit its lead mine; he also acts as guardian to young Audubon, who becomes engaged to Lucy Green Bakewell; quarrel with Dacosta follows.

1805

January 12-15 (?).—Walks to New York, where Benjamin Bakewell supplies him with passage money to France.

January 18 (about).—Sails on the *Hope* for Nantes, and arrives about March 18.

A year spent at "La Gerbetière," in Couëron, where he hunts birds with D'Orbigny and makes many drawings, and at Nantes, where plans are made for his return, with Ferdinand Rozier, to America.

1806

Enters the French navy at this time, or earlier, but soon withdraws.

March 23.—A business partnership is arranged with Ferdinand Rozier, and Articles of Association are signed at Nantes.

April 12.—Sails with Rozier on the *Polly*, Captain Sammis, and lands in New York on May 26.

They settle at "Mill Grove" farm, where they remain less than four months, meanwhile making unsuccessful attempts to operate the lead mine on the property.

September 15.—Remaining half interest in "Mill Grove" farm and mine acquired by Francis Dacosta & Company, conditionally, the Audubons and Roziers holding a mortgage.

1806-1807

Serves as clerk in Benjamin Bakewell's commission house in New York, but continues his studies and drawings of birds, and works for Dr. Mitchell's Museum.

1807

With Rozier decides to embark in trade in Kentucky.

August 1.—They purchase their first stock of goods in New York.

August 31.—Starts with Rozier for Louisville, where they open a pioneer store.

Their business suffers from the Embargo Act.

1808

June 12.—Married to Lucy Bakewell at "Fatland Ford," her father's farm near Philadelphia, and returns with his bride to Louisville.

1809

June 12.—Victor Gifford Audubon born at Gwathway's hotel, the "Indian Queen," in Louisville.

1810

March.—Alexander Wilson, pioneer ornithologist, visits Audubon at Louisville.

Moves down river with Rozier to Redbanks (Henderson), Kentucky.

December.—Moves with Rozier again, and is held up by ice at the mouth of the Ohio and at the Great Bend of the Mississippi, where they spend the winter.

1811

Reaches Sainte Geneviève, Upper Louisiana (Missouri), in early spring.

April 6.—Dissolves partnership with Rozier, and returns to Henderson afoot.

Joins in a commission business with his brother-in-law, Thomas W. Bakewell.

December.—Meets Vincent Nolte when returning to Louisville from the East, and descends the Ohio in his flatboat.

1812

The *annus mirabilis* in Kentucky, marked by a series of earthquakes, which begins December 16, 1811, and furnishes material for "Episodes."

Commission house of Audubon and Bakewell is opened by the latter in New Orleans, but is quickly suppressed by the war, which breaks out in June.

Spring.—Starts a retail store, on his own account, at Henderson.

November 30.—John Woodhouse Audubon, born at "Meadow Brook" farm, Dr. Adam Rankin's home near Henderson.

1812-1813

Storekeeping at Henderson, where he purchases four town lots and settles down.

1816

March 16.—Enters into another partnership with Bakewell; planning to build a steam grist- and sawmill at Henderson, they lease land on the river front.

1817

Thomas W. Pears joins the partnership, and the steam mill, which later became famous, is erected. (After long disuse or conversion to other purposes, "Audubon's Mill" was finally burned to the ground on March 18, 1913.)

1818

Summer.—Receives a visit from Constantine Samuel Rafinesque, who becomes the subject of certain practical jokes, at zoölogy's future expense, and figures in a later "Episode."

1819

After repeated change of partners, the mill enterprise fails, and Audubon goes to Louisville jail for debt; declares himself a bankrupt, and saves only his clothes, his drawings and gun. Resorts to doing crayon portraits at Shippingport and Louisville, where he is immediately successful.

1819-1820

At Cincinnati, to fill an appointment as taxidermist in the Western Museum, just founded by Dr. Daniel Drake; settles with his family and works three or four months, at a salary of $125 a month; then returns to portraits, and starts a drawing school.

1820

Decides to publish his "Ornithology," and all his activities are now directed to this end.

October 12.—Leaves his family, and with Joseph R. Mason, as pupil-assistant, starts without funds on a long expedition down the Ohio and Mississippi Rivers, to New Orleans, hoping to visit Arkansas, and intending to explore the country for birds, while living by his talents: from this time keeps a regular journal and works systematically.

1821

January 7.—Enters New Orleans with young Mason without enough money to pay for a night's lodging.

February 17.—Sends his wife 20 drawings, including the famous Turkey Hen, Great-footed Hawk, and White-headed Eagle.

Obtains a few drawing pupils; is recommended by John Vanderlyn and Governor Robertson, but lives from hand to mouth until June 16, when Audubon and Mason leave for Shippingport; a fellow passenger, Mrs. James Pirrie, of West Feliciana, offers Audubon a position as tutor to her daughter, and with Mason he settles on her plantation at St. Francisville, Bayou Sara, where he remains nearly five months; some of his finest drawings are made at this time.

October 21.—Leaves abruptly and returns with Mason to New Orleans, where he again becomes a drawing teacher, and resumes his studies of birds with even greater avidity.

December.—Is joined by his family, and winter finds them in dire straits.

1822

March 16.—To Natchez with Mason, paying their passage by doing portraits of the captain and his wife; while on the way finds that many of his drawings have been seriously damaged by gunpowder; teaches French, drawing and dancing at Natchez, and Washington, Mississippi.

July 23.—Parts with Mason, after giving him his gun, paper and chalks, with which to work his way north.

September.—Mrs. Audubon, who was acting as governess in a family at New Orleans, joins him at Natchez, where she obtains a similar position.

Receives his first lessons in the use of oils from John Stein, itinerant portrait painter, in Natchez, at close of this year.

1823

January.—Mrs. Audubon is engaged by the Percys, of West Feliciana parish, Louisiana, and starts a private school at "Beechwoods," belonging to their plantation, in St. Francisville, where she remains five years.

March.—Audubon leaves Natchez with John Stein and Victor on a painting tour of the South, but meeting with little success, they disband at New Orleans; visits his wife, and spends part of summer in teaching her pupils music and drawing.

Adrift again; both he and Victor are taken ill with fever at Natchez, but when nursed back to health by Mrs. Audubon, they return with her to "Beechwoods."

September 30.—Determined to visit Philadelphia in the interests of his "Ornithology," he sends on his drawings and goes to New Orleans for references.

October 3.—Starts with Victor for Louisville, walking part of the way.

1823-1824

Winter spent at Shippingport, where Victor becomes a clerk to his uncle, Nicholas A. Berthoud.

Paints portraits, panels on river boats, and even street signs, to earn a living.

1824

To Philadelphia, to find patrons or a publisher; thwarted; is advised to take his drawings to Europe, where the engraving could be done in superior style; befriended by Charles L. Bonaparte, Edward Harris, Richard Harlan, Mr. Fairman, and Thomas Sully, who gives him free tuition in oils.

August 1.—Starts for New York, with letters to Gilbert Stuart, Washington Allston, and Samuel L. Mitchell; is kindly received and made a member of the Lyceum of Natural History.

August 15.—To Albany, Rochester, Buffalo, Niagara Falls, Meadville, and Pittsburgh, taking deck passage on boats, tramping, and paying his way by crayon portraits.

September.—Leaves Pittsburgh on exploring tour of Lakes Ontario and Champlain for birds; decides on his future course.

October 24.—Returns to Pittsburgh, and descends the Ohio in a skiff; is stranded without a cent at Cincinnati; visits Victor at Shippingport, and reaches his wife in St. Francisville, Bayou Sara, November 24.

1825-1826

Teaches at St. Francisville, and gives dancing lessons at Woodville, Mississippi, to raise funds to go to Europe.

1826

May 17.—Sails with his drawings on the cotton schooner *Delos*, bound for Liverpool, where he lands, a total stranger, on July 21.

In less than a week is invited to exhibit his drawings at the Royal Institution, and is at once proclaimed as a great American genius.

Exhibits at Manchester, but with less success.

Plans to publish his drawings, to be called *The Birds of America*, in parts of five plates each, at 2 guineas a part, all to be engraved on copper, to the size of life, and colored after his originals. The number of parts was at first fixed at 80, and the period of publication at 14 years; eventually there were 87 parts, of 435 plates, representing over a thousand individual birds as well as thousands of American trees, shrubs, flowers, insects and other animals of the entire continent; the cost in England was £174, which was raised by the duties to $1,000 in America.

Paints animal pictures to pay his way, and opens a subscription book.

October 26.—Reaches Edinburgh, where his pictures attract the attention of the ablest scientific and literary characters of the day, and he is patronized by the aristocracy.

November, early.—William Home Lizars begins the engraving of his first plates at Edinburgh, and on the 28th, shows him the proof of the Turkey Cock.

Honors come to him rapidly, and he is soon elected to membership in the leading societies of science and the arts in Great Britain, France and the United States.

1827

February 3.—Exhibits the first number of his engraved plates at the Royal Institution of Edinburgh.

March 17.—Issues his "Prospectus," when two numbers of his *Birds* are ready.

April 5.—Starts for London with numerous letters to distinguished characters and obtains subscriptions on the way.

May 21.—Reaches London, and exhibits his plates before the Linnæan and Royal Societies, which later elect him to fellowship.

Lizars throws up the work after engraving ten plates, and it is transferred to London, where, in the hands of Robert Havell, Junior, it is new born and brought to successful completion eleven years later.

Summer.—Affairs at a crisis; resorts to painting and canvasses the larger cities.

December.—Five parts, or twenty-five plates, of *The Birds of America* completed.

1828

March.—Visits Cambridge and Oxford Universities; though well received, is disappointed at the number of subscribers secured, especially at Oxford.

September 1.—To Paris with William Swainson; remains eight weeks, and obtains 13 subscribers; his work is eulogized by Cuvier before the Academy of Natural Sciences, and he receives the personal subscription, as well as private commissions, from the Duke of Orleans, afterwards known as Louis Philippe.

1829

April 1.—Sails from Portsmouth on his first return to America from England, for New York, where he lands on May 1.

Summer.—Drawing birds at Great Egg Harbor, New Jersey.

September.—To Mauch Chunk, and paints for six weeks at a lumberman's cottage in the Great Pine Woods.

October.—Down the Ohio to Louisville, where he meets his two sons, one of whom he had not seen for five years; thence to St. Francisville, Bayou Sara, where he joins his wife, from whom he had been absent nearly three years.

1830

January 1.—Starts with his wife for Europe, first visiting New Orleans, Louisville, Cincinnati, Baltimore, and Washington, where he meets the President, Andrew Jackson, and is befriended by Edward Everett, who becomes one of his first American subscribers.

April 1.—Sails with Mrs. Audubon from New York for Liverpool. Settles in London; takes his seat in the Royal Society, to which he was elected on the 19th of March; resumes his painting, and in midsummer starts with his wife on a canvassing tour of the provincial towns; invites William Swainson to assist him in editing his letterpress, but a disagreement follows.

Changes his plans, and settles again in Edinburgh; meets William MacGillivray, who undertakes to assist him with his manuscript, and together they begin the first volume of the *Ornithological Biography* in October.

1831-1839

The *Ornithological Biography*, in five volumes, published at Edinburgh, and partly reissued in Philadelphia and Boston.

1831-1834

In America, exploring the North and South Atlantic coasts for birds.

1831

March.—First volume of the *Ornithological Biography* published, representing the text of the first 100 double-elephant folio plates.

April 15.—Returns with his wife to London.

May-July.—Visits Paris again in the interests of his publications.

August 2.—Starts with his wife on his second journey from England to America, and lands in New York on September 4.

Plans to visit Florida with two assistants, and obtains promise of aid from the Government.

October-November.—At Charleston, South Carolina, where he meets John Bachman and is taken into his home.

November 15.—Sails with his assistants in the government schooner *Agnes* for St. Augustine.

1832

April 15.—In revenue cutter *Marion* begins exploration of the east coast of Florida; proceeds to Key West, and later returns to Savannah and Charleston.

Rejoins his family at Philadelphia, and goes to Boston; there meets Dr. George Parkman, and makes many friends.

August.—Explores the coasts of Maine and New Brunswick, and ascends the St. John River for birds.

Returns to Boston, and sends his son Victor to England to take charge of his publications.

1832-1833

Winter.—In Boston, where he is attacked by a severe illness induced by overwork; quickly recovers and plans expedition to Labrador.

1833

June 6.—Sails from Eastport for the Labrador with five assistants, including his son, John Woodhouse Audubon, in the schooner *Ripley* chartered at his own expense.

August 31.—Returns to Eastport laden with spoils, including few new birds but many drawings.

September 7.—Reaches New York and plans an expedition to Florida.

September 25.—Visits Philadelphia and is arrested for debt, an echo of his business ventures in Kentucky; obtains subscribers at Baltimore, and in Washington meets Washington Irving, who assists him in obtaining government aid; finds patrons at Richmond and at Columbia, South Carolina.

October 24.—Reaches Charleston and changes his plans; with his wife and son passes the winter at the Bachman home, engaged in hunting, drawing and writing.

1834

The number of his American subscribers reaches 62.

April 16.—Sails with his wife and son on the packet *North America* from New York to England with large collections.

Settles again in Edinburgh, and begins second volume of his *Biography*, which is published in December.

1835

Many drawings, papers and books lost by fire in New York.

Part of summer, autumn and winter in Edinburgh, where the third volume of his *Ornithological Biography* is issued in December.

1836

Audubon's two sons, who have become his assistants, tour the Continent for five months, traveling and painting.

August 2.—Sails from Portsmouth on his third journey from England to the United States; lands in New York on Sept. 6 and canvasses the city.

September 13.—Hurries to Philadelphia to obtain access to the Nuttall-Townsend collection of birds, recently brought from the Rocky Mountains and Pacific Coast; is rebuffed, and bitter rivalries ensue; Edward Harris offers to buy the collection outright for his benefit.

September 20.—Starts on a canvassing tour to Boston, where he meets many prominent characters, and obtains a letter of commendation from Daniel Webster, who writes his name in his subscription book. Visits Salem, where subscribers are also obtained; meets Thomas M. Brewer, and Thomas Nuttall, who offers him his new birds brought from the West.

October 10.—Is visited by Washington Irving, who gives him letters to President Van Buren and recommends his work to national patronage.

October 15.—Returns to Philadelphia, where attempts to obtain permission to describe the new birds in the Nuttall-Townsend collection are renewed; he is finally permitted to purchase duplicates and describe the new forms under certain conditions.

November 10.—To Washington, to present his credentials, and is promised government aid for the projected journey to Florida and Texas.

1836-1837

Winter.—Spent with Bachman at Charleston, in waiting for his promised vessel; makes drawings of Nuttall's and Townsend's birds, and plans for a work on the *Quadrupeds of North America*.

1837

Spring.—Starts overland with Edward Harris and John W. Audubon for New Orleans; there meets the revenue cutter *Campbell*, and in her and her tender, the *Crusader*, the party proceeds as far as Galveston, Texas; visits President Sam Houston.

May 18.—Leaves for New Orleans, and on June 8 reaches Charleston. John Woodhouse Audubon is married to Bachman's eldest daughter, Maria Rebecca.

To Washington, and meets President Martin Van Buren.

July 16.—Sails with his son and daughter-in-law on the packet *England* from New York; reaches Liverpool on August 2d, and on the 7th is in London.

The panic of this year causes loss of many subscribers, but Audubon decides to extend *The Birds of America* to 87 parts, in order to admit every new American bird discovered up to that time.

1838

June 20.—Eighty-seventh part of *The Birds of America* published, thus completing the fourth volume and concluding the work, which was begun at Edinburgh in the autumn of 1826.

Summer.—By way of a holiday celebration tours the Highlands of Scotland with his family and William MacGillivray.

Autumn.—To Edinburgh, where, with the assistance of MacGillivray, the fourth volume of his *Biography* is issued in November.

1839

May.—Fifth and concluding volume of the *Ornithological Biography* is published at Edinburgh. *A Synopsis of the Birds of North America*, which immediately follows, brings his European life and labors to a close.

Late summer.—Returns with his family to New York, and settles at 86 White Street. Victor, who preceded his father to America, is married to Mary Eliza Bachman.

Projects at once a small or "miniature" edition of his *Ornithology*, and begins work on the *Quadrupeds*. Collaboration of Bachman in this project is later secured.

1840-1844

First octavo edition of *The Birds of America* is published at Philadelphia, in seven volumes, with lithographic, colored plates and meets with unprecedented success; issued to subscribers in 100 parts, of five plates each with text, at one dollar a part.

1840

June.—Begins a correspondence with young Spencer F. Baird, which leads to an intimate friendship of great mutual benefit, Baird discovering new birds and sending him many specimens.

1841

Purchases land on the Hudson, in Carmansville, at the present 157th Street, and begins to build a house.

July 29.—Writes to Spencer F. Baird that he was then as anxious about the publication of the *Quadrupeds* as he ever was about procuring birds.

1842

April.—Occupies his estate, now included in the realty section of upper New York City called Audubon Park, which he deeded to his wife and named for her "Minnie's Land."

September 12.—Starts on a canvassing tour of Canada, going as far north as Quebec, and returns well pleased with his success, after spending a month and traveling 1,500 miles.

Plans for his western journey nearly completed.

1843

March 11.—At fifty-eight, sets out with four companions for the region of the Upper Missouri and Yellowstone Rivers, but is unable to attain his long desired goal, the Rocky Mountains.

November.—Returns with many new birds and mammals.

1845-1846

The Viviparous Quadrupeds of North America, in collaboration with the Rev. John Bachman, issued to subscribers in 30 parts of five plates each, without letterpress, making two volumes, imperial folio, at $300.00.

John W. Audubon, traveling in Texas, to collect materials for his father's work.

1845

Engrossed with drawings of the *Quadrupeds*, in which he receives efficient aid from his sons.

July 19.—Copper plates of *The Birds of America* injured by fire in New York.

December 24.—Bachman, his collaborator, issues ultimatum through Harris, but work on the *Quadrupeds*, which had come to a stand, is resumed.

1846-1847

John W. Audubon in England, painting subjects for the illustration of the *Quadrupeds of North America.*

1846-1854

The Viviparous Quadrupeds of North America, in collaboration with John Bachman, published in three volumes, octavo, text only, by J. J. and V. G. Audubon; volume i (1847) only appeared during the naturalist's lifetime.

1847

Audubon's powers begin to weaken and rapidly fail.

1848

February 8.—John W. Audubon joins a California company organized by Colonel James Watson Webb, and starts for the gold fields, but his party meets disaster in the valley of the Rio Grande; he leads a remnant to their destination and returns in the following year.

1851

January 27.—Jean Jacques Fougère Audubon dies at "Minnie's Land," before completing his sixty-sixth year.

Chapter 1

INTRODUCTION

It is more than three-quarters of a century since Audubon's masterpiece, *The Birds of America*, was completed, and two generations have occupied the stage since the "American Woodsman" quietly passed away at his home on the Hudson River. These generations have seen greater changes in the development and application of natural science and in the spread of scientific knowledge among men than all those which preceded them. Theories of nature come and go but the truth abides, and Audubon's "book of Nature," represented by his four massive volumes of hand-engraved and hand-colored plates, still remains "the most magnificent monument which has yet been raised to ornithology," as Cuvier said of the parts which met his astonished gaze in 1828; while his graphic sketches of American life and scenery and his vivid portraits of birds, drawn with the pen, can be read with as much pleasure as when the last volume of his *Ornithological Biography* left the press in 1839. This appears the more remarkable when we reflect that Audubon's greatest working period, from 1820 to 1840, belonged essentially to the eighteenth century, for the real transition to the nineteenth century did not begin in England before 1837; then came the dawn of the newer day that was to witness those momentous changes in communication and travel, in education, democracy and ideas, which characterize life in the modern world.

When Audubon left London for Paris on September 1, 1828, it took him four days by coach, boat and diligence to reach the French capital, a journey which in normal times is now made in less than eight hours. Mail then left the Continent for England on but four days in the week, and to post a single letter cost twenty-four sous. Writing at Edinburgh a little earlier (December 21, 1826), Audubon recorded that on that day he had received from De Witt Clinton and Thomas Sully, in America, letters in answer to his own, in forty-two days, and added that it seemed absolutely impossible that the distance could be covered so rapidly. This was indeed remarkable, since the first vessel to cross the Atlantic wholly under its own steam, in 1838, required seventeen days to make the passage from New York to Queenstown.

"Walking in Paris," said Audubon in 1828, "is disagreeable in the extreme; the streets are paved, but with scarcely a sidewalk, and a large gutter filled with dirty black water runs through the middle of each, and people go about without any kind of order, in the center, or near the houses." The Paris of that day contained but one-fourth the number of its present population. Having reaped the fruits of the Revolution, it was enjoying peace under the Restoration; moreover, it was taking a leading part in the advancement of natural science, of which Cuvier was the acknowledged dean. It was but a year before the death of blind and aged Lamarck, neglected and forgotten then, but destined after the lapse of three-quarters of a century to have a monument raised to his memory by contributions from every part of Europe and America, and to be recognized as the first great evolutionist of the modern school.

Audubon had not seen his ancestral capital for upwards of thirty years, not since as a young man he was sent from his father's home near Nantes to study drawing in the studio of David, at the Louvre. Though in the land of his fathers and speaking his native tongue, his visit was tinged with disappointment. At the age of forty-three he was engaged in an enterprise which stands unique in the annals of science and literature. But fifty plates, or ten numbers, of his incomparable series had been engraved, and this work had then but thirty subscribers. That he was bound to sink or swim he knew full well. On August 30 he wrote: "My subscribers are yet far from enough to pay my expenses, and my purse suffers severely from want of greater patronage." This want he had hoped to satisfy in France, but after an experience of eight weeks, and an expenditure, as he records, of forty pounds, he was obliged to leave Paris with only thirteen additional names on his list. Yet among the latter, it should be noticed, were those of George Cuvier, the Duke of Orleans and King Charles X, while six copies had been ordered by the Minister of the Interior for distribution among the more important libraries of Paris. Moreover, he had won the friendship and encomiums of Cuvier, which later proved of the greatest value. The savants who gathered about him at the meeting of the Royal Academy of Sciences, over which Cuvier presided, exclaimed, "Beautiful! Very beautiful! What a work!," but "What a price!," and acknowledged that only in England could he find the necessary support. Audubon concluded that he was fortunate in having taken his drawings to London to be engraved, for the smaller cost of copper on that side of the Channel was an item which could not be overlooked. Little did he dream that commercial greed for the baser metal would send most of his great plates to the melting pot half a century later. No doubt he was right also in concluding that had he followed certain advisers in first taking his publication to France, it would have perished "like a flower in October." It should be added that King Charles' subscription expired with his fall two years later, while that of Cuvier ended with his death in 1832.

Audubon was one of those rare spirits whose posthumous fame has grown with the years. He did one thing in particular, that of making known to the world the birds of his adopted land, and did it so well that his name will be held in everlasting remembrance. His great folios are now the property of the rich or of those fortunate institutions which have either received them by gift or were enrolled among his original subscribers, and wherever found they are treasured as the greatest of show books. The sale of a perfect copy of the *Birds* at the present day is something of an event, for it commands from $3,000 to $5,000, or from three to five times its original cost. All of Audubon's publications have not only become rare but have increased greatly in price; they are what dealers call a good investment, an experience which probably no other large, illustrated, scientific or semi-scientific works have enjoyed to a like degree.

As has been said of Prince Henry the Navigator, though in different words, John James Audubon was one of those who by a simple hearted life of talent, devotion and enthusiasm have freed themselves from the law of death. Audubon was a man of many sides, and his fame is due to a rare combination of those talents and powers which were needed to accomplish the work that he finally set out to do. His personality was most winning, his individuality strong, and his long life, bent for the most part to attain definite ends, was checkered, adventurous and romantic beyond the common lot of men.

Few men outside of public life have been praised more lavishly than Audubon during his active career. Though he had but few open enemies, those few, as if conscious of the fact, seemed to assail him the more harshly and persistently. In reading all that has been said about this strenuous worker both before and since his death, one is continually struck by the perverse or contrary opinions that are often expressed. He was not this and he was not that, but he was simply Audubon, and there has been no one else who has at all closely resembled him or with whom he can be profitably compared. One charges that he did not write the books which bear his name. Another complains that he was no philosopher, and was not a man of science at heart; that he was vain, elegant, inclined to be selfish, inconsequential, and that he reverenced the great; that he shot birds for sport; that he was a plagiarist; that he was the king of nature fakirs and a charlatan; that he never propounded or answered a scientific question; and, finally, that though at times he wrote a graphic and charming style and showed occasional glimpses of prophetic insight, he cannot be trusted; besides, he might have been greatly indebted to unacknowledged aid received from others.

These or similar charges were brought against Audubon during his lifetime, as they have been made against many another who has emerged quickly from obscurity into world-wide renown. Many attacks upon his character were assiduously repelled by his friends, though seldom noticed publicly by himself; as if conscious of his own integrity, he was content to await the verdict of time, and time in America has not been recreant to his trust. Some of these charges it may be necessary to examine at length, if found to be justified in any degree, while others may be brushed aside as unworthy of even passing consideration. Evidence of every sort is now ample, as it seems, to enable us to do justice to all concerned, to penetrate the veil that has hidden much of the real Audubon from the world, and to place the worker and the man in the fuller light of day.

The reader who follows this history may expect to find certain blemishes in Audubon's character, for the most admirable of men have possessed faults, whether conscious of them or not. The lights in any picture would lose all value were the shadows wholly withdrawn. If we blinded ourselves to every fault and foible of such a man, we might produce a sketch more pleasing to certain readers, but it would lack the vitality which truth alone can supply. The more carefully his character is studied, however, as Macaulay said of Addison, the more it will appear, in the language of the old anatomists, "sound in the noble parts, free from all taint of perfidy, of cowardice, of cruelty, of ingratitude, of envy."

In this attempt to present a true and unbiased estimate of Audubon in relation to his time, we have the advantage of dealing with a well rounded and completed life, not with a broken or truncated one. He impressed many of his contemporaries in both Europe and America with the force of his contagious enthusiasm and prolific genius, and their opinions have been recorded with remarkable generosity. On the other hand, "if a life be delayed till interest and envy are at an end," said an excellent authority,[1] "we may hope for impartiality, but must expect little intelligence," because the minute details of daily life are commonly so volatile and evanescent as to "soon escape the memory, and are rarely transmitted by tradition." Such details, which often reveal character while they add color and life to the narrative, have been amply supplied, as the reader will find, by Audubon himself, not only in his journals and private letters already published but in the numerous documents of every sort that are now brought to light.

---

[1] Samuel Johnson, *The Rambler*, No. 60.

If "the true man is to be revealed, if we are to know him as he was, and especially if we are to know the influences that molded him and so profoundly affected him for good or evil, we must begin at the beginning and follow him through his struggles, his temptations, his triumphs." It might be better to start "in the cradle," or even forty years before he was born, for, as modern biology teaches us, nature is stronger than nurture and race counts for much. Certainly this man can never be understood if removed from the environment which time and circumstance gave him; he needs the historical background, furnished in part by his contemporaries, some of whom were rivals with whom he had often to struggle to make his way. In recounting this history, in many cases hitherto unwritten, we must recognize the proverbial difficulty of tracing human motives to their proper source, and endeavor to form no harsh judgments without ample basis in documentary or other evidence.

No more ardent and loyal American than John James Audubon ever lived. His adopted country, which he would fain have believed to have been that of his actual birth, was ever his chief passion and pride, and for him the only abode of sweet content. Few have seen more of it, of its diversified races, climates and coasts, its grand mountains, its noble lakes and rivers, its virgin forests and interminable prairies, with all the marvelous stores of animal and plant life which were first truly revealed to the pioneer woodsman, artist and naturalist. None has been more eager to hand down to posterity, ere it be too late, a true transcript of its wild and untameable nature while, as he would say, still fresh from the Creator's own hand. Audubon's beneficent influence during his long enforced residence abroad, as a representative of American energy and capacity, can hardly be measured, while in his own land few were more potent in bringing the nation to a consciousness of its unique individuality and power.

Audubon, as has been said, saw nature vividly colored by his own enthusiasm, and he never looked at her "through the spectacles of books." His writings, however unpolished or written with whatever degree of speed, have the peculiar quality of awakening enthusiasm in the reader, who, like the youth poring over *Robinson Crusoe*, feels within him a new ardor, in this instance, for hunting and studying birds and for leading a life of adventure in the wilderness. It would be as unjust to judge of Audubon's rare abilities as a descriptive writer from the letters, journal jottings and miscellaneous extracts given in this work, as to weigh his accomplishments as an artist from his itinerary portraits or his early sketches of animals in crayon point and pastel. Those cruder products of his pen and brush, however, as the reader will find, possess a high degree of interest from the light which they throw on the development of his character and art, as well as from their personal and historical associations. His best and only finished literary work, the *Ornithological Biography*, in five large volumes, with the revisions and additions which later appeared, abound in animated pictures of primitive nature and pioneer life in America as well as vivid portraits of the birds and other characteristic animals.

A good illustration of Audubon's habit of blending his own experiences with his biographies of birds is found in the introduction to his account of the Common Gannet:

> On the morning of the 14th of June 1833, the white sails of the Ripley were spread before a propitious breeze, and onward she might be seen gaily wending her way towards the shores of Labrador. We had well explored the Magdalene Islands, and were anxious to visit the Great Gannet Rock, where, according to our pilot, the birds from which it derives its name bred. For several days I had observed numerous files proceeding northward, and marked their mode of flight while thus travelling. As our bark dashed

> through the heaving billows, my anxiety to reach the desired spot increased. At length, about ten o'clock, we discerned at a distance a white speck, which our pilot assured us was the celebrated rock of our wishes. After a while I could distinctly see its top from the deck, and thought that it was still covered with snow several feet deep. As we approached it, I imagined that the atmosphere around was filled with flakes, but on my turning to the pilot, who smiled at my simplicity, I was assured that nothing was in sight but the Gannets and their island home. I rubbed my eyes, took up my glass, and saw that the strange dimness of the air before us was caused by the innumerable birds, whose white bodies and black-tipped pinions produced a blended tint of light-grey. When we had advanced to within half a mile, this magnificent veil of floating Gannets was easily seen, now shooting upwards, as if intent on reaching the sky, then descending as if to join the feathered masses below, and again diverging toward either side and sweeping over the surface of the ocean. The Ripley now partially furled her sails, and lay to, when all on board were eager to scale the abrupt side of the mountain isle, and satisfy their curiosity.[2]

Audubon's accounts of the birds are copious, interesting and generally accurate, considering the time and circumstances in which they were produced. When at his best, his pictures were marvels of fidelity and close observation, and in some of his studies of mammals, like that of the raccoon, in which seemingly every hair is carefully rendered, we are reminded of the work of the old Dutch masters and of Albrecht Dürer; notwithstanding such attention to microscopic detail, there is no flatness, but the values of light and shade are perfectly rendered. In his historical survey of American ornithology, Elliott Coues was fully justified in designating the years 1824-1853 as representing the "Audubonian Epoch," and the time from 1834 to its close as the "Audubonian Period." "The splendid genius of the man, surmounting every difficulty and discouragement of the author, had found and claimed its own.... Audubon and his work were one; he lived in his work, and in his work will live forever."[3]

There is no doubt that Audubon regarded an honest man as the quintessence of God's works, and though he sometimes set down statements which do not square with known facts, this was often the result of lax habits, or of saying what was uppermost in his mind without retrospection or analysis. When memory failed or when more piquancy and color were needed, he may have been too apt to resort to varnish, but for everything written on the spot his mind was as truth-telling as his pictures. In considering the good intent of the man, his extraordinary capacity for taking pains, and his vast accomplishments, criticism on this score seems rather captious. On the other hand, when it came to dealing with his own early life, that was a subject upon which he reserved the right to speak according to his judgment, and in a way which will be considered later.

Audubon left England to settle his family finally in America in the autumn of 1839, when he was fifty-four years old, and since he lived but twelve years longer, probably few are now living who retain more than a childish memory of his appearance in advanced age. Many Londoners will recall an odd character, an aged print dealer who used to sit alone, like a hoary spider in its web, in his little shop in Great Russell Street, close to the British Museum, and another of similar type, who may still haunt a better known landmark, the old "naturalist's shop" in Oxford Street, not far from Tottenham Court Road and but a minute's walk from the

[2] *Ornithological Biography* (Bibl. No. 2), vol. iv, p. 222.
[3] Elliott Coues, *Key to North American Birds*, 4th ed., p. xxi (Boston, 1890).

spot where most of Audubon's *Birds* were engraved. Both had seen the naturalist walk the streets of London and had known him in business relations. He occasionally strolled into the old naturalist's shop, which has been occupied by father and son for nearly a century. The son, then a young clerk, is now (1913) the crabbed veteran who still waits on customers but never waits long; should you hazard a question before making a purchase, he will roar like the captain of a ship and leave you to your own devices; but show him money and the change in his demeanor is wonderful; his hearing improves, his tone softens, and he may recount for you what he remembers of times long past, which is not much. Audubon in the thirties seemed to him like an aged man, an impression quite natural to a youth. He also remembered seeing Charles Waterton, Audubon's declared enemy and supercilious critic, William Swainson, his one-time friend, and William MacGillivray, his eminent assistant; that they were great rivals expressed the sum of his reflections. He recalled the time when Oxford Street was filled, as he expressed it, with horses and donkeys, and of course knew well the old Zoölogical Gallery, No. 79 Newman Street, in which for a time Robert Havell & Son conducted a shop in connection with their printing and engraving establishment. The latter, when moved by Robert Havell, Jr., to No. 77 Oxford Street, was nearly opposite the old Pantheon, which still lingers, and not far from the corner of Wrisley Street, the present site of Messrs. Waring & Gillow's large store.

We already possess several biographies of Audubon, and many of his letters of a personal or scientific interest and most of his extant journals, though but a fraction of those which originally existed, have been published. "America, my Country," has not forgotten him. Mount Audubon rises on the northerly bound of Colorado as an everlasting reminder of the last and grandest of all his journeys, that to the Missouri River in 1843. American counties and towns,[4] as well as parks and streets in American cities, bear his name. At least four of his beloved birds have been dedicated to him. In 1885, thirty-four years after his death, the New York Academy of Sciences began a popular movement through which a beautiful cross in marble was raised in 1893 above his grave in Trinity Cemetery.[5] The "one hundred and twenty-fifth anniversary"[6] of the naturalist's birth was celebrated in New York in 1905, and at

---

[4] Audubon, in Audubon County, Iowa, in Beeker County, Minnesota, and in Wise County, Texas, as well as Audubon, Montgomery County, Pennsylvania, in which his old farm, "Mill Grove," is situated. Audubon Avenue is the first of the subterranean passages which lead from the entrance of Mammoth Cave, and is noted for its swarms of bats. Audubon Park, New York City, between the Hudson River and Broadway and extending from 156th, to 160th Streets, embraces a part of "Minnie's Land," the naturalist's old Hudson River estate, but is a realty designation and is now almost entirely covered with buildings.

[5] The Audubon Monument Committee of the New York Academy of Sciences was appointed October 3, 1887, and made its final report in 1893, when this beautiful memorial was formally dedicated. Subscriptions from all parts of the United States amounted to $10,525.21. The monument is a Runic cross in white marble, ornamented with American birds and mammals which Audubon has depicted, and surmounts a die bearing a portrait of the naturalist, modeled from Cruikshank's miniature, with suitable inscriptions, the whole being supported on a base of granite; the total height is nearly 26 feet, and the weight 2 tons. It was presented to the Corporation of Trinity Parish by Professor Thomas Eggleston, and received by Rev. Dr. Morgan Dix. The cemetery has since been cut in two by the extension of Broadway; the monument is in the northerly section, close to the parish house of the Chapel of the Intercession.

The monument at New Orleans, mentioned below, was erected under the auspices of the Audubon Association, at a cost of $10,000, most of which was secured through the efforts of Mrs. J. L. Bradford, $1,500 having been contributed by residents of the Crescent City. The figure is in bronze, and stands on a high pedestal of Georgia granite.

The beautiful bust of Audubon at the American Museum of Natural History is by William Couper, of Newark, N. J.

[6] As will later appear, this was in reality the 120th anniversary.

the American Museum of Natural History an admirable marble bust of Audubon was unveiled on a notable occasion, December 29, 1906, when similar honors were paid to Louis Agassiz, Spencer Fullerton Baird, Edward Drinker Cope, James Dwight Dana, Benjamin Franklin, Joseph Henry, Joseph Leidy, John Torrey, and Alexander von Humboldt. On November 26, 1910, a statue of Audubon, after an admirable design by the veteran sculptor, Edward Virginius Valentine, of Richmond, Virginia, was unveiled in Audubon Park, New Orleans, where the naturalist, with pencil in hand, is represented in the act of transferring to paper the likeness of a favorite subject. He also occupies a niche in the Hall of Fame at New York University.

Figure 2. Statue of Audubon by Edward Virginius Valentine in Audubon Park, New Orleans. Published by courtesy of Mr. Stanley Clisby Arthur.

Figure 3. The Audubon Monument in Trinity Cemetery, New York, on Children's Day, June, 1915. Published by courtesy of the Corporation of Trinity Parish, New York City.

In recent times Audubon's name has become a household word through the medium of the most effective instrument which has yet been devised for the conservation of animal life in this or any country, the National Association of Audubon Societies for the Protection of Wild Birds and Animals. This has become the coordinating center for the spread and control of a great national movement that received its first impulse in 1886.[7] Launched anew ten

[7] The first Audubon Society, devoted to the interests of bird protection, was organized by Dr. George Bird Grinnell, editor of *Forest and Stream*, in 1886, and 16,000 members were enrolled during the first year; Dr. Grinnell was also the father of the *Audubonian Magazine*, which made its first appearance in January, 1887; by the

years later, it has advanced with ever increasing momentum, until now it is the governing head of twenty-nine distinct State societies, as well as eighty-five affiliated clubs and similar organizations. In 1916 it counted a life membership of 356, with 3,024 sustaining members, and realized a total income of over $100,000. It should be added that during the past six years over 2,900 Junior Audubon Clubs have been formed in the schools, through which nearly 600,000 children have been instructed in the principles of the Audubon Society. Well may it be that this admirable organization, with its successful efforts for remedial legislation in state and nation; its initiative, with the aid of the National Government, in establishing Federal reservations or sanctuaries for the perpetuation of wild life; its educational activities through the extension of its influence to the pupils of the public schools; and its watch and ward over all the varied interests of its cause, will keep the name of Audubon greener to all future time than the most cherished of his works.

Of Audubon's works the public now sees but little and knows even less, all without exception having been long out of print. His admirable plates of birds and mammals have been widely copied and still serve for the illustration of popular books, but most of his publications were projected on too large and expensive a scale for general circulation, having been first sold to subscribers only and often at great cost. No complete reprint, revision or abridgment of his principal volumes has been made for half a century. No complete bibliography of Audubon has ever been prepared, and none will remain completed long, for it is hard to imagine a time when comment on his life, his drawings, and his adventures will altogether cease.

In May, 1834, William MacGillivray, who was assisting him in the technical parts of the *Ornithological Biography*, suggested that Audubon write a biography of himself, and predicted a wide popularity for such a work. Audubon entertained the idea but was then too deeply immersed in *The Birds of America* to give it much attention; yet in 1835 he wrote out a short sketch, entitled *Myself*, addressing it in the fashion of that day to his two sons, and then laid it aside. Mrs. Audubon evidently had access to this manuscript when the life of her husband, to be referred to later, was in course of preparation, and thus it has furnished, directly or indirectly, nearly all that has been published concerning the naturalist's early life. This fragment, which extends to about thirty printed pages, was characterized by Audubon as a "very imperfect (but perfectly correct) account of my early life," and though written with an eye to its possible publication, which was clearly sanctioned, it was evidently never revised. The manuscript was long lost but eventually was "found in an old book which had been in a barn on Staten Island for years"; it was first published by the naturalist's granddaughter, Miss Maria R. Audubon, in 1893, and again in 1898. As will later appear, this account is inaccurate in many important particulars.

Audubon expressed the intention of extending his personal history, which he promised to delineate with a faithfulness equal to that bestowed on the birds, but the task was never resumed. Yet more than most writers have done, he wove the incidents of his own career into

---

middle of that year the membership in the new society had increased to 38,000, but with the disappearance of the *Magazine* in 1889 the movement languished and came to a speedy end. In 1896 a fresh start was taken by the inauguration of State societies in Massachusetts and Pennsylvania, and the movement gathered greater force through the inauguration in 1899 of the admirably conducted magazine, *Bird-Lore*, as its official organ. The State societies were federated in 1902, and the National Committee then created gave place in 1905 to the National Association. See Gilbert Trafton, *Bird Friends*, for an excellent summary of the work of the Audubon Societies, and the "Twelfth Annual Report of the National Association of Audubon Societies," *Bird-Lore*, vol. xviii (1916).

the pages of his principal works, and this strong personal flavor added much to their charm. Unfortunately, in giving such personal or historical details he is most vulnerable to a critic, who insists first upon accuracy, for errors of various sorts and confused and conflicting statements are far too common.

Of the more formal biographies of Audubon, the first to appear was a slender volume entitled *Audubon: the Naturalist of the New World*, by Mrs. Horace Stebbing Roscoe St. John, published in England in 1856.[8] In the same year this work was expanded and reissued by the publishers who at that time had charge of the sale of Audubon's works in America.[9] The American publishers explained in their edition that inasmuch as "the fair authoress in preparing her interesting sketch of Audubon ... appears not to have been aware of the publication of his second great work, the *Quadrupeds of North America* (which had not been advertised, we believe, in Europe) they have taken the liberty of giving some account of it and making numerous extracts from its pages."[10] Perhaps the most interesting or valuable things in this little volume at the present day are the woodcut on the title page showing Audubon's house on the Hudson as it then appeared, surrounded by tall trees, and, inserted on a flyleaf, a list of all of Audubon's published works and the prices at which they could be procured in New York just prior to the Civil War (see Note, Vol. II).

In 1868 there appeared in England a work of combined and confused authorship, commonly referred to as "Buchanan's *Life of Audubon*," the "sub-editor," as he called himself, having since become better known as an original, skilled and prolific writer of verse, drama, fiction and literary criticism. At that time Robert Buchanan was twenty-six years old, and had published five volumes of poems in rapid succession, some of which had been received with favor by the public. A second and third edition of this *Life* followed in 1869. Finally the work was resurrected and again sent to press, unrevised, in 1912, when it appeared in "Everyman's Library," at a shilling a copy, with an introduction which had served as a review of the work in 1869.

A recent biographer of Alexander Wilson speaks of Buchanan as "commissioned by Mrs. Audubon to write her husband's life," but the lady herself, as well as Buchanan, has told a different story. It seems that in about the year 1866, Mrs. Audubon prepared, "with the aid of a friend," an extended memoir of her husband, which was offered to an American publisher but without success. The "friend," at whose home Mrs. Audubon was then living, was the Rev. Charles Coffin Adams,[11] rector of St. Mary's Church, Manhattanville, now 135th, Street, New York. The Adams manuscript, which consisted chiefly of a transcript from the naturalist's journals, then in possession of his wife, was completed presumably in 1867. In the

[8] In this year Charles Lanman, writer, and at a later time librarian of the Library of Congress, wrote to Victor Audubon as follows: "Are not you and your family willing *now* to let me write a book about your illustrious father? I feel confident that I could get up something very interesting and which would not only help the big work, but make money. I could have it brought out in handsome style, and should like to have well engraved a portrait and some half dozen views in Kentucky, Louisiana, and on the Hudson. Write me what you think about it." Lanman's letter is dated "Georgetown, D. C., Oct. 8, 1856"; on November 1 Victor Audubon replied, declining the proposal.

[9] Messrs. C. S. Francis & Company, of 554 Broadway, New York.

[10] The publishers in this instance do not appear to have been better informed, for the text of the *Quadrupeds*, from which they quote, was written by John Bachman, and the first volume of it was issued in London in 1847.

[11] Rev. Dr. Adams was rector of this parish for twenty-five years, from 1863 to his death in February, 1888; he was the author of three volumes on religious subjects and various smaller tracts; from 1855 to 1863 he had charge of a church in Baltimore, Maryland, and while there published an anonymous pamphlet entitled "Slavery by a Marylander; Its Institution and Origin; Its Status Under the Law and Under the Gospel" (8 pp. 8vo. Baltimore, 1860).

summer of that year it was placed in the hands of the London publishers, Messrs. Sampson Low, Son, & Marston, who without any authority turned it over to one of their hard-pressed, pot-boiling retainers, Robert Buchanan, poet and young man of genius. Buchanan boiled down the original manuscript, as he said, to one-fifth of its original compass, cutting out what he regarded as prolix or unnecessary and connecting "the whole with some sort of a running narrative."[12] Mrs. Audubon was unable to recover her property from either publishers or editor or to obtain any satisfaction for its unwarranted use. Whatever defects the Adams memoir may have possessed, this is much to be regretted, since, as her granddaughter has said, Mrs. Audubon had at her command many valuable documents, the originals of which have since been destroyed.

Buchanan, like Audubon, had been reared in comparative luxury, "the spoiled darling of a loving mother." After the failure of his father in various newspaper enterprises about four years before this time, he had gone up to London with but few shillings in his pocket and had begun life there literally in a garret. The reflection that Audubon had fought a similar but much harder battle in that same London thirty years before, and won, should possibly have awakened in him a somewhat friendlier spirit than was then displayed. It must be admitted, however, that Buchanan produced a very readable story, although there was not a word in his whole book which showed any real sympathy with Audubon's lifelong pursuits, any knowledge of ornithology, or any interest in natural science. Though expressing unbounded admiration for the naturalist, his foibles and faults seem to have hidden from this biographer the true value of his distinguished services. In respect to a knowledge of natural history it should be added that Buchanan laid no claims, and of Audubon's accomplishments in this field comparatively little was said, the book, like the Adams' manuscript from which it was drawn, being mainly composed of extracts from the naturalist's private journals and "Episodes," as he called his descriptive papers. It was here that Audubon made the strongest appeal to this literary editor, who concluded his preface with the following words of praise: "Some of his reminiscences of adventure ... seem to me to be quite as good, in vividness of presentment and careful colouring, as anything I have ever read."

Buchanan dilated on Audubon's pride, vanity and self-conceit, faults which may have belonged to his youth but which were never mentioned by his intimate friends and contemporaries except under conditions which reflected rather unfavorably upon themselves. Complaints on this score were spread broadcast by reviewers of this work, seventeen years after the naturalist's death and with the suddenness of a new discovery. They were undoubtedly based on the unconscious and allowable egotisms of such personal records as Audubon habitually made for the members of his family when time and distance kept them asunder. Vanity and selfishness could have formed no essential parts of a character that merited the eulogy which follows:

> Audubon was a man of genius, with the courage of a lion and the simplicity of a child. One scarcely knows which to admire most—the mighty determination which

[12] Buchanan said that the manuscript submitted to him was inordinately long and needed careful revision; he added that "while he could not fail to express his admiration for the affectionate spirit and intelligent sympathy with which the friendly editor discharged his task, he was bound to say that his literary experience was limited." After copying a passage from one of Audubon's journals, this editor had the unfortunate habit of drawing his pen through the original; in this way hundreds of pages of Audubon's admirable "copper-plate" were irretrievably defaced.

> enabled him to carry out his great work in the face of difficulties so huge, or the gentle and guileless sweetness with which he throughout shared his thoughts and aspirations with his wife and children. He was more like a child at the mother's knee, than a husband at the hearth—so free was the prattle, so thorough the confidence. Mrs. Audubon appears to have been a wife in every respect worthy of such a man; willing to sacrifice her personal comfort at any moment for the furtherance of his great schemes; ever ready to kiss and counsel when such were most needed; never failing for a moment in her faith that Audubon was destined to be one of the great workers of the earth.[13]

No one will deny, however, that Buchanan was right in saying that in order to get a man like Audubon understood, all domestic partiality, the bane of much biography, must be put aside; but it is equally important to make such allowances as the manifold circumstances of time and place demand, and to be a reasoner rather than a fancier. This work abounds in errors, but it is not clear to what extent they were due to carelessness on Buchanan's part.

It was certainly a mistake to attribute Buchanan's attitude to partiality for Alexander Wilson, who, like himself, was a Scotchman. It was a case of temperament only, for gloom and poverty had embittered his life. As his sister-in-law and biographer[14] said of him, "he was doomed to much ignoble pot-boiling.... He had few friends and many enemies," and "had received from the world many cruel blows," while "no man needed kindness so much and received so little." Perhaps the best key to the sad history of this able writer was given by himself when he said: "It is my vice that I must love a thing wholly, or dislike it wholly." His wife, we are told, was much like himself, and "like a couple of babies they muddled through life, tasting of some of its joys, but oftener of its sorrows." Undoubtedly Robert Buchanan was a genuine lover of truth and beauty; he has written numerous sketches of birds and outdoor scenes, but with no suggestion of nature as serving any other purpose than that of supplying a poet with bright and pleasing images.

It was with the purpose of correcting the false impressions created by animadversions in Buchanan's *Life* that Mrs. Audubon, with the aid of her friend, James Grant Wilson, revised this work and published it in America under her name as editor, in 1869. The changes then made in Buchanan's text, however, were of a minor character and most of its errors remained uncorrected. The naturalist's granddaughter, Miss Maria R. Audubon, was inspired in part by similar feelings in preparing, with the aid of Dr. Elliott Coues, her larger and excellent work in two volumes, entitled *Audubon and His Journals*, which appeared in 1898. To her all admirers of Audubon owe a debt of gratitude for giving to the world for the first time a large part of his extant journals, as well as many new facts bearing upon his life and character. Other briefer biographies of Audubon which have appeared have been taken so completely from the preceding works, and have repeated and extended their errors to such an extent, as to call for little or no comment either here or in the pages which follow.

Through the discovery in France of new documentary evidence in surprising abundance we are obliged to draw conclusions contrary to those which have hitherto been accepted, and the new light thus obtained enables us to form a more accurate and just judgment of Audubon the man, and of his work.

---

[13] Robert Buchanan, *The Life of Audubon*, p. vi.

[14] See Harriet Jay, *Robert Buchanan: Some Account of His Life, His Life's Work, and His Literary Friendships* (London, 1903). Robert Williams Buchanan was born at Caverswell, Lancashire, August 18, 1841, and died in London, June 10, 1901.

Chapter 2

JEAN AUDUBON AND HIS FAMILY

Few names of purely Gallic origin are today better known in America, or touch a more sympathetic chord of human interest, than that of Audubon, and few, we might also add, are so rare. John James Audubon first made his family name known to all the world, and though he left numerous descendants, it has become well nigh extinct in America, and is far from common in France. The great Paris directory frequently contains no entry under this head; Nantes knows his name no longer, and it is rare in the marshes of La Vendée, where at some remote period it may have originated.

The lists of the army of five thousand which Rochambeau's fleet brought to our aid in the American War of Independence show but a single variant of this euphonious patronym, in Pierre Audibon,[15] a soldier in the regiment of Touraine, who was born at Montigny in 1756; but in the fleet of the Count de Grasse which coöperated with our land-forces at the Battle of Yorktown, on October 19, 1781, a ship was commanded by an officer with whom we are more intimately concerned. This was Captain Jean Audubon, who was later to become the father of America's pioneer woodsman, ornithologist and animal painter.

By birth a Vendean, at the age of thirty-seven Jean Audubon had plowed the seas of half the world, and in the course of his checkered career, as sailor, soldier, West Indian planter and merchant, had met enough adventure to furnish the materials for a whole series of dime novels. Short of stature, with auburn hair and a fiery temper, he was then as stubborn and fearless an opponent as one could meet on the high seas, and one of the gamest fighting cocks of the French merchant marine. How much Jean Audubon's son owed to his French Creole mother will never be known, but to this self-taught, thoroughly capable, and enterprising sailor we can surely trace his restless activity, his versatile mind and mercurial temper, as well as an inherent capacity for taking pains, which father and son possessed to a marked degree.

The true story of Jean Audubon's career has never been told, but even at this late day it will be found an interesting human document; and what is more to our purpose, it throws into sharp outline much that has hitherto remained obscure in the life of his remarkable son. The first Audubon to leave any imprint, however faint, upon the history of his time, this honest, matter-of-fact sailor, would have been the last to wish to appear in the garb of fiction, and we shall base our story solely upon the unimpeachable testimony of public and private records,

---

[15] For similar spelling of the name by John James Audubon.

which researches in France had happily brought to light before the beginning of the war in 1914.[16]

Jean Audubon came by his sailor's instincts and fighting prowess naturally, for his father, Pierre Audubon of Les Sables d'Olonne, was a seaman by trade. Like his son he captained his own vessel, and for years made long voyages between French ports in both the old and the new worlds. Pierre Audubon, the paternal grandfather of John James Audubon, and the first of that name of whom we have found any record,[17] lived at Les Sables d'Olonne, where with Marie Anne Martin, his wife, he reared a considerable family in the first half of the eighteenth century.

Les Sables, at the time of which we speak, was a small fishing and trading port on the Bay of Biscay, fifty miles to the southwest of Nantes, but is now become a city of over twenty thousand people. Lying on the westerly verge of the Marais, or salt marshes and lakes of La Vendée, the inhabitants of the district, and more particularly of the *Bocage*, or plantations, to the north and northeast, were noted from an early day for their conservatism, as shown in a firm adherence to ancient law and custom, as well as for their unswerving loyalty to the old nobility and to the clergy. Like their Breton neighbors on the other side of the Loire, the Vendeans were honest, industrious, and faithful to their civic obligations; they were also independent, resourceful, and knew no fear. When the neighboring city of Nantes planted trees of liberty and displayed the National colors in 1789, the Vendeans were stirred to indignation and later to arms, while the Chouans on the right bank of the river were quick to follow their example; in short, the rebels of La Vendée raised such a storm that for months the very existence of the infant Republic was threatened. This spirit of revolt to the newer order, the *Chouanerie*, as it came to be called, was stamped out for the time, but a few smoldering embers always remained, ready to burst into flame at the slightest provocation; recrudescent symptoms of this tendency had to be suppressed even as late as 1830, when Charles X, the last Bourbon king, lost his crown. Pierre Audubon's family, no doubt, shared many characteristics of their Vendean and Breton neighbors, but as the sequel will show, one at least did not approve of their political course, for he took up arms against them, and presumably against many of his own kith and kin.

Jean Audubon was born at Les Sables on October 11, 1744, and was christened on the same day, his godfather being Claude Jean Audubon, in all probability an uncle after whom he was named, and his godmother, Catharine Martin, presumably an aunt. Twenty-one children, according to the naturalist, blessed the union of Pierre Audubon and his wife, and were reared to maturity. Whether this statement is strictly accurate, or what became of so large a family cannot now be ascertained.[18]

---

[16] For notice of these records of Jean Audubon and his family, see the Preface, and for the most important documents, Appendix I.

[17] Pierre Audubon's service in the merchant marine of France is undoubtedly recorded in the archives of the Department of Marine in Paris, but all researches in that direction were suddenly halted by the war.

[18] Jean Audubon had a brother Claude, and on February 27, 1791, he wrote to him, asking for 4,000 francs, which he needed for the purchase of a boat. It was probably this brother who lived at Bayonne, and left three daughters, Anne, Dominica, and Catherine Françoise, who married Jean Louis Lissabé, a pilot (see Vol I.). If this inference be correct, and the sum referred to was demanded in payment of a debt, it may explain a statement of the naturalist that his father and his uncle were not on speaking terms.

Another brother is said to have been an active politician at Nantes, La Rochelle and Paris from 1771 to 1796, when he dropped out of sight for a number of years. When heard of again he was living at La Rochelle in affluence and piety. This was apparently the Audubon to whom the naturalist referred in certain of his journals

Pierre Audubon was engaged by the French Government to transport the necessities of war to Cape Breton Island in 1757, when the world-wide struggle between France and Great Britain for supremacy in the New World was at hand. The French were determined at all hazards to hold their great fortress of Louisburg, which had been taken by the English but again restored to the French not many years before. This was the strongest and most costly fortress on the American continent, as well as a great center for the valuable trade in salted fish. By a coincidence, or possibly out of compliment to his wife, Pierre's ship bore the name of *La Marianne*, and when he sailed from his home port of Les Sables d'Olonne on April 15, 1757, he took with him his own son, Jean, as cabin-boy, when the lad was but thirteen years old. In the following May Great Britain threw down the gauntlet to France, and the terrific seven years' struggle began. The great fortress of Louisburg fell in the following year to the English fleet, and was left a heap of ruins. His father's ship, the *Mary Ann*, was involved, and young Jean Audubon, who thus began his fighting career at fourteen, was wounded in the left leg and made a prisoner. With many of his compatriots he was taken to England, landing on November 14, 1758, where he remained in captivity for five years; he was released but a short time before the treaty of peace was signed at Paris, February 10, 1763. Apart from her interests in the West Indies, France was stripped at this time of all her vast possessions in America, save only the two little islands of Saint Pierre and Miquelon.

Whether Pierre Audubon shared the fate of his son we are unable to say, for at this point he drops out of our records and we do not hear of him again. It is certain that he never made another voyage with Jean, who returned to his native town with his passion for the sea unabated, and at nineteen reentered the merchant marine as a novice. His next voyage, on the ship *La Caille*, Captain Pigeon, was to execute a governmental commission at the Island of Miquelon. Five golden years of his youth had been spent in captivity; if productive of nothing else they had given him a knowledge of the English tongue, but they had also engendered bitter hatred of the English race, a feeling which his son confessed to have shared in his youthful days.[19]

The period from 1766 to 1768 was occupied in four voyages to Newfoundland, probably in the interest of the codfish trade, first as sailor before the mast in *Le Printemps*, and then as lieutenant in a ship called also *La Marianne*, with alternate sailings from, and to, La Rochelle and Les Sables d'Olonne. On his third voyage to Newfoundland, which was made in 1767, when he was twenty-three years old, Jean Audubon ranked as lieutenant of his vessel, but in the summer of 1768 he shipped again from Les Sables as sailor before the mast for a short

---

and private letters as one who, possessing the secret of his birth and early life, had done both him and his father an irreparable injury (see Vol. I).

A sister, Marie Rosa Audubon, was married in 1794 to Pierre de Vaugeon, a lawyer at Nantes; their only son, Louis Lejeune de Vaugeon, was living at Nantes as late as 1822, when he deeded his former home to Henri Boutard. (The substance of this and the preceding paragraph is based partly upon data furnished by Miss Maria R. Audubon.)

Jean Audubon gave his daughter, Rosa, the name of her aunt, but in later life seems to have broken off all relations with his brothers. Upon his death his will was immediately attacked by Mme. Lejeune de Vaugeon, of Nantes, and by the three nieces from Bayonne. The naturalist does not give the name of the aunt who, as he said, was killed during the Revolution at Nantes, but I have found no reference to any other.

[19] This was recalled by the naturalist on March 5, 1827, when he wrote: "As a lad I had a great aversion to anything English or Scotch, and I remember when travelling with my father to Rochefort in January, 1800, I mentioned this to him.... How well I remember his reply.... 'Thy blood will cool in time, and thou wilt be surprised to see how gradually prejudices are obliterated, and friendships acquired, towards those that we at one time held in contempt. Thou hast not been in England; I have, and it is a fine country.'" (See Maria R. Audubon, *Audubon and His Journals*, vol. i, p. 216).

trading cruise on the coast of France; in this instance the vessel, called *Le Propre*, was captained by Pierre Martin, who was possibly an uncle. At this juncture Jean Audubon enlisted in the French navy (service for the State) as a common sailor, and made two voyages on governmental business from the port of Rochefort, serving altogether nearly nine months (1768-9). After the termination of this last engagement nothing is heard of Jean for over a year, when in 1770 he makes his first appearance at Nantes, the city that was to know him in many capacities for nearly half a century. There he reentered the merchant marine, and on November 1, 1770, began a series of eight voyages, lasting as many years, to the island of Santo Domingo, the western section of which was then in possession of France.

Since much of the mystery which hitherto has shrouded the early life of John James Audubon is involved in the West Indian period of his father's career, we shall now trace this history in considerable detail.

The great export trade of French Santo Domingo in those days was in brown and white sugar, then known as the "Muscovado" and "clayed" sorts, which for the year 1789 amounted to over 141,000,000 pounds, valued at more than 122,000,000 francs; and in coffee, which in the same year totaled nearly 77,000,000 pounds, estimated to be worth nearly 52,000,000 francs.[20] While all such estimates were no doubt very crude, they serve to illustrate the richness of the prize that attracted Frenchmen by hundreds to the colony, an island that to many seemed a paradise in prospect, but which proved to be a purgatory in disguise.

Jean Audubon's voyages were all made in the interest of this valuable trade. Since they commonly lasted from six months to nearly a year, they became doubly hazardous to a French sailor after the outbreak of the American Revolution, for if he escaped his Scylla, the inveterate pirate, he might expect to encounter an equally formidable Charybdis in an English privateer. Though the northwestern corner of Santo Domingo was the center of their forays, Jean never lost a ship to the buccaneers, and though sometimes caught by the English, he never surrendered. He made three successive voyages from 1770 to 1772 in *La Dauphine*, commanded by Jean Pallueau, first as lieutenant and later as captain of the second grade, but on his last five voyages to the West Indies he captained his own ships, known as *Le Marquis de Lévy* (1774), *Les Bons Amis* (1775-6), and *Le Comte d'Artois* (1777-8).

---

[20] In 1789 over 7,000,000 pounds of cotton and 758,628 pounds of indigo were exported from the French side of the island, while further products of that year, including smaller amounts of cocoa, molasses, rum, hides, dye-woods, and tortoise shell, swelled the grand total of exports to 205,000,000 livres or francs. Bryan Edwards, however, whose deductions were based on official returns, placed the average value of all exports from French Santo Domingo for the years 1787, 1788, and 1789, at 171,544,000 livres in Hispaniola currency, or £4,765,129 sterling; this would be equivalent to about $23,158,426, and imply a purchase value of the French livre or franc of about 13½ cents in American money.

The number of plantations of every kind in the French colony was estimated by Edwards in 1790, at the outbreak of the Revolution, at 8,536; there were over 800 sugar plantations, over 3,000 coffee estates, to mention two such resources. If to these items we add nearly half a million slaves, the total valuation of the movable and fixed property of the French planters and merchants of this period would reach 1,557,870,000 francs. In 1788, 98 slave ships entered the six principal ports on the French side, and landed 29,506 negroes; Les Cayes received 19 of these ships, which delivered at that port 4,590 blacks. These slaves were sold for 61,936,190 livres, or at the rate of 2,008.37 livres each; according to Edwards this was equivalent to £60 sterling, or to about $291.60 in American money, at the rate of 14½ cents to the livre or franc. See particularly Francis Alexander Stanilaus, Baron de Wimpffen, *A Voyage to Santo Domingo in the Years 1788, 1789, and 1790*, translated by J. Wright (London, 1817); and also Bryan Edwards, *An Historical Survey of the French Colony in the Island of San Domingo* (London, 1797).

Captain Audubon was married on August 24, 1772, at Paimbœuf, to Anne Moynet,[21] a widow of some property, who had been born at Nantes in 1735 and was thus nine years his senior. Her married name was Ricordel. She possessed several houses at Paimbœuf, and acquired one in 1777, which was rented to the Administration at the time of the Revolution (see Vol. I), as well as a dwelling at Nantes, where she lived while her roving sailor of a husband was in Santo Domingo or the United States. Madame Audubon was a woman of simple tastes, devoted to culture, and, as we shall see, possessed of a kind heart.

When Captain Audubon left Les Cayes, Santo Domingo, on his last trading voyage, in the spring of 1779, bound for Nantes with a valuable cargo, his ship, *Le Comte d'Artois*, was attacked by four British corsairs and two galleys. With the odds overwhelmingly against him, he fought until his crew were nearly all killed or disabled, and after an abortive attempt to blow up his vessel, tried to escape in his shallop. For the second time he was made a prisoner by the English, who in this instance took him to New York, then in the possession of British troops. He was landed in that city on May 12, 1779, and was held there as a prisoner of war for thirteen months. If our inference be correct, he finally owed his release to the efforts of the French Ambassador, Monsieur de la Luzerne, the same, we believe, who had been a Governor of Santo Domingo, and who in 1790 became its Minister of Marine. As will be seen presently, this diplomat again exerted himself in Captain Audubon's behalf.

It is interesting to find that on this occasion Jean Audubon was fighting not only for his life, but for his property. His vessel, *Le Comte d'Artois*, was very heavily armed. Though of only 250 tons, she carried no less than ten cannon, four of which were mounted on gun carriages, and ten bronze pivot guns, which might imply that she was originally designed as a privateer. The ship was not destroyed when her captain was made prisoner, but was taken by the English to Portsmouth, New Hampshire (?), and burned there before December 15 of the following year.[22] Before starting on this disastrous voyage Captain Audubon had sold the vessel and his interest in her cargo to the Messrs. Lacroix, Formon de Boisclair and Jacques, with whom later he had extensive dealings in slaves; but he was not paid, and though an indemnity seems to have come from the British Government, he was never able to obtain a satisfactory settlement of the Formon claim.[23]

---

[21] As signed by herself, but variously spelled "Moinet," or "Moynette" in family documents of the period. On August 28, four days after their marriage, they drew up and signed a mutual contract regarding the disposition of their property in case children should be born to them.

[22] The destruction of *Le Comte d'Artois* is noticed in a document bearing date of January 19, 1782; the name of the town only is given, but it is probable that it refers to the United States.

[23] For repeated reference to this unsettled claim, see his letter of 1805 to Francis Dacosta, where the name is written "Formont."

The bill of sale of *Le Comte d'Artois* was drawn on February 21, 1779, when Jean Audubon appeared "before the notaries of the king in the seneschal's court of Saint Louis," and was described as "resident at Les Cayes, opposite the Isle à Vaches." The document, which in my copy is incomplete, reads in part as follows:

"The present M. Jean Audubon, captain-commander of the ship *Le Comte d'Artois*, of Nantes, armed for war and now laden with merchandise, anchored in this roadstead of Les Cayes, dispatched, and at the point of departure for France; armed by the Messrs. Coirond Brothers, merchants at the said city of Nantes, acting in his own name as one interested in the armament and cargo of the vessel, as well as in his capacity as captain; [he] acting as much also for the said furnishers of arms as for the others interested in the said armaments, and the merchandise, which will be hereafter mentioned, in consideration of the rights of each, promises to have these presents accepted and approved in due time; which said person, appearing in said names, in the quality aforesaid, by these presents has sold, ceded, given up, transferred, and relinquished all his legal rights in the aforesaid ship, to the business-associates Lacroix, Formon de Boisclair & Jacques, three merchants in partnership, living in this town, purchasers conjointly and severally, for themselves and the assigns of each, to the extent of one third; To wit: the said ship *Le Comte d'Artois*, of the said port of Nantes, of about two

Jean Audubon's release from captivity in New York, in June, 1780, probably marks the period of his first intimate acquaintance with the United States. We know only that he did not return immediately to either Santo Domingo or France, but became an enthusiast for the American cause, and sought the earliest opportunity to avenge his wrongs at the hands of the British. He did not have long to wait, for through the exertions of the Ambassador de la Luzerne, he was placed in command of the corvette *Queen Charlotte*. With her, in October, 1781, he joined the fleet of the Count de Grasse before Yorktown,[24] where he soon witnessed the surrender of Cornwallis and the humiliation of his enemies. After this turning point of the war Captain Audubon remained in the United States, and in April, 1782, commanded a merchantman called *L'Annette*,[25] in which he was also personally interested, and delivered a cargo of Virginia tobacco at the port of Nantes. Shortly after his return to America in the same year he was placed in command of an American armed vessel *The Queen* and sent on another mission to France. Near the Chaussée des Saints he was attacked by a British privateer, but after a stubborn fight at close quarters he sank his enemy and entered the port of Brest. Nothing is said of the taking of prisoners on such occasions, and there were doubtless few survivors among the defeated crew. This command Jean Audubon held until peace was concluded between Great Britain and her former colonies in America, probably until the close of 1783. The hostile army was disbanded in the spring of that year, the treaty of peace was made definitive in September, and on November 25, 1783, the last British troopers left the city of New York.

---

hundred and fifty tons, at present anchored in this roadstead of Les Cayes, dispatched, and at the point of departure for France, with all its rigging, outfit, and dependences, which consist among other things of two sets of sail, complete, and newly fitted out, all the tools, and the reserve sets of these, with the munitions of war, consisting of ten cannon, four of them mounted on gun carriages, and all that goes with them...." (Translated from the French original in possession of Monsieur Lavigne.)

24 The fact that Captain Audubon did not accompany Rochambeau's fleet which assembled at Brest in April, 1780, and reached Newport in mid-July, may account for the omission of his name from the lists that have been recently published. See *Les Combattants François de la Guerre Américaine*, 1778-1783 (Paris, 1903).

25 Others interested in this vessel were Messrs. David Ross & Company, with whom Captain Audubon later had financial difficulties.

Chapter 3

JEAN AUDUBON AS SANTO DOMINGO PLANTER AND MERCHANT

After the American struggle for liberty had been finally won, Captain Audubon resigned his commission held in the United States and returned to his home at Nantes, but town or country could not hold him long. Lured by the prospects of great wealth which Santo Domingo offered to the merchant of those days, and having learned by long experience in her ports the devious methods by which fortunes were attained, he decided to give up the sea and embark in colonial trade. For six years, from 1783 to 1789, he lived almost continuously in the West Indies, and as merchant, planter, and dealer in slaves amassed a large fortune. Meanwhile his wife, who had seen little of him since their marriage in 1772, remained at Nantes.

Captain Audubon traveled through the United States early in 1789, and again late in that year when on his way to France, probably in the first instance returning to Santo Domingo by way of the Ohio and the Mississippi. Symptoms of unrest were already prevalent in the northern provinces of the island but had caused no serious alarm in the south. Jean Audubon's aim seems to have been to collect debts due him in the United States and to leave the capital invested there. At all events it was on this occasion that he purchased the farm of "Mill Grove," near Philadelphia, the history of which will be given a little later. He had no intention, however, of living in Pennsylvania, for he immediately leased this estate to its former owner and hurried away.

July 14, 1789, found the elder Audubon enlisted as a soldier in the National Guards at Les Cayes. These colonial troops, which were originally militia organizations modeled after similar bodies in France, were reorganized at this time to meet any possible emergencies. Affairs in the southern provinces of Santo Domingo had followed, up to this moment, their normal course, and Jean Audubon, who could have learned nothing of what had transpired at home, decided to entrust his various interests to the hands of agents and return to France. This was probably in late August or early September, 1789, as we know that he first returned to the United States and visited Richmond, Virginia, at the close of that year.[26] Strangely enough, on the twentieth day of the former month the National Assembly at Paris had voted the celebrated Declaration of Rights, which was destined to upturn the whole social system of Santo Domingo and to convert that island into a purgatory of the direst anarchy, strife, and

[26] See letter to Dacosta, Vol. I.

bloodshed which the world had ever known, or at least remembered; but fully six weeks must have elapsed before news of this grave decision could have reached the colony.

At this time Jean Audubon was no doubt regarded as a very rich man, and though he happened to leave Les Cayes at a critical moment, little could he have dreamed of the disaster that awaited him there as well as in his beloved France. His personal affairs during this eventful period, involving as they necessarily do the early life of his distinguished son, have hitherto been shrouded in the dark and sinister history of that ever smiling but ever turbulent island. Now, however, the veil of mist that has settled over the page can be penetrated at the most important points. In this and subsequent chapters we shall follow the life of father and son through the course of events which has been thus briefly summarized.

To return to the earlier threads of our narrative, at about the close of 1783, Captain Audubon was engaged by the Coirond brothers, colonial merchants at Nantes, to take charge of their foreign trade, which centered chiefly at Les Cayes,[27] Santo Domingo, then a most thriving and populous town, as it is today the largest seaport on the southern coast of the Republic of Haiti. Their ships brought sugar, coffee, cotton and other West Indian products to France, and laden with fabrics, wines and every luxury known to the colonists of that day, returned to Les Cayes, as well as to Saint Louis, an important port a little farther to the east, where these merchants also possessed warehouses and stores.

In a short time Jean Audubon had acquired an independent business of his own, both as a planter and merchant. He made his home at Les Cayes, but extended his enterprises to Saint Louis and possibly to other points. From this time onward he commonly described himself as *négotiant*,[28] or merchant, and his son, when writing to his father from America, addressed him in this way. His business letters and other documents of the period refer to his house at Les Cayes, his plantations of cane and his sugar refinery, his exportation of colonial wares, his purchases of French goods, particularly at Nantes, and to his trade in black slaves which eventually assumed large proportions. How important his sugar plantations may have been is not known, but a tax-receipt shows that at one time he possessed forty-two slaves.[29] The naturalist said that his father acquired a plantation on the Ile à Vaches, an island of considerable importance at the southern bound of the roadstead of Les Cayes and nine miles from the town, but we have found no other reference to it.

Great numbers of negroes must have passed through Jean Audubon's hands, as shown by his bills of sale, which strangely reflect the customs of a much later and sadder day on the North American continent. In one of these bills, dated at Les Cayes, September 16, 1785, Jean is credited with one-half the net proceeds of the sale of forty negroes, bought originally of M. Th. Johnston for the sum of 60,000 francs, and sold by Jean Audubon and Messrs. La Croix,

---

[27] The proper name of this seaport town, as given by all French cartographers and writers, is *Les Cayes*, meaning "the cays" or "keys" (small islands, Spanish *cayos*); omitting the article it is often simply written "Cayes." French residents on the island, however, when dating or addressing a letter or receipting a bill would naturally write "*aux Cayes*," meaning of course "at The Cays," where the document was signed or where the person to whom the letter was addressed resided (see the Sanson bill, and bills of sale of negroes. It was thus an easy step for Englishmen, in ignorance or disregard of the French usage, to call the town "Aux Cayes"; even as early as 1797, Bryan Edwards, though giving the name correctly on his map, which doubtless had a French source, wrote "Aux Cayes" in his text; the corruption has survived, and is occasionally found in standard works, but is too egregious to be tolerated.

[28] And sometime as *marchand*, more strictly a retailer.

[29] Since a colonist's wealth was estimated upon the number of slaves he could afford, and since a slave was regarded as equivalent to a return of 1,500 francs a year, Jean Audubon's income on this basis would have been 63,000 francs.

Formon & Jacques for 71,552 francs; after deducting 183 francs for food and treatment, the net returns became 71,369 francs, and Jean's profits, on a half-interest basis, 5,684 francs, or about 142 francs per head. The prices of these slaves, which were sold to planters on the island when not retained for their own use, ranged from 1,500 to 2,100 francs, or from $300 to $420, at the present rate of exchange. It is interesting to notice that while these negroes were held for sale, the exact period of which is not stated, they received as food eighty bunches of bananas and three beef heads; though under the care of a physician, it is not surprising to find that one of them died. Another bill, bearing date of August 7, 1785, records the sale to Jean Audubon of ten negroes and three negresses for a total sum of 26,000 francs; 16,000 francs of this amount was paid in sugar, but what is particularly interesting now is the fact that a balance of 2,000 francs was finally cancelled on June 9, 1788, a year or more after Jean Audubon, according to the accepted accounts, is supposed to have lost his wife and his property and to have fled from the island. Mme. Anne Moynet Audubon never visited America, and her husband, as we have seen, left Santo Domingo in 1789, before the outbreak of the revolution. His property remained substantially intact until after 1792, and in some years, it is believed, yielded him in rents 90,000 francs, which at present rates in American money would be equivalent to $18,000. In giving his certificate of residence at Nantes in that eventful year, Captain Audubon publicly declared that he possessed a dwelling, a sugar refinery, and warehouses or stores at both Les Cayes and Saint Louis. Moreover, his West Indian estate was not completely settled until 1820, two years after his death.

Figure 4. Les Cayes, Haiti: The Wharf and Post Office; at the Left Is Seen a Pile of Logwood Awaiting Shipment.

Figure 5. Les Cayes, Haiti: The Market and Church of Sacré Cœur. After Photographs Made at Les Cayes in June, 1917, and Obtained through the Kindness of Mr. Ferdinand Lathrop Mayer, Secretary of Legation, Port-au-Prince, Haiti.

Slaves were regarded in Santo Domingo as an indispensable commodity, as they had been in Virginia and the Carolinas for a century past, and were still to be for three-quarters of a century to come; the "friends of the blacks" as the abolitionists were called, were considered by most planters as the enemies of the whites. Degradation and cruelty, ever attendant upon a system that drew its chief support from the self-interest of a class, were all too common in the island, yet there were many who earnestly strove to soften the lot of their slaves. Though a born fighter, Jean Audubon was humane, and the evidence, so far as it goes, shows that his own slaves were treated with kindness and consideration.

This period in Santo Domingo, particularly from the year 1785 to 1789, not only is important for our story, but happened to mark a crisis in French sovereignty in America. It will be necessary, therefore, to follow certain events in a history which can serve only as a warning to mankind, for it contains little to satisfy the understanding and nothing to excite the fancy or gladden the heart. It is to be noticed first, however, that according to the accepted accounts, John James Audubon was born of a Spanish creole mother, in Louisiana, in 1780. Shortly after his birth, his mother is said to have gone to Santo Domingo, where she perished in a local uprising of the blacks, when Jean Audubon's plantations and property were totally destroyed; Jean managed to escape with only his two children, a few faithful slaves, and a part of his money and valuables, to New Orleans, whence he subsequently went to France. Investigation of existing records has proved that these statements are not in accord with the facts, but before entering into further personal details it will be well to examine those conditions on the island of Santo Domingo which led many into easy fortune only to involve them later in a ruin as complete and irretrievable as it was unforeseen and unnecessary.

For nearly a hundred years the western half of Santo Domingo had been held by France, and to every outward appearance it had enjoyed such unbounded and steadily increasing prosperity that it was regarded with envy on every side; in fine, it seemed to be one of the richest and most desirable colonies in the whole world. Historians, said an observer of a later day,[30] were "never weary of enumerating the amount of its products, the great trade, the warehouses full of sugar, cotton, coffee, indigo and cocoa; its plains covered with splendid estates, its hillsides dotted with noble houses; a white population, rich, refined, enjoying life as only a luxurious colonial society can enjoy it." Few could then see the foul blot beneath so fair a surface, or realize that what had been bought by the misery and blood of a prostrate race would demand an equivalent, and that a settlement might be forced.

Negroes had been imported into Santo Domingo from the African coasts in incredible numbers, first by Spain after she had succeeded in exterminating the inoffensive native Caribs, and later by France. One hundred thousand blacks of all ages were entering the colonies each year, and to secure this number of *bossals*, as the native Africans were called, involved the death of nearly as many more, either through the fighting that preceded their capture on land, or from the terrors of pestilence or shipwreck that awaited them at sea. By 1790 the blacks of Santo Domingo outnumbered the whites sixteen to one, and the number of blacks then in the island was estimated at 480,000, in contrast to 30,800 whites, and about 24,000 free mulattoes or "people of color."

Under French rule the blacks had been subjected, as many believed, to a system of slavery unsurpassed for cruelty and barbarity. No doubt there were Frenchmen who, in their fierce struggle to become rich, worked their slaves beyond human endurance and did not hesitate to terrorize them with the severest punishment upon the first symptoms of revolt; but, on the whole, such sweeping denunciations were probably unjust. An impartial observer and historian of that day, himself an Englishman,[31] declared that the French treated their slaves quite as well as the English did theirs, and clothed them better. He believed that the lot of the Santo Domingo blacks at the period of which we speak would compare favorably with that of the peasantry of Europe, a comment made familiar to American ears when applied to the slave population of the South. The real trouble came from the more enlightened disaffection of the mulattoes and free negroes, fanned by the fanatic zeal of abolitionists abroad, particularly of those who formed the society of *Les Amis des Noirs* in France, who were determined to carry out their policies by any means and at whatever cost.

The mulattoes were really in worse plight than the actual slaves, for they were virtually slaves of the State and had no master to whom they could appeal, being subject to military service without pay, to the *corvée* or labor upon the highways, the hardships of which were insupportable, as well as to a constant and galling tyranny. The law was invariably framed in favor of the white man, who, if he struck a mulatto, was subject to a trivial fine, while retaliation by the man of color might cost him his right hand. It should be added, however, that custom was usually more lenient than the law, and that such atrocious enactments were generally a dead letter.

---

[30] See Sir Spencer St. John, *Hayti, or the Black Republic*, 2d ed. (New York, 1889).

[31] Bryan Edwards, Esq., M.P., F.R.S., &c., *An Historical Survey of the French Colony in the Island of San Domingo* (London, 1797).

As might have been expected in the circumstances, the mulattoes took their revenge on the despised blacks, whom they were permitted to hold as slaves. They were notoriously the hardest taskmasters in the island, and in return they were naturally envied and hated by the ignorant mass of black humanity. The whites, to complete the discord, were divided among themselves, the Frenchmen from Europe affecting a superiority over the white Creoles, the seasoned natives of the island, a condition that never made for good feeling. Moreover, the white planter, who endeavored to gain a foothold by producing sugar, cotton or coffee, seems to have had a just grievance against the merchants whom the law favored and who set the price for negroes and all other commodities that had to be bought in exchange for produce. Such at least was the conviction and experience of a keen observer, Francis Alexander Stanislaus, Baron de Wimpffen,[32] who went to Santo Domingo in 1788, tried to establish himself as a coffee planter at Jaquemel, on the southern coast not far from Les Cayes, and after three years of fruitless effort, gave up the attempt in disgust, glad to escape, as from the flames of purgatory, to the United States, where he settled in Pennsylvania. Baron de Wimpffen's lack of success no doubt colored his impressions of the country to some extent, but after making due allowance on this score, we find in his letters, beyond a doubt, an essentially true picture of Santo Domingan society and plantation life at the very time and place with which our story is most intimately concerned. A sketch of the picture which the Baron has drawn, though in brief outline, will enable us better to understand the real condition of affairs.

The prevailing taste in Santo Domingo, according to this observer, was creolian tinctured with *boucan*, or with the characteristics of the buccaneers. White society on the island was divided into governmental or town officials, merchants, and planters, the several classes having their own interests, which were often conflicting. The planters were concerned only with negroes, their sugar, their cotton or their coffee, and could talk of nothing else; values were reckoned in negroes, or in sugar, for which slaves were commonly exchanged. The laxity of morals, the absence of schools, and the total lack of books were patent on every hand. After sunset dancing was the chief form of amusement in the towns, and handsome mulattoes were the acknowledged Bacchantes of the island. It was from this class that housekeepers were usually chosen by the greater part of the unmarried whites. They had "some skill," said Baron de Wimpffen, "in the management of a family, sufficient honesty to attach themselves invariably to one man, and great goodness of heart. More than one European, abandoned by his selfish brethren, has found in them all the solicitude of the most tender, the most constant, the most generous humanity, without being indebted for it to any other sentiment than benevolence."

Expense of cultivation at this time is said to have risen out of all proportion to the value of the product. While negro service was a prime necessity to the planter, the African mine was becoming exhausted; even then slave dealers were penetrating a thousand leagues or more from the Guinea coast. Added to the cost of slaves, which was yearly increasing and had already reached to 2,000 or even 3,000 francs per head, the Government exacted a ruinous capitation tax, which bore with special weight on the planter.[33] Physicians and lawyers, however ignorant, exacted exorbitant fees; masons and carpenters, however inefficient,

---

[32] See Note, Vol. I.

[33] The Superior Council, sitting at Port-au-Prince, in 1780 fixed the tax for the parish of Les Cayes at the rate of 2 francs, 10 centimes per head, which in this instance was certainly trifling. (Note furnished by M. L. Lavigne.)

demanded an unreasonable wage; they, we are told, with the merchant and official governmental class, were the only money makers on the island. The merchant whom we have seen taking the planter's produce at his own price, in exchange for slaves again at his own price, had the advantage in every business transaction; the planter, as a result, was his chronic debtor, and at usurious rates.

Subject to an enervating climate, which Europeans with their intemperate habits could seldom endure for long, the planter, though weak and sick himself, was often obliged to be overseer, driver, apothecary, and nurse to his negroes, the slave of his slaves. In spite of every care, out of one hundred imported negroes the mortality was nearly twenty per cent in the first year. Where less oversight was given to their food, the slightest scratch was likely to degenerate into a dangerous wound, while the most dreaded disease, then known in English as the "yaws" and in French as *la grosse vérole* (to distinguish it from the smallpox, *la petite vérole*), was a scourge for which no remedy had then been found. Every slave was branded with a hot iron on the breast, with both the name of his master and that of the parish to which he belonged, but notwithstanding such precautions desertions were far from uncommon.

The Santo Domingan blacks were put to work in the morning with a crack of the *arceau*, a short-handled whip, delivered on their backs or shoulders, and so accustomed had they become to the regularity of this stimulus that they could hardly be set in motion without it. How to manage the true *bossal*, as distinguished from the African creole, with humanity and success was a problem to which many considerate planters must have addressed themselves in vain, if, as this one declared, the black's ruling passion was to do nothing, and he was by nature a thief, to whom indulgence was weakness and injustice a defect of judgment that excited both his hatred and his contempt.

Stanilaus further observed that the soil of Santo Domingo was then already becoming exhausted, and he believed that the day of rapid fortunes for the planter had passed. "Calculate now," said he, "the privations of every kind, the commercial vicissitudes, the perpetual apprehensions, the disgusting details, inseparable from the nature of slavery; the state of languor or anxiety in which he vegetates between a burning sky, and a soil always ready to swallow him up, and you will allow with me that there is no peasant, no day-labourer in Europe, whose condition is not preferable to that of a planter of San Domingo." "I never met," he adds, "a West Indian in France who did not enumerate to me with more emphasis than accuracy, the charms of a residence at Saint Domingo; since I have been here, I have not found a single one who has not cursed both Saint Domingo, and the obstacles, eternally reviving, which, from one year to another, prolong his stay in this abode of the damned."

Having followed De Wimpffen to this point, the reader is entitled to hear his parting epigrams. "The more I know," he said, of the inhabitants of Saint Domingo, "the more I felicitate myself on quitting it. I came hither with the *noble* ambition of occupying myself solely in acquiring a fortune; but destined to become a master, and consequently to possess slaves, I saw, in the necessity of living with them, that of studying them with attention to know them, and I depart with much less esteem for the one, and pity for the other. When a person is what the greater part of the planters are, he is made to have slaves; when he is what the greater part of the slaves are, he is made to have a master."

Whether Jean Audubon's long experience would have confirmed all that has just been said is doubtful, for he was primarily a merchant or dealer and thus belonged to the favored class. But what especially interests us now is that both he and De Wimpffen were owners of plantations in the southern province of Santo Domingo at the same time. The one who wished to retain a valuable property followed the custom of the time by confiding the management of his affairs to an agent, either at a fixed salary or on a profit-sharing basis; while the other, who stayed long enough to discern the trend of events, was glad to sell his land and his slaves and shake the dust of the island from his feet forever.[34]

Before resuming the intimate details of our narrative, we must follow the whirlwind of political events already set in motion in the island colony. In the spring of 1789 the white colonists of Santo Domingo took administrative matters into their own hands, and without vestige of legal authority, elected and dispatched eighteen deputies to the States-General, then sitting in France. These men reached Versailles in June, a month after that body had declared itself the National Assembly, but only six were ever admitted to its counsels. For a long time opposition to the planters had been fomented in Paris by the "Friends of the Blacks," the abolition society to which we have referred; stories of cruelty to the slaves, colored and intensified in passing from mouth to mouth, as invariably happens when atrocity tales are used as partisan weapons, added to the arrogance and extravagant habits of many planters when resident in the mother country, did not tend to soften the prejudice of the public towards their class. The planters could get no consideration at home, and, as we have seen, the Declaration of Rights followed promptly in August, while a legislative Assembly was ordered in September. Meantime the mulattoes on the island were clamoring for the political rights which the decree had promised them, and, to make matters worse, some of the influential whites espoused their cause, even preaching the enfranchisement of the blacks, from whom up to this time little had been heard. In short, the whites were divided as effectually as were blacks and mulattoes.

The dominant party in Santo Domingo, led by the Governor-General, were determined to uphold the old despotic *régime*, while the General Assembly, which met at Saint Marc in obedience to orders from the mother country, on April 16, 1790, drafted a new constitution. The clash came in July of this year, and in the northern province, where the first blood of the revolution was drawn at Port-au-Prince. On October 12, 1790, James Ogé, a mulatto, inspired, financed and equipped by the "Friends of the Blacks" in Paris, landed secretly in Santo Domingo, established a military camp at Cap François and called all mulattoes to arms. His plan was to wage war on the whites as well as upon all mulattoes who refused to join his standard of revolt; but Ogé and his company were quickly suppressed, and this incompetent leader, who fled to Spanish territory, was later extradited and broken on the wheel. This episode naturally infuriated the whites against all mulattoes, who took up arms at Les Cayes and at other points. The whites also armed, and a skirmish occurred at Les Cayes, Jean Audubon's old home, where fifty persons on both sides lost their lives, but a temporary truce was immediately effected. This was the first serious incident in which the town of Les Cayes figured in the bloody revolution of Santo Domingo; it occurred, we believe, in the late autumn of 1790. Audubon's mother had then been dead four years, and her son, the future naturalist, had left the country in the fall of 1789; in order to bring out these facts clearly it has seemed necessary to enter into this detail.

Later events in Santo Domingo now moved in a direction and with a velocity which few then were able to comprehend. The danger and the potency of the volcano that had long been muttering beneath their feet needed but a few touches from without to reveal its full explosive

[34] Baron de Wimpffen sailed from Port-au-Prince for Norfolk, Virginia, in July, 1790, about a year after Jean Audubon had left the island.

power. These were furnished not only by the mulattoes, many of whom, after having fought under French officers in the American Revolution, had returned to the island and there spread wide the spirit of disaffection and revolt; but also by the National Assembly in France, which by its vacillating policies destroyed every hope of reconciliation. In March, 1790, this Assembly granted to the citizens of Santo Domingo the right of local self-government, but only a year later, on May 15, 1791, tore up this decree and emancipated the mulattoes. When the news reached the island six weeks later, the colony was thrown into the utmost consternation; the whites as a class refused point-blank to accept the decision and summoned an Assembly of their own, which met in August. The mulattoes again took up arms, and the blacks, who by this time had been won to their side, started a general revolt which had its origin on a plantation called "Noé," in the parish of Acul, nine miles from Cap François. They began by burning the cane fields and the sugar houses and murdering their white owners. Thenceforth Santo Domingan history becomes an intricate and disgusting detail of conspiracies, treacheries, murders, conflagrations, and atrocities of every description. The only ray of light comes from the first genuine leader of the blacks, the gallant but unfortunate Toussaint, in 1793.

As has already been intimated, Jean Audubon's Santo Domingo property suffered long after he left the island, and certainly after 1792 when, as we shall soon see, revolutions were demanding his attention and all his energies at home.

Chapter 4

AUDUBON'S BIRTH, NATIONALITY, AND PARENTAGE

Santo Domingo, though repeatedly ravaged by the indiscriminate hand of man, is a noble and productive land, which, for the diversity and grandeur of its scenery and the rare beauty of its tropical vegetation, was justly regarded as one of the garden spots of the West Indies and worthy to be in truth a "Paradise of the New World." For every lover of birds and nature this semi-tropical island, and especially Les Cayes, upon its south-westerly verge in what is now Haiti, will have a peculiar interest when it is known, that there, amid the splendor of sea and sun and the ever-glorious flowers and birds, the eyes of America's great woodsman and pioneer ornithologist first saw the light of day.

Jean Audubon met somewhere in America, and probably at Les Cayes, a woman whom he has described only as a "creole of Santo Domingo," that is, one born on the island and of French parentage, and who is now known only by the name of Mlle. Rabin.[35] To them was born, at Les Cayes, a son, on the twenty-sixth of April, 1785. This boy, who was sometimes referred to in early documents as "Jean Rabin, *créole de Saint-Domingue*," and who again was called "Fougère" (in English, "Fern"), received the baptismal name of Jean Jacques Fougère six months before his sixteenth birthday.

The bill of the physician, Doctor Sanson of Les Cayes, who assisted at young Audubon's birth still exists, and as the reader will perceive, it is a highly unique and interesting historical document.[36] Written in the doctor's own hand, it is receipted by him, as well as approved and signed by Jean Audubon himself. This tardy discovery, along with other pertinent records in the commune of Couëron, in France, finally resolves the mystery which has ever hedged the Melchizedek of American natural history. The child's name, of course, is not given in the bill, but authentic records of Audubon's subsequent adoption and baptism agree so completely in names and dates as to establish his identity beyond a shadow of doubt. Much other documentary evidence which also has recently come to light is all in harmony with these facts, and further shows that the natal spot and time as given in the Sanson bill can refer only to this talented boy. But before turning to these legal documents we must examine the personal record of Jean Audubon's physician.

---

[35] This was one of the commonest names among the French Creoles of Santo Domingo, and was possibly assumed, though the evidence is inconclusive.

[36] For photographic reproduction see below; and for transliteration and translation, Appendix I, Documents Nos. 1 and 1a; for "Fougère" see Appendix I, Documents Nos. 2 and 3; and for "Jean Rabin," Documents Nos. 14, 16, 17 and 18.

Dr. Sanson's carefully itemized account, to the amount of 1,339 francs, extends over a period of nearly two years, from December 29, 1783, to October 19, 1785; it was accepted and signed by Captain Audubon on October 12, 1786, and receipted by the doctor when paid on June 7, 1787. The bill is interesting as a commentary on the medical practice of an early day, as well as for the light which it throws on Jean Audubon's Santo Domingan career, his establishment at Les Cayes, and his treatment of black slaves and dependents. This quaint document, moreover, tends to confirm a remark of Baron de Wimpffen to the effect that every doctor in Santo Domingo grew rich at his profession, and also recalls what he said in regard to the household remedies of the period. "Every colonist," to quote this observer again, "is commonly provided with a small chest of medicines, of which the principal are manna, salts, and rhubarb; the country itself produces tamarinds, and the leaves of the cassia tree, a slight infusion of which, with a little orange juice, makes as good a purge as a mixture more scientifically composed."

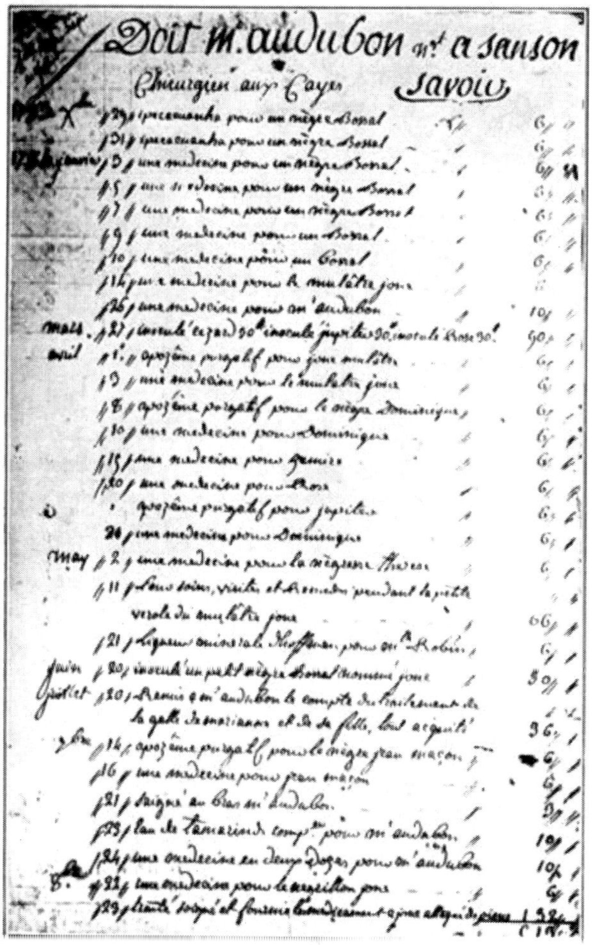

Doit Mr. Audubon … a Sanson
Chirurgien aux Cayes
Savoir

Figure 6. First Page of the Bill, Rendered by Dr. Sanson, of Les Cayes, Santo Domingo, to Jean Audubon for Medical Services from December 29, 1783, to October 19, 1785. After the Original Document in Possession of M. L. Lavigne, at Couëron. The Note in Upper Left-Hand Corner, "Très Curieux Mlle. Rabin & Son Enfant," Has Been Added by a Later Hand.

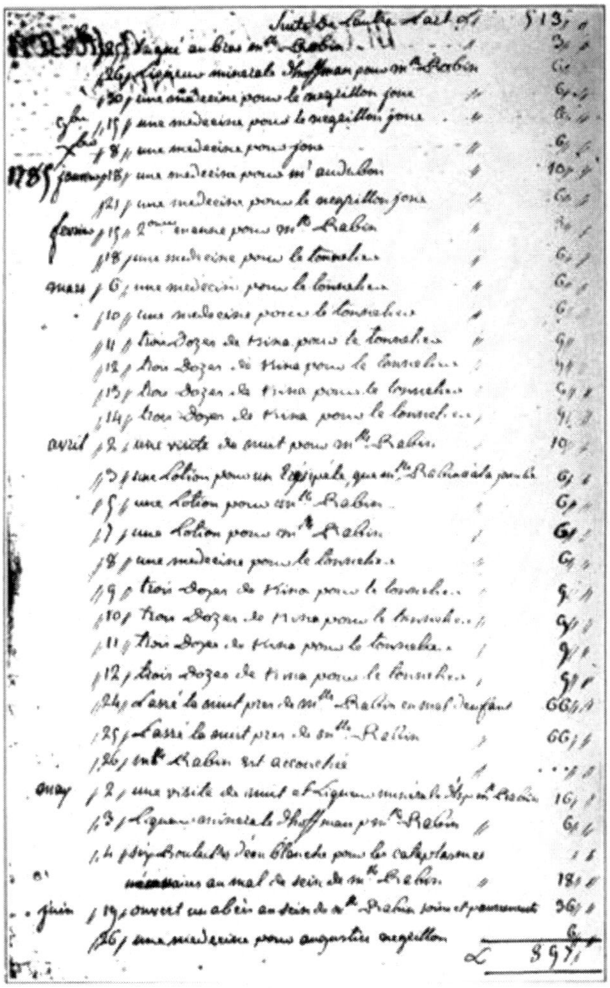

Figure 7. Second Page of the Sanson Bill, Bearing, in the Entry for April 26, 1785, the Only Record Known to Exist of the Date of Audubon's Birth.

This physician's chief resources are seen to have been ipecacuanha, purgative decoctions, including such as the tamarind tree provided, manna, mineral waters, lotions, plasters, and *kino*, an astringent juice derived from different leguminous plants, which gave a red color to the saliva, not to speak of "other medicines," the nature of which is not revealed, which were liberally supplied to whites and blacks, both old and young, alike. It will be noticed further that the slaves of African birth when not named are referred to as "*bossals*" though many young blacks and mulattoes are called "Joue";[37] that a cooper, attached presumably to the Audubon sugar refinery, was dosed thrice daily with *kino* on four days in succession; and that

[37] The word "*Joue*," which occurs eleven times in this document—as "mulatto Joue," "Joue mulatto," "negro bossal named Joue," and "little negro Joue"—suggests the English equivalent "Cheek," but no such usage appears to be authorized. It is evidently a proper name, and is more likely to prove the French rendering of a word common to one of the negro dialects of the island. On the other hand it might represent a corrupted pet name, like "*joujou*" or "*bijou*," bestowed by the French Creoles of Santo Domingo upon their favorite *négrillons* or *petits nègres*, which played a more or less ornamental *rôle* in many households, whether as footmen or servants. In any case the use of this word is doubtless purely local.

this favorite treatment was repeated a month later. A clerk in the establishment, Monsieur Aubinais, is mentioned as requiring frequent attention, as well as Jean Audubon himself, who was once bled at the arm.

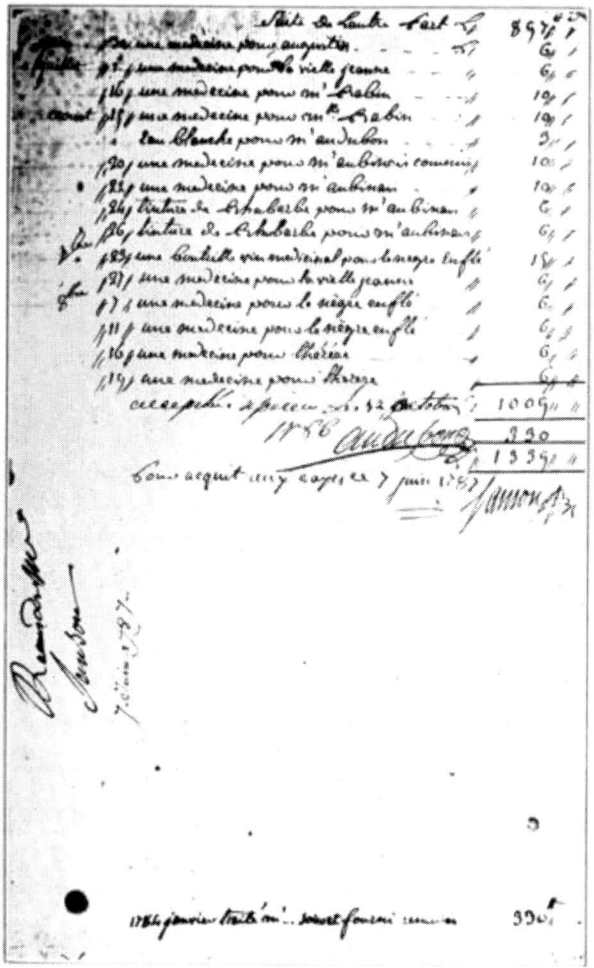

Figure 8. Third Page of the Sanson Bill, Signed as Accepted by Jean Audubon, October 12, 1786, and Receipted by the Doctor, When Paid, June 7, 1787.

In the entry for March 27, 1784, there is this interesting reference: "Inoculated Cæsar, Jupiter, and Rose, at thirty francs each, ninety francs"; and if there were any doubt why Cæsar had been inoculated, a hint is immediately given under May 11: "For attention, visits, and remedies, during the smallpox (*la petite vérole*) of the mulatto Joue, sixty francs"; again we read: "June 30, inoculated a little negro *bossal*, named Joue, thirty francs." Every fresh batch of negroes landed in the colonies led to a new outbreak of this terrible scourge, and but one other disease, *la grosse vérole*,[38] was more common or more fatal among the blacks. For a long period it had been a common practice to inoculate both whites and blacks directly with the smallpox in order to secure some degree of protection against its most virulent form, but

[38] See Vol. I, p. 46.

this method of fighting the devil with fire had its disadvantages. By the end of the eighteenth century opinion was about equally divided upon the advisability of continuing the measure, since induced variola or smallpox was apt to be virulent, and was often quite as infectious as when manifested in the usual and natural way. Then came Edward Jenner's grand discovery, made twelve years before this date but not announced until 1798, that vaccinia would prevent variola. Almost immediately vaccination spread like wild fire over Europe, and it has never been appreciated more fully or more highly lauded by the best representatives of the medical profession everywhere than at the present day.

The most interesting references in this historic document are to "Mlle. Rabin," whose name occurs no less than seventeen times, beginning May 21, 1784, and closing with the entry for the seventeenth of August, 1785. We learn that the physician spent the nights of April 24 and 25, 1785, at the woman's bedside, and that her child was born on the twenty-sixth day of that month, probably in the morning. It will be noticed further that she had been bled previously at the arm, that she had suffered also from the erysipelas, and that later she was treated for abscesses. These frequent attentions of the physician, extending over several months, the last record being for August 17, show only too clearly that at this time Audubon's mother was in feeble health. All that is further known about her is that she died either at the close of 1785 or in 1786, when her infant son was probably less than a year old.[39]

A daughter of Jean Audubon, Rosa, who was first called Muguet (in English, "Lily of the Valley"), was also born in Santo Domingo, and probably at Les Cayes, on April 29, 1787. Her mother, Catharine Bouffard, "*créole de Saint-Domingue*," who subsequently went to France, had another daughter, born also at Les Cayes, named Louise, who was living at La Rochelle in 1819.[40]

When Captain Audubon finally left the West Indies in the autumn of 1789, he took with him, in the care of trustworthy slaves, these two children, Fougère or Jean Rabin, aged four and a half years, and Muguet or Rosa, an infant of less than two. We know that he visited Richmond, Virginia, to collect a long outstanding claim against David Ross, then engaged in an iron industry near that city, and it is possible that he traveled by way of New Orleans and the Mississippi and Ohio rivers. After spending some time at the close of this year in the United States, he went to France and made a home for his children at Nantes. This city became essentially their permanent abode until their father's retirement from the navy on January 1, 1801, when he finally settled in the little commune of Couëron, on the north bank of the Loire. The storm that burst over Nantes soon after their arrival revealed the true colors of Jean Audubon's patriotism, and the man was seen at his best, as will be related in the following chapter.

---

[39] It was stated in the act of adoption, which was drawn up in March, 1794, that Audubon's mother had then been dead "about eight years," and the testimony of the Sanson bill shows that she was alive as late as October, 1785.

[40] The following letter of inquiry concerning Louise was written by Rosa's husband when Jean Audubon's will was being attacked in the courts at Nantes. It is dated at Couëron, June 26, 1819, and is addressed to "Monsieur Carpentier Chessé, engraver, place Royale, Nantes:"

"Following the friendly offer that you made me, I have the honor of asking you to undertake, at your next visit to La Rochelle, the following inquiries:

"1. There should be at La Rochelle (it is thought at the home of the widow Scipiot) a Miss Louise Bouffard, born at Les Cayes, Santo Domingo, in America.

"What is her position? What is she doing? What is her conduct? In short I should like to know absolutely all about her, being charged by the Madame, her mother, to make all inquiries."

(Translated from original in French, Lavigne MSS.)

Madame Audubon, who had no children of her own, tenderly received the little ones, thus wafted from over the sea to her door in the Rue de Crébillon.[41] As the story proceeds we shall see that she was a most kind, if over-indulgent, foster mother, and became excessively proud of her handsome boy. "The first of my recollective powers," said the naturalist when writing of himself in 1835,[42] "placed me in the central portion of the city of Nantes ... where I still recollect particularly that I was much cherished by my dear stepmother ... and that I was constantly attended by one or two black servants, who had followed my father to New Orleans and afterwards to Nantes."

Jean Audubon, who spent a good part of his life at sea and in a country almost totally devoid of morals, must be considered as the product of his time. He was better, no doubt, than many who made greater professions, better certainly than a Rousseau, who gave excellent advice to parents upon the proper methods of rearing their children but sent his own offspring to orphan asylums. As most men have their faults, said the son, the father "had one that was common to many individuals, and that never left him until sobered by a long life"; but, he added, "as a father, I never complained of him; his generosity was often too great, and his good qualities won him many desirable friends." Whatever his faults, Jean Audubon was just, generous and possessed of a kind heart. He was in reality a truer father than many who give their children their name but deny them sympathy and a wise oversight. Jean Audubon not only cherished the two children but made them his heirs. On March 7, 1793, Fougère at the age of eight and Muguet at six were legalized by a regular act of adoption in the presence of witnesses at Nantes as the children of Jean and Anne Moynet Audubon.

This step was taken at the very moment when the storm had burst over La Vendée, when the fate of Nantes was trembling in the balance and the life of her citizens was most insecure. The act of adoption reads:[43]

> Extract from the registers of births of the sections of La Halle and Jean Jacques of the commune of Nantes, department of the Loire inférieure, on the seventh of March, 1749, the second year of the Republic, one and indivisible, at ten o'clock in the morning.
>
> Before us, Joseph Theulier, public officer, elected to determine the public status of citizens, have appeared in the town hall, Jean Audubon, commanding the war sloop Cerberus, vessel of the Republic, aged forty-nine years, native of Les Sables d'Olonne, department of La Vendée, and Anne Moinet his wife, aged fifty-eight years, native of the former parish of Saint-Leonard, of this commune, who, assisted by René Toussaint Julien Beuscher, manufacturer, aged twenty-five years, living in the section of La Halle, Rubens Street, and by Julien Pierre Beuscher, marine surgeon, aged twenty-four years, living in the section of La Fraternité, Marchix Street, and employed steadily in the said war sloop

---

[41] A principal street in the old quarters of Nantes, leading from the Place Royale to Place Graslin. Jean Audubon named this street as his place of residence in 1792, when he was living in a house belonging to Citizen Carricoule. He made his home also at No. 39, rue Rubens, a short street, with many of its houses still intact, in the same quarter; this was rented of Françoise Mocquard for five years, beginning June 24, 1799 (*le 6 Messidor, an 7*), at four hundred francs per annum. He also dwelt at various times at No. 5, rue de Gigant, and in the rue des Carmes, where his wife possessed a house, as well as in the rue des Fontenelles and the rue Saint-Leonard. Very likely "La Gerbetière" at Couëron was occupied intermittently, especially in summer, after the outbreak of the Revolution and his reverses in fortune; even after his retirement there in 1801, he still kept a lodging (*pied-à-terre*) at Nantes, where, as it chanced, he died, though it was not his usual stopping-place.

[42] See Maria R. Audubon, *Audubon and His Journals*, vol. i, p. 8.

[43] For the original text of this act, here given in translation, see Appendix I, Document No. 2.

Cerberus, have declared before me that they do adopt and recognize from this moment as their lawful children, to wit:

A male child named Fougère, born since their marriage, which took place on the twenty-fourth of August, 1772, in the commune of Paimbœuf, in this department, to him, Jean Audubon, and a woman living in America, who has been dead about eight years, and a female child, named Muguet, born also since the marriage aforesaid, to him and another woman living in America, named Catharine Bouffard, of whose fate he is ignorant.

The two children being present, the first aged nine years, that will expire on the 22d of next April, the second aged seven years, that will also expire on the 26th of April next, and both having been born in America, according to this declaration that the witnesses above mentioned have signified as true, I have drawn up the present act, which the natural father and the mother by adoption, as well as their witnesses have signed, together with myself in this said day and year.

It will be noticed that in this legally attested document, Bouffard, the true name of Muguet's mother, is given, while the name of the mother of Audubon is suppressed. It might therefore be inferred that the name Rabin, which appears later, was assumed, but as already remarked, such evidence is not conclusive.

Fougère, who was also called Jean Rabin, was baptized on October 23, 1800, by a priest of the church of Saint-Similien at Nantes. The archives of this church for the period in question have disappeared, but Jean Audubon's copy of the record has survived, and reads as follows:[44]

The Act of Baptism of Jean Audubon-Rabin
October 23, 1800

We, the undersigned, certify to have baptized on this day Jean Jacques Fougère Audubon, adoptive son of Jean Audubon, lieutenant of a frigate of the Republic, and of Anne Moinet, his legitimate wife, who being present bear witness that the adoption of the said Fougère, made by them, is in accordance with the present act.

[Signed] Tardiveau, priest of Saint-Similien, of the town of Nantes.

The act of adoption was drawn at a time when Captain Audubon could have had little leisure to consult records had he been disposed to do so, but the dates of birth which he then gave for these two children were correct both as to the year and month. Fougère, however, was born on the twenty-sixth, instead of the twenty-second of April, and Muguet, on the twenty-ninth, instead of the twenty-sixth, of that month. Audubon's mother's name is indicated in numerous legal documents of later date, and, as will appear, in every instance her son's identity is clearly established.

---

[44] Research at Nantes in 1915 revealed that the baptismal records of the parish of Saint-Similien were wanting for the period from 1792 to 1803, so it is probable that they were destroyed in the Revolution. The municipal archives of Nantes possess a book of baptismal records of the city without distinction of parishes, but this shows the names of neither "Fougère," "Rabin," nor "Audubon," for the year in question.
The Abbè Tardiveau was *un prêtre assermenté*, or one of those priests who had sworn in 1790 to recognize the civil constitution of the clergy.
For copy of the act of baptism in the French original, see Appendix I, Document No. 3. It is impossible to say whether the heading as given in my copy of this act was in the original or not.

Young Audubon, who disliked the names of Fougère and Rabin, and naturally wished to be rid of their early associations, adopted the fanciful name of "La Forest,"[45] but used it only sporadically and for a short time. Some of his drawings of birds made at Nantes or Couëron as early as 1805, and in New York in 1806 and 1807, and possibly others of slightly later date, are signed "J. L. F. A.," or "J. J. L. Audubon."[46]

Jean Audubon and his wife are said to have settled some property upon "Jean Rabin, *créole de Saint Domingue*," which he refused to accept, saying, "my own name I have never been permitted even to speak; accord me that of Audubon, which I revere, as I have cause to do."[47] The reference in this instance was, I believe, to the final will of Lieutenant Audubon,[48] according to which his property, after being held in usufruct by his wife during her lifetime, was to be equally divided between their two adopted children. In his first will the son was referred to as "Jean Audubon," but in the second and last document, executed in 1816, two years before the testator's death, he appears as "Jean Rabin." Madame Audubon drew four wills; in the first, dated December 4, 1814, her adopted son is called "Jean Audubon"; in the next, of 1816, he is "Jean Rabin, *créole de Saint-Domingue*," while in a draft written December 26, 1819, he is styled simply "Jean Rabin"; finally, in her fourth and last testament of July 16, 1821, the wording is "Jean Audubon, called 'Jean Rabin.'" It is thus very plain that Audubon's foster parents considered it advisable to have his identity clearly set forth in legal documents. In one of his autobiographical sketches Audubon remarked that his own mother was said to have been as wealthy as she was beautiful, and if this were true, such caution might be explained and a key found to certain other enigmatical conditions which seemed to hedge his early life. But to such possibilities it will be necessary to revert at a later point of our story.[49]

This dual personality was set forth by the naturalist himself, but in a more curious form, in a power of attorney[50] executed at Henderson, Kentucky, on July 26, 1817, in favor of his

---

45 An English writer once gave the name of Audubon's mother as Mlle. La Forêt.

46 Audubon's signature underwent frequent variations during the first twenty-five years of his life, but after 1820 he almost invariably signed himself "John J.," or "J. J. Audubon." In the record of the civil marriage of his sister, at Couëron in 1805, his name appears as "J. J. L. Audubon;" in the "Articles of Association" with Ferdinand Rozier, signed at Nantes in 1806, it is "Jean Audubon," and in the release given on the dissolution of this partnership, at Ste. Geneviève, in 1811, the English form, "John Audubon," appears.

47 This statement was made to me by Miss Maria R. Audubon in 1914.

48 For full text of the six wills drawn at different times by Jean Audubon and his wife see Appendix I, Documents Nos. 13-18.

49 See Chapter 17.

50 This unique document reads as follows:

"To all to whom these presents may come: know ye that I, John Audubon, having special trust and confidence in my friend, G. Loyen Du Puigaudeau, of the Department of Loire and [*sic*] Inférieure, and Parish of Couëron, near Nantes, in the kingdom of France, [do constitute him] my true and lawful attorney, and the true and lawful attorney in fact of Jean Rabin, husband of Lucy Bakewell, of the County of Henderson and State of Kentucky, in the United States of America, for us [?], the said Jean Rabin, and in our name to our use and benefit, to ask, demand, sue for, recover, and receive all and every part of the Real and Personal Estate, that is to say Lands, Tenements, Grounds, Chattels, and credits, which I have, or either of us, in the Department of Loire and [*sic*] Inferieure in the kingdom of France, aforesaid, and to make sale of the same, either at auction, or by contract of the said Lands and Tenements, Goods, Chattells, and Credits, to receive the money arising from said sale, to give any Receipt, acquittance, or other discharge for the said money or any part thereof, if money or specie shall be received, or for any property he may receive in exchange or barter for said Real and personal Estate, and our said attorney, or the attorney of Jean Rabin aforesaid, is hereby authorized and empowered to make, give, execute, and deliver any Deed, Covenant, or transfer of said Real and Personal Estate to the purchaser of all or any part thereof for us, or for the said Jean Rabin, in as full and ample a manner as he, the said Jean, could do, was he personally present in said Department, in the Kingdom. In testimony whereof the said John

brother-in-law, Gabriel Loyen du Puigaudeau. This measure was taken more than a year after Audubon's father had drawn up his last will, in which the son was referred to as "Jean Rabin," and was evidently designed to facilitate any settlement of this will which events in France might render necessary. The naturalist was then engaged in his famous but disastrous financial enterprises on the Ohio River,[51] but whether any intimation had come to him of possible legal troubles, which later actually ensued in France, cannot be stated.

Figure 9. Audubon's Signature at Various Periods from 1805 to 1847. The First, Fourth and Sixth Are from Early Drawings; the Second from Audubon and Rozier's "Articles of Association"; the Fifth from a Release Given to Rozier; and the Remainder from Letters.

Audubon has hereunto set his hand and affixed his seal the Twenty Sixth day of July, Anno Domini One thousand & Eight hundred and Seventeen.

John J. Audubon [Seal within]

On the back of the preceding is the notary's certificate that Jean Audubon appeared before him; seal affixed, and dated July 26, 1817.

Signed, "A[mbroze] Barband,

Notary of Henderson County, Kentucky."

[51] See Chapter 16.

In reading the published accounts of Audubon's early life many have been puzzled by the absence of definite dates, as well as by the numerous contradictions in which they abound. It is needless to burden this narrative with a tedious reference to all these errors or to attempt to trace their origin, which no doubt had many sources, but since we have given the first true account of the naturalist's birth, we cannot pass these matters without a word of comment. The situation is somewhat involved, since we should possibly differentiate between what Audubon at different times believed to be true, and what he wished to make known to his family or to the public; possibly also we should discriminate between what he actually published over his own signature during his lifetime and the material which has appeared since his death, even though originally written by his own hand.

The first definite date which Audubon ever gave concerning his own life was that of his marriage in 1808, when he was twenty-three years of age, and all that he ever published of a biographical nature is to be found in his *Ornithological Biography*.[52] In the introduction to this work he simply said that he had "received light and life in the New World," and further that he returned to America from France, whither he had gone to receive the rudiments of his education, at the age of seventeen. Since Audubon's first return to America was in the autumn of 1803, when he was actually about eighteen and one-half years old, this statement is not so wide of the mark as to imply that the date of his birth was not then well understood. Moreover, the record of his adoption, which was certified to at the time of his baptism in 1800, was carefully preserved among the family documents, and there is no reason to suppose that knowledge of his age was ever withheld from him. Nevertheless, Audubon was inclined to overestimate his years, a characteristic rare in these days; when at Oxford in 1828 he was asked for his autograph, and was begged to inscribe also the date of his birth; "that," he said in recording the incident, "I could not do, except approximately," and his hostess was greatly amused that he should not know.

While going down the Ohio River in 1820, bound for New Orleans, Audubon took advantage of a rainy day to write in his journal something about himself that he thought his children at some future time might desire to know. This brief record may or may not have been at hand when in 1835 he wrote the more extended version that finally saw the light in 1893.[53] Since the manuscript of the later sketch was presumably in possession of Mrs. Audubon when the biography of her husband was prepared in New York about the year 1866, that account in its various versions has furnished biographers with practically all of the available material, not purely conjectural, concerning the naturalist's early life. Such additions as were made subsequently have proved to be very inaccurate.

In the first of these sketches, which, so far as it goes, is more in strict accord with facts, Audubon said nothing of his birth, and of his mother remarked only that he had been told that she was "an extraordinary beautiful woman," who died shortly after he was born. His father, he added, saw his wealth torn from him, until there was left barely enough to educate his two children, all that remained of the five, his three elder brothers[54] having been "killed in the wars." He then believed, as he said, that his first journey to France was made when he was two years old.

---

[52] Vol. i, p. v.

[53] Published by Maria R. Audubon in *Scribner's Magazine*, vol. xiii (1893).

[54] Whether Jean Audubon had other sons born in Santo Domingo is not recorded, and this reference of the naturalist, which was repeated in his later sketch, cannot be verified.

The later and fuller biography, referred to above as written in 1835 and published in 1893, begins with these words:[55]

> The precise period of my birth is yet an enigma to me, and I can only say what I have often heard my father repeat to me on this subject, which is as follows: It seems that my father had large properties in Santo Domingo, and was in the habit of visiting frequently that portion of our Southern States called, and known by the name of, Louisiana, then owned by the French Government.
>
> During one of these excursions he married a lady of Spanish extraction, whom I have been led to understand was as beautiful as she was wealthy, and otherwise attractive, and who bore my father three sons and a daughter,—I being the youngest of the sons and the only one who survived extreme youth. My mother, soon after my birth, accompanied my father to the estate [*sic*] of Aux Cayes,[56] on the island of Santo Domingo, and she was one of the victims during the ever-to-be-lamented period of the negro insurrection of that island.
>
> My father, through the intervention of some faithful servants, escaped from Aux Cayes with a good portion of his plate and money, and with me and these humble friends reached New Orleans in safety. From this place he took me to France, where having married the only mother I have ever known, he left me under her charge and returned to the United States in the employ of the French Government, acting as an officer under Admiral Rochambeau. Shortly afterward, however, he landed in the United States and became attached to the army under La Fayette.

The true history of Jean Audubon's commercial, naval, and civic career is given in the preceding and following chapters.

The naturalist, in his letters and journals, made frequent allusions to his age, but, as his granddaughter remarked, with one exception, no two agree; hence, his granddaughter concluded that he might "have been born anywhere from 1772 to 1783." In the face of such uncertainty she adopted the traditional date of May 5, 1780, adding that the true one was no doubt earlier. Audubon was thus five years younger than his biographers supposed, and twenty-one years were added to the age of his father, who actually lived to be only seventy-four years old, while his son died in his sixty-seventh year.

Wherever there is mystery there tradition is certain to raise its head, and though the naturalist carried his "enigma" to the grave, others, building upon his story, have fixed upon the very house in Louisiana in which he is said to have been born. Indeed, advocates of more than one house in that state as the probable scene of Audubon's nativity have arisen in recent times. We are obliged, therefore, to examine somewhat farther the now universally received but thoroughly erroneous idea that John James Audubon was a native of Louisiana at a time when that Commonwealth was part of a province of France.

Upholding a tradition of rather recent growth, Audubon's granddaughter has expressed the belief that the naturalist was born in a house belonging to the famous Philippe de Marigny and known as "Fontainebleau." This was a sugar plantation on the north side of Lake Ponchartrain, three miles east of what is now the village of Mandeville and twenty-five miles due north of New Orleans.

---

[55] See Maria R. Audubon, *Audubon and His Journals*, vol. i, p. 7.

[56] See Vol. I.

Pierre Enguerrand Philippe de Mandeville, Ecuyer Sieur de Marigny,[57] at one time owner of vast estates in and about New Orleans, was born in that city in 1750, and served as its alcade or mayor for two years. A lavish dispenser of hospitality, in 1798 he entertained in great state the Duke of Orleans, later known as Louis Philippe of France, together with his two brothers who accompanied him. He died at New Orleans, leaving five sons, of whom the third, Bernard Marigny, later became the owner of "Fontainebleau," which it has been mistakenly assumed was inherited from his father. At the time of the Duke of Orleans' visit just mentioned Jean Audubon had been out of the country nine years; there is no evidence of his ever having owned property at New Orleans, or ever having sustained any relations with the Marigny family.

Before following the Marigny myth further, it will be interesting to notice a late echo of the "Fontainebleau" story. In 1910 the Reverend Gordon Bakewell, then in his eighty-ninth year, gave some interesting reminiscences of Audubon, and spoke very definitely concerning both the time and place of his birth. Dr. Bakewell was a nephew of Mrs. Audubon, and as a youth, in 1834, had passed some time at her home in London. John W. Audubon, with his father's assistance, painted at that time a portrait of young Bakewell, who at a later day was welcomed in their home on the Hudson. Dr. Bakewell's contribution was as follows:[58]

> The uncertainty as to the place of Audubon's birth has been put to rest by the testimony of an eye witness in the person of old Mandeville Marigny now dead some years. His repeated statement to me was, that on his plantation at Mandeville, Louisiana, on Lake Ponchartrain, Audubon's mother was his guest; and while there gave birth to John James Audubon. Marigny was present at the time, and from his own lips, I have, as already said, repeatedly heard him assert the above fact. He was ever proud to bear this testimony of his protection given to Audubon's mother, and his ability to bear witness as to the place of Audubon's birth, thus establishing the fact that he was a Louisianian by birth.

We do not doubt the candor and sincerity of the excellent Dr. Bakewell, but are bound to say that the incidents as related above betray a striking lapse of memory and an even greater misunderstanding of recorded facts. Singularly a footnote to the paragraph quoted shows that the Marigny to whom he refers was, as must have been the case, Bernard Mandeville de Marigny, who was born in 1785, the same year as the naturalist. Since both were in the cradle at the same time, he is hardly available as a witness. Moreover, the official records of the United States Government prove that the estate called "Fontainebleau" was not in possession of the Marigny family at the time of Audubon's birth. The land in question was granted to a creole named Antonio Bonnabel, on January 25, 1799, by Manuel Goyon de Lemore, Governor-General of the Province of Louisiana and West Florida. Bonnabel sold his tract to Bernard Marigny in 1800, and Congress confirmed his title to it by a special act in 1836.[59]

---

[57] See J. W. Crozat, "Bibliographical and Genealogical Notes Concerning the Family of Philippe de Mandeville, Ecuyer Sieur de Marigny, 1709-1800," *Louisiana Historical Society Publications*, vol. v (New Orleans, 1911). The portrait referred to below now hangs in the H. Sophie Newcomb Memorial College, New Orleans.

[58] Gordon Bakewell *ibid.*, p. 31.

[59] See *Laws of the United States, Treaties, Regulations, and Other Documents Respecting the Public Lands*, vol. i, p. 301 (Washington, 1836). In Number 756, entitled "An Act for the Relief of Bernard Marigny, of the State of Louisiana," Marigny is mentioned as assignee of Antonio Bonnabel, and his claim, which was confirmed, is described as follows: a tract of land of 4,020 superficial arpents, in the State of Louisiana, parish of St.

Bernard Marigny served in the French army towards the close of the Napoleonic period, and his return to the United States from France, about 1818, is said to have been hastened by a duel which he fought with one of his superior officers. On his return he named Bonnabel's old tract on Lake Ponchartrain "Fontainebleau," in remembrance of the place where his regiment had been assigned for duty in France, and eventually built upon the estate a sawmill and a sugar-house, and planted sugar cane, living meanwhile on another plantation two and one-half miles away. The latter estate was allotted by him in 1832, when he gave it the name of Mandeville; the settlement thus started has since grown to a village of some 1,500 people. Here a summer house which belonged to Bernard's father still exists, although in altered form; it has been raised to accommodate a lower story, and is now known as the "Casino." According to those who have most carefully investigated existing records, this is the only house in Mandeville which belonged to the elder Marigny at the time of which we speak.

Bernard Marigny was one of those who befriended Audubon when he was in desperate straits at New Orleans in 1821, by advancing him money in return for portraits or drawings of birds. He died in that city in 1868, when in his eighty-third year, a poor and honest man.

Tammany, "bounded on the southwest by Lake Ponchartrain, and on the northwest by lands formerly owned by the heirs of Lewis Davis."

I am informed by Mr. Gaspar Cusachs, president of the Historical Society of Louisiana, who has carefully investigated the titles of this property and to whom I am indebted for much information concerning it and its owners, that the tract described above included the estate of "Fontainebleau." Marigny's claim included also a smaller tract of 774 arpents in the same parish. This land was bounded on the southwest by Lake Ponchartrain, on the north by Castin Bayou, and on the south by the tract acquired from Bonnabel; it was granted to the heirs of Lewis Davis in 1777, and certain of them filed a claim for it in 1812.

Chapter 5

Lieutenant Audubon as Revolutionist

The ancient city of Nantes, long famed for the beauty of its situation on the banks of a noble river, within easy reach of the sea, as well as for its importance in the arts of war and peace, numbered at the time of the Revolution 70,000 souls. The modern visitor to this favored spot will find quiet and orderly streets adorned with monumental statues (one of these representing Guépin, the revered historian of the city), the old buildings nearly all replaced by better, the Loire spanned by handsome bridges, and the ancient bounds of the town extended until it has become the sixth city of the Republic. Since Nantes formed a somber background to Audubon's youth, we shall follow in brief some of the ordeals through which his family, in common with thousands of other Nantais, were destined to pass during those eventful years which witnessed the close of the eighteenth century in France.

When Captain Audubon reached Nantes presumably not far from the beginning of 1790, he found the city in a state of the greatest turmoil and agitation. The commons, or third estate, included hundreds of its rich and influential citizens, and their demands for a fair hearing and a representation equal to that of the other orders had then passed the stage of open revolt, for they had planted their "liberty tree" and were sworn to defend it. In August of 1789 a permanent Committee of Public Safety had been constituted at Nantes, and by the end of that month 1,200 had volunteered for service in the National Guard. There were many loyalists in the city but they could not crush the ardent spirit of this revolt, and when in September money was needed to equip the revolutionary soldiery, young school children raised large sums for the popular cause. Jean Audubon immediately cast his lot with the revolutionists and joined the National Guard, but how much service he saw in the field cannot now be determined; it is known, however, that he was with these troops in the spring of 1792.[60]

In March, 1793, the loyalists of La Vendée rose to arms, and marching on Nantes under the able leadership of Charette, threatened to put its garrison to the sword if it were not surrendered within six hours. The National Guard met these invaders outside the walls and left the citizens to shift for themselves. Thus thrown upon their own resources, the Nantais showed that they could help themselves. They requisitioned and used for defense everything at hand; they exhumed the leaden coffins in their grand cathedral and appropriated waterspouts for ammunition, while their church bells were molded into cannon. Though held in check, the Vendeans laid siege to the city, and but for the resolution of its mayor, Baco,

[60] One period of this service bears date of May 31.

Nantes would probably have fallen—in which event Audubon would have had a different history and would probably never have become a pioneer naturalist in America. Baco, disregarding the advice of his military chiefs, immediately placarded the walls of Nantes decreeing death to any who should suggest capitulation, and called all the inhabitants to arms, sparing neither woman nor child. The Vendeans had met their match, for they were dealing with many of their own blood, but though the siege began in early March, they were not effectually dispersed until the end of June, and then only after much bloodshed without the walls. When the immediate crisis had passed, the Constitution of the Republic was unanimously accepted by the eighteen sections of Nantes, on the twenty-first day of July, 1792.

A few months later in that fateful year a more terrible calamity befell the city, when the reign of terror under the notorious ultra-revolutionist, Jean B. Carrier, began. Carrier reached Nantes on October 8 and at once proposed to exterminate both the Vendean royalists and their Nantais sympathizers. He reorganized the entire administration to suit his purposes, and to carry out his plans recruited from the lowest classes a revolutionary army to spy upon, denounce and arrest private citizens, many of whom were sent to Paris for trial when not secretly dispatched. The whole district was soon paralyzed by the barbarity of the crimes then committed, and the unhappy Vendeans were dragged to Nantes, to be shot, guillotined or drowned, in such numbers that the city was unable to bury its dead or the river to discharge them to the sea. Thus perished thousands, uncounted if not unknown, and the pestilence of typhoid fever that immediately followed claimed another heavy toll regardless of political sympathies. While these dire scenes were being enacted, Jean Jacques Fougère Audubon, then a lad of eight years, was living in the heart of Nantes, and his father was one of its leading revolutionists. An aunt of the future ornithologist, according to his account, who was one of these wretched victims of revolutionary fury, was dragged through the streets of Nantes before his eyes, but apparently she did not actually meet her death at that time.[61]

That Jean Audubon moved his family out of Nantes during the revolutionary crisis is possible, and Couëron would have been available as a place of refuge. Many Nantais are known to have fled to Lorient on the coast of Brittany, where they found in the heroic youth Julien the ardent and fearless patriot who was destined to become the real savior of their stricken city. Young Julien denounced Carrier in his letters to Robespierre, and when one of these was intercepted, defied him in person. When his stirring appeals finally reached the Tribunal at Paris, its misnamed representative was recalled, and left Nantes under cover of night on February 14, 1794. During his mad reign of four months, Carrier had gone far towards carrying out his theory of republican government, that should begin, as he openly avowed, by "suppressing" half of the population of France. The records show that nearly nine thousand bodies were buried in Nantes in a little over three months, from January 15 to April 24, 1794. The plague of fever no doubt accounted for many of these, but the wide reaches of the Loire never told their full story.

Though the most grievous affliction of Nantes passed with the recall of Carrier, the city had no lasting peace until the execution of the Vendean leader, Charette, in March, 1796; "Poor Charette," said Audubon, writing in his journal at Liverpool, December 24, 1827, "whom I saw shot on the place de Viarme at Nantes." This virtually ended the war in the Vendée, but the Chouans, under their intrepid chief, Dupré, the miller, called "Tête-Carrée,"

---

[61] See Vol. I.

managed to furnish considerable excitement, and raided Nantes in 1799. Dupré's followers stole in secretly at three o'clock on the morning of October 19 and left before daylight, after liberating fifteen royalists from the prison, which seems to have been their chief purpose. The cannon of alarm was fired from the Chateau; the tocsin sounded, calling the city to arms; there was much street fighting, but it was too foggy and dark to distinguish friend from foe, and when the National Guard was finally assembled, the enemy had vanished. This brief attack cost the city twenty-one deaths and wounds for twice the number,[62] but it was only a passing incident in comparison with events that had gone before. Thenceforth the history of the town is blended with that of the nation.[63]

Lieutenant Jean Audubon — Anne Moynet Audubon

Figure 10. After Oil Portraits Painted between 1801 and 1806, Now in Possession of M. L. Lavigne, at Couëron.

We have only slight indications of Jean Audubon's activities from the close of 1789, when, according to his own statement, he was in the United States, to the period of his service in the National Guard at Nantes in the spring of 1792; he was then living in the house of Citizen Carricoule, rue de Crébillon, and the lease of his "Mill Grove" farm, which was renewed in October, 1790, was dated at Nantes. We may safely assume that he was engaged in revolutionary business during most of this interval: his name begins to appear in the written

[62] The mayor, Saget, at the moment he was crossing the Place Egalité (the Place Royale of today) received point-blank a ball in his right thigh and another in his left leg, and lost both limbs.

[63] For the revolutionary history of Nantes I am chiefly indebted to M. A. Guépin's excellent *Histoire de Nantes*, 2d ed. (Nantes, 1839); Hipp. Etiennez, *Guide du Voyageur à Nantes, et aux Environs* (Nantes, 1861); A. Lescadien et Aug. Laurent, *Histoire de la Ville de Nantes*, t.2 (Nantes, 1836); F. J. Verger, *Archives curieuses de la Ville de Nantes et des Départments de l'Ouest*, t. 5 (Nantes, 1837-41); and to a scholarly monograph by Dugast-Matifeux, entitled *Carrier à Nantes: Précis de la Conduite patriotique et révolutionnaire des citoyens de Nantes* (Nantes, 1885).

records of Nantes and of the department of the Lower Loire in January, 1793, and existing documents[64] show that he was engaged as a commissioner and member of the Department and as a member of the Council of the Navy until the twenty-fifth of June, when he enlisted for active service in the navy of the Republic. Jean Audubon served also on various republican committees, his duties comprising the enlistment of recruits, organizing the National Guard, soliciting funds and food supplies for Nantes, finding cannon and other military or naval materials, posting proclamations, administering the oath of allegiance, and watching the movements of loyalist troops in the district. We have seen that the father of the naturalist was a game and determined fighter, and there is ample written testimony to prove that in the commune of Nantes he was regarded as an ardent patriot, who could be relied upon to act with tact, and if necessary with force.

Figure 11. Jean Audubon. After an Oil Portrait Painted by the American Artist Polk, at Philadelphia, About 1789, Now in Possession of Mrs. Morris Frank Tyler. Published by Courtesy of Miss Maria R. Audubon.

[64] The unpublished documents of this Department are preserved in the archives of the Préfecture at Nantes, and through the courtesy of their custodians I was enabled to examine them freely. These documents deal with all the revolutionary changes in church and state consequent upon the breaking down of the old *régime*, and with the enrollment of volunteers and the dispatch of armed forces to centers of disturbance throughout that district. The present manuscripts are said to represent but a fraction of those which originally existed, the archives having been subjected to repeated raids, thefts, and wanton destruction by fire and other means. The most important have been listed and published by the Government in summary form under the title, *Les Archives du Département de la Loire Inférieure, 1790-1799*, Série L. (Nantes, 1909).

Having been appointed a Civil Commissioner by the Directory of the Department on January 17, 1793, Citizen Audubon was sent to Savenay, a town of some importance twenty-five miles to the northwest of Nantes. His instructions on this mission were to gather useful information on the civil, moral and political state of the district, "in order to bring a remedy," and to administer the oath of allegiance to all administrative and judicial bodies. Jean began operations without delay, and his report, which was kept in journal form and embraces the period from January 19 to September 10, 1793, is an interesting document; it covers fifty-one large foolscap pages, written now in a fine and again in a bold, regular hand, in the course of which his characteristic signature[65] occurs no less than twenty-two times, each section of the report having been signed as completed. In one section of this journal he wrote: "Our operations having been finished, we assembled around the tree of liberty, and there sang the hymn of the Marseillaise, which was interrupted with frequent shouts of '*Vive la république!*,' '*Vive la nation!*,' and more than one charge of musketry."

Figure 12. One of Jean Audubon's Signatures in His Report to the Directory, 1793. From the Original in the Archives of the Préfecture at Nantes.

Jean Audubon with eight others was charged with organizing the National Guard in the canton of Pellerin, and ordered to accompany the detachment that marched to the relief of Pornic, March 27, 1793. The Citizen was busy also in other directions. He said in his report:

> In virtue of the power conferred upon us by the Central Committee, on the ninth of April we were transported to the parish of Couëron, where we arrived at seven o'clock in the morning. Proclamations were posted both at Couëron and at Port Launay close by, while some were sent across the river to Pellerin. We availed ourselves on this occasion of the services of two officers of a corsair, who demanded that we aid in removing from Pellerin four cannon with four-pound balls, and we succeeded in putting to flight a small barque and four men, who an hour later returned with cannon.... The parish of Couëron appears very tranquil, and is in a better mood than [at first] seemed to us.

A little later Jean proceeded to Paimbœuf on a similar errand. His letters to the citizen-administrators of that commune are dated at Nantes on the seventeenth of April and the fourteenth of May; in one of these he refers to "the sum of four hundred francs" due from the

[65] During the Revolution Jean Audubon always added to his signature the cabalistic sign of three dots between parallel lines, which possibly stood for the three watchwords of the Republic—"*Liberté, Egalité, Fraternité*."

Administration "for one year's rent of my house in calle Rondineau (*à la calle rondino*), which you have taken for a *corps de garde*".

In July and August of this second year of the Republic, Citizen Audubon was sent to his native town of Les Sables d'Olonne to follow the movements of the loyalist generals Westermann and Boulart,[66] a mission which could hardly have been agreeable if, as seems to have been the case, some of his own people were loyal to the old *régime*. Correspondence by sea between Les Sables and Nantes, which was open before the siege, was not broken at this time, for the royalists had named one of their representatives, Benoit, as a delegate "to fraternize with the citizens of Nantes, to invite the authorities to correspond, and beg them to send food if they had more than they required." Four of Jean's letters, dated at Les Sables on the fifth and eighth of July and the sixth of August, besides one from La Rochelle on the fourteenth of July, all addressed to the Administration of the Loire inférieure, have been preserved.

In the manuscript records of the Department for 1793 is found also a notice of Jean's appointment as Special Commissioner, with a memorandum of all the money paid to reimburse him for the expenses of his numerous journeys. Thus, it is noted that he had been paid 145 francs for a service of twenty-nine days, which would represent the modest allowance of a dollar a day. Another item shows that he had received 100 francs for a tour of ten days; a note which was added to this item to explain the Directory's sanction for the payment of another forty-five francs and ten sous reads as follows: "by its order of the sixth of March last, the Council had, in effect, named Citizen Audubon as its Commissioner, to visit the coasts and to secure signatures, with full power to treat with all people, to acquire materials for the navy and other objects of his mission; if this mission did not prove successful, it was solely through force of circumstances, and not from any lack of zeal on his part."[67]

On the twenty-fifth of June, 1793, while engaged in duties to which we have just referred, Jean Audubon was appointed, with rank of ensign, to command the Republican lugger named the *Cerberus*.[68] During this charge, which lasted until the twenty-second of November of the following year, he fought one of the stiffest engagements of his career. On the twelfth of July he encountered the *Brilliant*, an English privateer of fourteen cannon which had captured an American ship laden with flour; and after a desperate battle which lasted three hours, in the course of which Jean was wounded in the left thigh, the Englishman, beaten and obliged to surrender his prize, was glad to escape under cover of night. Jean towed the American into the port of La Rochelle, and afterwards sent to the Administration a full account of the

[66] In the published orders and correspondence of the royalist General Boulart the following letter, given here in translation, is addressed to Citizen Audubon: "I give you notice, Citizen, that my aide-de-camp will arrive immediately from Niort. I beg you to do all in your power to come this evening to confer with me, since I have something to ask you of the utmost importance. I also inform you that there has arrived at Les Sables Citizen Anguis, the people's representative. Perhaps it would be more advantageous that you should see him this evening, and that tomorrow early we attempt to bring all three together. You could depart in the morning for Nantes." [Signed] "The General Boulart." Jean Audubon filed this letter from the enemy with his Department, but his answer is not given. See Ch. L. Chassin, *Etudes Documentaires sur La Révolution Française: La Vendée Patriote, 1793-1800*, vol. ii, p. 306, t. 1-4 (Paris, 1894-1895).

[67] *Délibérations-Arrêtés de Directoire du Département.* In MSS. pp. 107-108.

[68] Jean was actually in command of this war vessel in March of that year, as shown by a document given in full in Chapter 4.

engagement.[69] Ensign Audubon's next command was a dispatch boat called *L'Eveillé* ("The Awakened"), on which he served for nearly nine months, from November 23, 1794, to August 14, 1795. He was then detailed for port duty at La Rochelle from August 15, 1795, to January 24, 1797. His last ship was *L'Instituteur* ("The Institutor"), which he commanded with the rank of *lieutenant de vaisseau*, January 25 to October 3, 1797, while he was engaged in governmental business between the ports of La Rochelle and Brest.

The financial losses which Lieutenant Audubon sustained at Les Cayes in consequence of the revolution in Santo Domingo were a crushing blow to him; he never recovered his fortune, later estimated by his son-in-law at a sum which at that day would have been fabulous.[70] The business house in which he was interested failed; his plantations, refinery, houses and stores, the rents from which, as we have seen, in certain years after 1789, had yielded 90,000 francs, were presumably ravaged and partially destroyed. When the news of this misfortune reached him after 1792, his hands were tied by revolutions at home. Though he applied to his Government for relief, as undoubtedly did a host of other losers, he was eventually granted only a small indemnity, not exceeding 30,000 francs.

Friends of Jean Audubon at Nantes had made repeated demands of the Ministry of Marine that he be given a rank more in accord with his patriotism and efficient service to the State, and on October 11, 1797, he was commissioned lieutenant-commander (*lieutenant de vaisseau*),[71] one grade below that of captain. He held this rank for three years, during which he was engaged in vigilance service at Les Sables d'Olonne and in military duty at Rochefort, or until he was retired from the navy for disability, January 1, 1801 (*le 11 nivose, an 9*), at the age of fifty-seven.[72] He had served the State for over eight years, and his total period of active duty on sea and land when employed in the merchant marine and navy of France, as estimated from port to port, amounted to nineteen years, nine months and twelve days, while it had extended with interruptions over more than forty years.[73] After this long period of service, when, suffering from a pulmonary affection, he applied to his Government for a pension, he received the paltry annuity of 600 francs or $120.

[69] These records are on file in the archives of the Department of Marine at Paris, but access to them will doubtless be denied until peace is restored in Europe.

[70] M. L. Lavigne writes that he possesses a copy of a letter addressed by M. G. L. du Puigaudeau to a lawyer in Paris, in which it is stated that Lieutenant Audubon's losses amounted to 1,500,000 francs. After making due allowance for the psychological tendency to overestimate losses, especially when sustained in remote and romantic lands, the true amount was no doubt large.

[71] Or "lieutenant of a frigate," and corresponding to "mate" in the merchant marine.

[72] The certificate which Lieutenant Audubon received at the time of his discharge is preserved among the Lavigne manuscripts and documents at Couëron, and is headed:
Port de Rochefort.
ETAT des Services du Citoyen *Jean Audubon* natif des Sables d'Ollonne Département de La Vendée âgé de 58 ans.
It is signed by the Chief of Administration, Daniel, the Naval Commander-in-Chief of the District, Martin, and by the naval commissioner and clerk, February 26, 1801 (*le sept Ventose, an 9 de la République*).

[73] Jean Audubon was 11 years, 6 months and 25 days in the service of the merchant marine of France (*service au commerce*), in the course of which he rose to the rank of captain of the first grade in 1774. He served in the French navy (*service à l'état*) 8 years, 2 months and 17 days, ranking successively as sailor, ensign-commander, and lieutenant-commander (*lieutenant de vaisseau*); 8 months and 22 days of this period (1768-1769) were in intervals of peace, and 7 years, 5 months and 25 days (1793-1801), in times of war. Any conflict which may seem to occur in titles must be attributed to this double service.

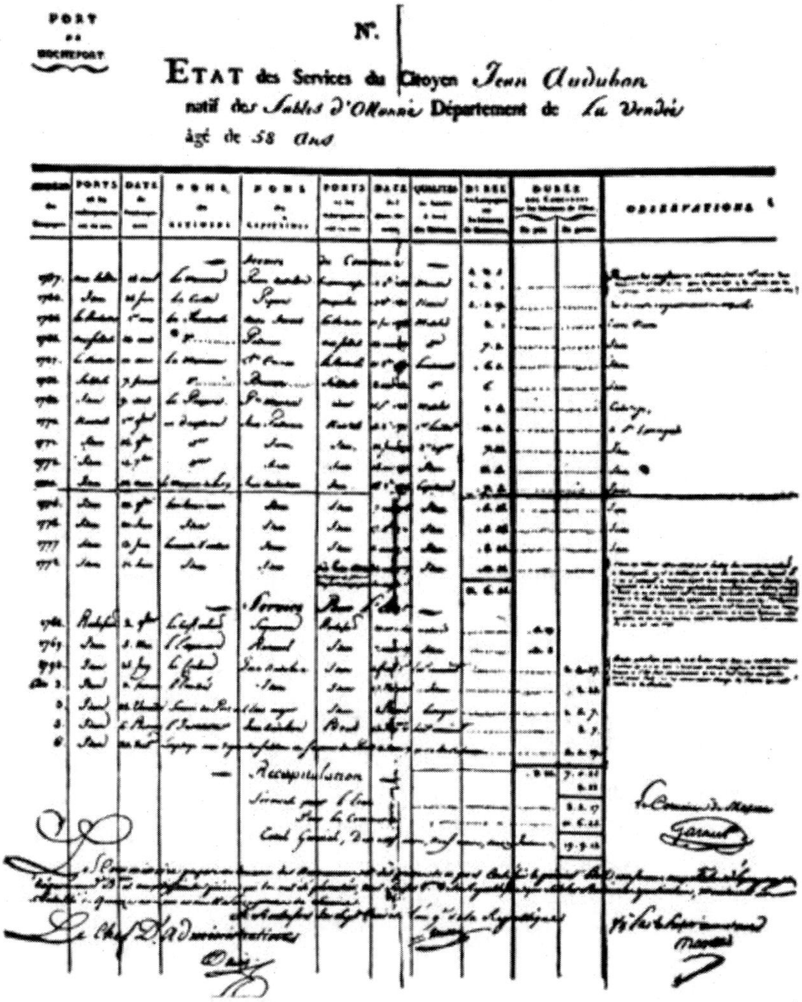

Figure 13. Certificate of Service of Lieutenant Jean Audubon, February 26, 1801.
From a Photograph of the Original in the Lavigne MSS.

With this modest pension and a property yielding an income not above $2,000 a year[74], Lieutenant Audubon retired to his quiet villa of "La Gerbetière," at Couëron, where he could indulge his taste for country life and for raising his favorite fruits and flowers; he is said to have kept some livestock, but could have been a farmer only on a modest scale. Meanwhile he continued to maintain a house, or at least rooms, at Nantes, whither he went periodically to conduct his correspondence and business affairs. The following letter of attorney, issued by Lieutenant Audubon a year after he had retired from the navy, shows that he still had interests in Santo Domingo, and was endeavoring to collect rents, long overdue, from houses and stores that belonged either to himself or to his clients. Whether through the dishonesty of agents or from what other cause, this property which the elder Audubon held in his own right seems gradually to have melted away:

[74] This property was evidently encumbered to a considerable extent, for he repeatedly filed with the Department letters for the removal of restrictions placed upon it (*lettres pour obtenir la main levée*). I cannot give the dates of these letters, but believe that they were drawn in 1801 or shortly after.

The 19th pluviose, in the eleventh year of the Republic, one and indivisible [January 7, 1802], before the public notaries of the department of Loire inférieure, who reside in Nantes and Doulon, the undersigned have seen present the citizen Jean Audubon, lieutenant of frigate, retired, and proprietor at Santo Domingo, aged 59 years, infirm and unable in consequence of his infirmities to go himself to attend to his business affairs in Santo Domingo, living in Rubens Street, in the Mocquard house,[75] No. 39, in the city and commune of Nantes, department of Loire inférieure:

Who has made and constituted for his general and special attorney Jean François Blanchard, merchant, and originally from the commune of Chateaubriand, department of Loire inférieure, living at the town of Les Cayes, in the southern section of the island of Santo Domingo, opposite Ile à Vaches, to whom he gives full and complete powers to revoke for him, and in his name, every preceding bill of attorney, for the purpose of managing the stores [*magazins*] at Les Cayes, in the southern part of Santo Domingo, opposite Ile à Vaches: To demand and obtain all accounts from the holders of said properties, who have had or still have charge of them there; to examine the said accounts, to debate, close up and stop them ... to lease the said properties, without the power of making any extensive repairs to them whatsoever, about which he had not informed the constituent in France, and that he has not authorized him there to do, at least by a special letter, it being understood that the actual tenant is obliged to make all the necessary repairs to the said houses and stores to the extent of 15,000 francs, and he should not use more than 4,000 francs yearly for the space of five years, counting from the month of thermidor, year 8 [July 19-August 17, 1800].

It is demanded of citizeness Fauveau, or of her assigns, to know the reason why she has failed, to the present moment, to pay to the constituent in France for the domicile of the citizeness Coyron,[76] the twelve thousand six hundred francs that she should annually pay to him, according to the act of July 15, 1788, as given by Domergue, notary at Les Cayes. You will satisfy them with the state of the dwelling house in the plain of Jacob, opposite Ile à Vaches.

This was sold by the said act to the said citizeness Fauveau and to her late husband by the said constituents, to whom he will report regularly on the state of affairs, at least twice in the year....

[Signed at Nantes] J. Royer [one of the undersigned notaries]

Lieutenant Jean Audubon died at Nantes,[77] when on a visit to that city, on February 19, 1818, at the age of seventy-four, "regretted most deservedly," said his son, "on account of his

---

[75] This house was rented at the time to Françoise Mocquard, but it is probable that Lieutenant Audubon had reserved rooms which were occupied during his visits to the city while his permanent home was at Couëron. In the power of attorney issued by Jean Audubon, his wife, and Claude François Rozier, at Nantes, April 4, 1806, the senior Audubon gave his residence as "rue Rubens, No. 39."

[76] Presumably a widow of one of the Coyrons (or Coironds), merchants at Nantes, whose business interests in Santo Domingo were entrusted to Jean Audubon's hands in 1783.

[77] The following extract from the registry of deaths at Nantes, which is here given in translation, indicates that Lieutenant Audubon passed away suddenly, since his death did not occur in his own apartment:

"In the year 1818, on the 19th day of February, at eleven o'clock in the morning, in the presence of the undersigned, deputies and officers of the civil service, delegates of Monsieur the Mayor of Nantes, have appeared the Messrs. Gabriel Loyen du Puigaudeau, gentleman of leisure, son-in-law of the deceased, residing hereafter at Couëron, and Francis Guillet, grocer, living on the Quai de la Fosse, of legal age, who have certified in our presence that on this day, at six o'clock in the morning, Jean Audubon, retired ship-captain,

simplicity, truth, and perfect sense of honesty"; "his manners," he continues, "were those of a most polished gentleman ... and his natural understanding had been carefully improved both by observation and by self education." Jean Audubon's means in France had been reduced partly by bad debts, for he seems to have been generous in lending money to his friends; Madame Audubon found herself greatly hampered by lack of ready money, although, as her son-in-law remarked, her hands were full of notes.

When Jean Audubon applied for nomination to the naval service of the Republic in 1793, we find a description of his previous life and habits recorded as a part of the information required by the Committee of Public Safety. The commune of Nantes at that time gave a flattering testimonial to his patriotism, in which he was described as an officer of merit, who had acquired through long experience at sea an extensive knowledge of navigation, who was a man of honor, and devoid of any inclination to vice or gambling; his nautical experience had been chiefly gained in American waters, the voyages of his choice being those to Santo Domingo and the United States.

At the age of forty-eight the elder Audubon thus briefly described himself: short in stature, measuring five feet, five inches; figure, oval; eyes, blue; nose and mouth, large; eyebrows, auburn; hair and beard turned gray. Contrary to the naturalist's expressed belief, there seems to have been little or no physical resemblance between father and son. At a corresponding age, John James Audubon, according partly to his own account, stood five feet, ten inches in stockings; his hair was dark brown; he had sunken, hazel eyes, flecked with brown, and of remarkable brightness; while his clean-cut profile showed an aquiline nose. "In temper," said the son, to continue the comparison, "we much resembled each other, being warm, irascible, and at times violent, but it was like the blast of a hurricane, dreadful for a time, when calm almost instantly returned."

Though passionate at times, Jean Audubon was a man of force and decision, as his career amply shows. If he does not loom large in the history of his time or was but little known beyond the limits of his province, it must be remembered that the time called forth thousands of the ablest men of his nation.

---

pensioner of the State, born at Les Sables d'Olonne, department of La Vendée, husband of Anne Moinet, died in the house of Mlle. Berthier, in the Chaussée de le Madeleine, No. 24, 4th Canton.

"The witnesses have signed with us the present act, after it was read to them. The deceased was 74 years of age."

{ Gabriel Loyen du Puigaudeau,
"Signed in the register:- { Guillet, and Joseph de la
{ Tullaye, deputy

The Audubons and Du Puigaudeaus were probably buried in one of the large cemeteries at Nantes, since no trace of their graves has been found at Couëron by M. Lavigne.

Chapter 6

School Days in France

It is now commonly believed that of the three great factors which mold character—environment, training and heritage, the last is the most important, since it alone is predetermined and unalterable. Environment may be uncertain or unsuitable, training defective or deferred, but blood is the one possession of which the child cannot be robbed; and since it sets the limits to possibility, in no small degree must it determine the acquisitions and accomplishments of a lifetime. This, however, is not the whole truth. Race may account for much, but it does not account for everything; the child is effectually robbed whenever it is not permitted to realize to the full upon its inheritance. To be able to convert possibilities into actualities it must receive fit training and right incentives, and if at critical times the proper spur is wanting, its patrimony may be sadly wasted. The "good environment" for the youth, too often thought to be the soft conditions of an easy life, is in truth that only which provides the proper and necessary stimulus. This may be now fear or pride, now hard necessity or bitter want; again, an awakened sense of responsibility or ambition to excel may be induced by concrete examples and fostered, as it often is, by lofty purposes and the uplift of a high ideal.

Audubon's life affords a striking proof of the power which environment can exert in awakening dormant capacity, in developing talents to their full and calling into use every force held in reserve. When we consider what his life work finally became, and what he eventually accomplished in a field for which he had no training, except in drawing, we find it easier to wonder at the man than to criticize him. With a formal schooling in France of the slenderest sort, in which the writing of his own language was never completely mastered, at eighteen he came to America and adopted a new tongue, which he first heard from the Quakers. Twenty years more were to elapse before he had a definite plan,—during which his environment was mainly that of a trader and storekeeper in the backwoods, never remote from the white man's frontier, hardly the soil one would seek for the development of budding talents in art, literature or science. Failure in trade was one of the spurs which started Audubon on his ultimate career, for it led to the immediate development of the talents which he possessed; the encouragement which he received from his wife was undoubtedly another. When he finally emerged, like a somewhat wild but well ripened fruit, at the age of forty, rich in experience, ready to absorb what from lack of earlier motives or opportunities he had failed to acquire, and with the determination to succeed, he won recognition as much through his personality and enthusiasm as by his extraordinary versatility and talents.

In an early sketch of his life Audubon said that his father had given both him and his sister an education appropriate to his purse; his teachers were possessed of agreeable talents, and he might have stored up much had not the continental wars in which France was then engaged forced him from school at an early age, when, much against his will, he entered the navy as midshipman, at Rochefort. This naval experience terminated, as he then recorded, in 1802, during the short peace between England and France; he was then seventeen years of age.[78] This was the year following his father's retirement, and the year previous to his first independent visit to the United States.

More details of this early period were given later, when the naturalist spoke with great affection of his foster mother, to whom his education had been mainly entrusted. "Let no one speak of her as my step-mother," said he; "I was ever to her as a son of her own flesh and blood, and she was to me a true mother." His every idle wish was gratified, he tells us, and his every whim indulged, in accordance with the notion that fine clothes and full pockets were all that were needed to make the gentleman: "She hid my faults, boasted to every one of my youthful merits, and, worse than all, said frequently in my presence, that I was the handsomest boy in France."

If Madame Audubon broke the prevailing tradition and by going to the other extreme did her best to spoil this affectionate boy, some allowance must be made for parental over-indulgence. In 1793, when the future naturalist was eight years old, the public buildings of his city had been converted into prisons and its streets were both unsanitary and unsafe, while in the following year, as we have seen, a mortal plague began to rob the prisons and the guillotine. Many had lost their all in the tempest that swept over them; many more had fled, and public schooling at Nantes must have been at a stand or disorganized for a considerable period.

Young Audubon could not have tasted much schooling before the outbreak of the Revolution, when he was seven years old, and but little after it, since this discipline practically terminated in 1802. His passionate love of nature, which was undoubtedly innate, was manifested at an early day. Living things of every description which he found by the banks of the Loire or along the stonewalls and hedgerows of Couëron gave him the greatest pleasure, but birds were his early favorites. These he soon began to depict with pencil and crayon, but to the dryer discipline of the school he ever turned with laggard feet.

When the versatile Lord Avebury, who became one of the greatest modern students of the powers of ants and other social insects, was four years old, his mother made this record in her diary: "His great delight is in insects. Butterflies, Caterpillars or Beetles are great treasures, and he is watching a large spider outside my window most anxiously." The same boy at eight, when writing home from school, added this postscript to a letter: "I am a favorite with most of the boys because I do not care about being laughed." The boy who has a good inheritance, follows his own bent, and does "not care about being laughed," may be on the road to success and with talents may achieve distinction. John James Audubon was one of those boys, although his path was never strewn with the roses that many have imagined.

---

[78] Audubon said that he was at the time fourteen years old, which could not have been the case, but when writing in 1835 he placed this experience at shortly before his return to America, which would have been in the winter of 1805-6; "I underwent," to quote this later account, "a mockery of an examination, and was received as a midshipman in the navy, went to Rochefort, was placed on board a man-of-war, and ran a short cruise. On my return, my father had in some way obtained passports for Rozier and me, and we sailed for New York."

The naturalist tells us that his father hoped that he would follow in his footsteps, or else become an engineer, and he saw that his son was instructed in the elements of mathematics, geography, fencing and music. But as Lieutenant Audubon was continually on the move, supervision in those matters fell to the over-indulgent stepmother, with the result that, instead of doing his duties at school, young Audubon took to the fields. Every night, he said, he would return with his lunch basket well laden with the spoils of the day—birds' nests, eggs, and curiosities of every sort destined for the museum into which his room had already been transformed. He was then in the "collecting stage," when that sense of possession dominates the heart of the boy, which, if well directed, can be turned to excellent account.

Lieutenant Audubon encouraged his son's taste for natural history and for drawing, but did not regard such accomplishments as a substitute for what he considered more serious subjects. He himself had suffered too much from lack of a formal education and was resolved to give his children the best opportunities within their reach. "Revolutions," he once remarked, according to his son, "were not confined to society, but could also take place in the lives of individuals," when they were all "too apt to lose in one day the fortune they had before possessed; but talents and knowledge, added to sound mental training, assisted by honest industry," could "never fail, nor be taken from any one when once the possessor of such valuable means."

When the elder Audubon returned from one of his periodic cruises, "my room," said the naturalist, "made quite a show," and the father complimented him on his good taste; but upon being questioned in regard to the progress made in his other studies, he could only hang his head in silence. His sister Rosa, on the contrary, who was also called to account, was warmly commended upon the improvement shown in her musical exercises. The next morning at dawn a carriage was drawn up before the Audubon door, and with the father and son, together with the latter's trunk and violin, was soon proceeding in the direction of Rochefort. The sailor had laid his plans and was about to execute them in his own way. Presently, said the son, his father drew forth a book and began to read, thus leaving him to his own resources. In this way they traveled for a number of days, not an unnecessary word being spoken during the entire journey, until the walls of Rochefort had been passed, and they alighted at the door of the father's house in that city. When they had entered, the naturalist continues, "my father bade me sit by his side, and taking one of my hands, calmly said to me: 'My beloved boy, thou art now safe. I have brought thee here that I may be able to pay constant attention to thy studies; thou shalt have ample time for pleasures, but the remainder *must* be employed with industry and care. This day is entirely thine, and as I must attend to my duties, if thou wishest to see the docks, the fine ships-of-war, and walk around the wall, thou mayest accompany me.'"

The youth accepted his father's proposal with good grace, and was presented to the officers whom they met, but he soon found that he was like a prisoner of war on parade. He was enrolled at once in the military school, where he was placed under the immediate care of Gabriel Loyen du Puigaudeau, his future brother-in-law. It was not long, however, before young Audubon gave his guardian the slip; he jumped from the window of his prison and made for the gardens of the Marine Secrétariat, but a corporal, whom he had recognized as a friend, suddenly nipped his plans in the bud; he was ordered, he said, aboard a pontoon, then lying in port, and there was obliged to remain until his father, who was absent at the time, finally released him, "not without a severe reprimand." The following record, written long after, is reminiscent of this period: "This day twenty-one years since I was at Rochefort in

France. I spent most of the day at copying letters of my father to the Minister of the Navy.... What has happened to me since would fill a volume.... This day, January first, 1821, I am on a keel boat going down to New Orleans, the poorest man on it."

Audubon's stay at Rochefort, the date of which is no doubt correctly given in the journal just quoted, was destined to be short. After a year he returned to Nantes, and later to "La Gerbetière," whereas before he spent all of his leisure in roaming the fields and looking for birds, their nests, their eggs and their young. At about this time, when fifteen years of age, Audubon began to make a collection of his original drawings of French birds, which was greatly extended in 1805 and 1806.

He has recorded that at the behest of his foster mother, who was an ardent Catholic, he was confirmed in that Church when "within a few months of being seventeen years old"; he was surprised and indifferent, but "took to the catechism, studied it and other matters pertaining to the ceremony, and all was performed to her liking." Since no record of this act has been found, it is probable that the ceremony in question was confused with that of his baptism, which, as we have noticed, occurred on October 23, 1800, six months before he attained his sixteenth birthday.

After having seen something of the character of Audubon's early training in France, it will not be surprising to find that when, at the age of forty-five, he first seriously began to write for publication and in English, which was not his mother tongue, he found himself handicapped in many ways. In after life he wrote that the only school which he had ever attended was that of Adversity, and that his tuition there had been of a prolonged and elaborate character. Though this statement was made under the stress of present feeling, it was not wholly devoid of truth.

Chapter 7

FIRST VISIT TO THE UNITED STATES, AND LIFE AT "MILL GROVE"

If there were ever a time when Lieutenant Audubon wished to see his son following the victorious eagles of Napoleon, whom he is said to have idolized, the hated conscription of that day, which was robbing every home in France of its best blood, might well have brought counsels of prudence. Little could the father have thought that by following other eagles of his own choice, his son was destined to add a far greater luster to the family name. Whatever may have turned the scale, in 1803 a decision was quickly reached, and the issue was fortunate for the future of natural science in America; it was decided that young Audubon should emigrate at once to the United States, with what end in view we shall soon see expressed in the sailor's own words. Accordingly, to his "intense and indescribable pleasure," the future naturalist, who had now passed his eighteenth birthday, eagerly prepared for the journey, the first of many that were later to become memorable in the annals of American science. No record of this voyage has been preserved, but from evidence derived from a variety of sources we can fix the time as the autumn of 1803.[79]

Audubon's introduction to the country of his adoption proved most inauspicious, for, to follow his account, when walking to Greenwich in Connecticut, some thirty miles from New York, to cash the letter of credit that his father had given him, he was seized with the yellow

[79] Audubon, writing in 1820, described himself at this time as "a young man of seventeen, sent to America to make money (for such was my father's wish), brought up in France in easy circumstances;" but in the same journal he said that he did not reach Philadelphia until three months after landing, and that "shortly after" his arrival at "Mill Grove" the Bakewell family moved to "Fatland Ford." Mr. G. W. Bakewell, the historian of his family, states that in the spring of 1804, William Bakewell, Audubon's future father-in-law, with his son, Thomas, traveled through Pennsylvania, Virginia and Maryland in search of a farm; they purchased "Fatland Ford," which was then the property of James Vaux. Audubon's account of the Pewee (*Ornithological Biography*, vol. ii, p. 124) shows that he was at "Mill Grove" before April 10, when "the ground was still partially covered with snow, and the air retained the piercing chill of winter." If these various statements are correct, they would indicate that Audubon left Nantes about the middle of November, 1803, and that he finally reached "Mill Grove" not far from the end of March, 1804. On the other hand, Mr. W. H. Wetherill, the present owner of "Mill Grove," informs me that his records indicate that the Bakewells occupied "Fatland Ford" in January, 1804. If this were the case, young Audubon could not have left France later than August, 1803. Too much weight, however, should not be attached to such references of a biographical character in Audubon's own writings; for in the account referred to above Audubon said that after his first visit to the United States he remained two years in France and returned to America "early in August;" while we know that his sojourn in France lasted but little more than a year and that he landed in New York on the 28th of May.

fever.[80] Fortunately at this critical moment his captain came to his aid, and placed him in the care of two Quaker ladies who kept a boarding house at Morristown in New Jersey. To the faithful ministrations of these kindly sisters the naturalist believed that he owed his life.

When Jean Audubon finally left the United States not far from the beginning of 1790, he placed his business interests in America in charge of an agent, named Miers Fisher, "a rich and honest Quaker of Philadelphia," and to the hands of this trustworthy man he now confided his son. Accordingly, when young Audubon had been nursed back to health, word was sent to his father's friend, who came in his carriage and drove the lad to his own home in the outskirts of Philadelphia. To follow the account which the naturalist gave, when writing of this visit a quarter of a century later, his host, finding his charge to be a comely youth, and having a daughter "of no mean appearance," proposed that he should remain with them and become one of the family. Audubon seems to have suspected that this was a premeditated scheme to entangle him in marriage, and as he had no liking for the severity of Quaker manners, determined to make his escape. This, he said, was finally accomplished by appealing to his own rights and to the honest Quaker's sense of duty in seeing him established on the estate which his father had designed for him.

Though effective for the time, as will presently appear, this appeal was quite fanciful, for Jean Audubon's ideas concerning the future of his son were of a more practical character, and he had no intention at this time of establishing him at "Mill Grove," which was soon to be sold. The friend to whom the following letter was addressed is implored to aid in finding a good American family in which his son could acquire the English language as a step to entering trade:[81]

> This will be handed to you by my son, to whom, I request you will render every service in your power, wishing that you should join Mr. Miers Fisher to procure him a good and healthy place where he might learn English. I come to point out to you Morristown, and look for a good and decent family in that place to recommend him to her as your own Son. This service from you will deserve my everlasting gratitude. I am Sir, with consideration.
>
> Yr Mo ob Ser—.

Mr. Miers Fisher, who evidently received a copy of this letter, no doubt considered his own family as good as the best, and in detaining young Audubon at his home, we must credit him with the desire of following the instructions thus received.

"Mill Grove," which was finally reached in the spring of 1804,[82] was a new-found paradise to the young naturalist. Here, however, he was destined to spend but little over a year, though it was doubtless the happiest year of his life. The farm was then conducted by a

[80] A plague of genuine yellow fever had visited New York in 1795, but in 1804 and 1805 the city suffered from a malignant fever of another type, and to such an extent that 27,000 persons, or one-third of the entire population, are said to have fled to escape the pestilence. This was possibly the malady which seized young Audubon not far from the beginning of the former year.

[81] The rough draft of a letter in English, evidently written by Lieutenant Audubon to be delivered by his son to the ship's captain, and probably in duplicate to his agent, Miers Fisher, but bearing no name or date, (Lavigne MSS.)

[82] See Vol. I.

Quaker, named William Thomas, who was installed as tenant with his wife and family. It was arranged, said Audubon, that he should receive from them a quarterly allowance in ready money, in a sum that "was considered sufficient for the expenditure of a young gentleman."[83]

Well might any youth fond of wild life in the country have fallen in love with this secluded spot, the beauty and charm of which are suddenly revealed to the visitor of today as he approaches it from the old Philadelphia road. Standing high on the rugged banks of the Perkioming Creek, which empties into the Schuylkill River just below this point, the old house, facing west, commands a wide and diversified scene, extending from the living waters below, over bottom lands of the valley, to the dim, undulating lines of the Reading hills in the farther distance. This old landmark[84] of Colonial times remains today in perfect preservation, thanks to the never-failing care and interest of the present owner,[85] who has done all in his power to maintain its historic associations, and to keep the memory of the naturalist green in one of the few spots in America where material landmarks of his career have not been completely effaced. The place has had an interesting history, and though Audubon's occupancy was brief, it affected, as we shall see, his whole after-life.

Audubon thought nothing of walking to and from Philadelphia when no conveyance was at hand, but to-day the railroad brings the traveler within a mile and a half of his old farm. Not far to the south, beyond the present railway station of Protectory, lies Valley Forge and the wooded hills where Washington's ragged veterans passed in log huts the ever memorable winter of 1777-8. Audubon fancied that his father had made the acquaintance of General Washington at that date, but this was eleven years before the place had come into the possession of his family, and at that time Captain Audubon was sailing the seas. Equally fanciful also was the idea that his mother had once lived there, which he expressed in a letter written from New York on February 10, 1842, to young Spencer F. Baird, at Carlisle, Pennsylvania. The naturalist was assuring his young friend that the slow but beautiful "Little Carlisle" was to be preferred to "Great New York, with all its humbug, rascality, and immorality," and added: "It is now a good long time since I was young, and resided near Norristown in Pennsylvania. It was then and is now a very indifferent place as compared with New York; but still my heart and my mind oftentime dwell in the pleasure that I felt there, and it always reminds me that within a few miles of that village, my Mother did live."

---

[83] The yearly rent of "Mill Grove" in 1804, according to the accounts of Francis Dacosta, who had then acquired a half interest in it, amounted to $353.34.

[84] "Mill Grove" farm is in Montgomery County, twenty-four miles northwest of Philadelphia, in the town known, after 1823, as Shannonville, but in 1899 rechristened "Audubon;" Norristown is five miles to the east.

[85] Mr. William H. Wetherill of Philadelphia, whose hospitality I have enjoyed and to whom I am indebted for many interesting facts and records pertaining to "Mill Grove." Samuel Wetherill, Mr. W. H. Wetherill's grandfather, was one of the first to bring "black rock," or coal, from Reading to Philadelphia. Samuel Wetherill, Junior, who is said to have started the first woolen mill in the country and to have produced the first white lead made in the United States, purchased "Mill Grove" for the sake of its minerals in 1813, the war having put a stop to all importations from England at that time. He actually succeeded in extracting several hundred tons of lead from the "Mill Grove" mines, doing better, it is thought, than any who preceded or followed him. Samuel Wetherill, Junior, died in 1829, and was succeeded in the lead and drugs industry by his four sons, of whom Samuel Price Wetherill became the owner of "Mill Grove" in 1833. The farm remained in the hands of the Wetherill family until 1876, and returned to them again, when the present owner came into possession, in 1892.

Figure 14. "Mill Grove" in 1835, Showing The Mills on the Bank of Perkioming Creek, the Farmhouse, and the Old Smelting Works (Built by Samuel Wetherill), Then in Disuse. After a Water-Color Painting by Charles Wetherill, Son of Samuel Wetherill, and Uncle of William H. Wetherill, the Present Owner of the Estate.

Figure 15. "Mill Grove" as It Appears to-Day. The above from Photograph by, and This Published by Courtesy of, Mr. W. H. Wetherill.

The soil of this farm region is of a dark red color, owing to a friable shale which outcrops everywhere. The high, wooded bank of the Perkioming abounds in caves, scooped out by the hand of nature or man, as well as in great pits and shafts, for deep down under its shale, "Mill

Grove" was rich in minerals, particularly the sulphide of lead, associated with copper and zinc, to reach which many excavations have been made. The lead mines of this farm are said to have been famous in Revolutionary times, and have been worked sporadically for a hundred years; if traditions are trustworthy, many a winged bullet that laid a Red-coat low in the War of Independence was a messenger from "Mill Grove." In some of the old conveyances, which go back to the time of Penn, the place was commonly designated as the "Mill Grove Mines Farm." It is recorded that the original tract of two thousand acres, extending from the Schuylkill to the Perkioming as far as the mouth of Skippack Creek, was sold to Tobias Collett by William Penn in 1699 for fifteen shillings. We shall soon see that the mineral wealth which "Mill Grove" was supposed to hide beneath its rugged slopes was a source of no little trouble to the Audubons, the Roziers, and their successors for many a year.

At the foot of the declivity towards the west, half hidden by foliage, stood a picturesque stone mill, at a point where a solid rampart had been thrown across the stream to divert its power to the use of man. Hard by was the miller's house, which antedates the mansion, and which was built and first occupied by James Morgan, who came into possession of the property in 1749. It was this old mill site, originally distinct from the farm, that gave the name to the place. Behind the gristmill an extensive sawmill, built over the mill race, was also in operation. Today the dam is broken through, and the great mill wheel of wood and iron, twelve feet in diameter and fifteen feet wide, has come to rest after turning for more than a century.

Like the mill, the original house on the hilltop was built of rough-hewn native stone, which is brown or red and very hard. It consists of two stories, with central hall, and a curiously divided attic with dormer windows, which Audubon is said to have converted into a museum. A marble slab in the south gable bears the date of 1762; an addition of the same rough stone was built on the north side, but at a considerably lower level, in 1763, and the commemorative tablet in this instance bears the initials "J. M.," proving that the construction of the buildings of "Mill Grove" was due to the old miller, James Morgan. The interior, with its odd chimney-corner, low ceilings, bold fireplace and hand-wrought iron-work, bears witness to a time when honest, substantial construction and pride in workmanship received the first consideration. The present owner of "Mill Grove" has added attractive porches at the front and back. Ampelopsis climbs over the walls, which are shaded by handsome trees; one of these, a fine black walnut at the easterly porch, which in August bore its great green balls in full clusters, must have been vigorous in Audubon's day, and possibly suggested the introduction of sprays of this full-fruited tree into some of his plates.

While on a visit from Santo Domingo in 1789, concerned with his business interests, Captain Audubon spent some time in Philadelphia. On March 28, 1789, he purchased the "Mill Grove" property, at that time consisting of 284½ acres of land, mansion house, mill, barns, furniture, tools and livestock, from Henry Augustin Prevost[86] and his wife, for the sum of 2,300 English pounds, in gold and silver. He never lived there, and that he never intended to make it his immediate residence is shown by the fact that in less than a fortnight he leased

[86] In 1761 James Morgan, the first miller and builder, conveyed one-half of the mill site of five acres to Roland Evans, who came into possession of the other half, with the adjoining farm, in 1771; the property was sold to Governor John Penn in 1776; it passed to Samuel C. Morris in 1784, and to the Prevosts in 1786.

the farm in its entirety, as already noticed, to its former owner, and gave him a mortgage which stood for seventeen years.[87]

Young Audubon lived at "Mill Grove" from the winter of 1804 to the spring of 1805, and again for a few months in the summer of 1806, the year of its final sale by the Audubons and Roziers. In his journal of 1820 the naturalist wrote that his father had once the honor of being presented to General Washington, and also to Major Crogan, of Kentucky, "who was particularly well acquainted with him." Jean Audubon left at "Mill Grove" oil portraits of himself and of Washington, both by an inferior American artist named Polk,[88] and it is probable that the one of himself was painted while he was at Philadelphia in the spring of 1789; the drawing is hard and flat, but the appearance of the face clearly indicates a man past middle life, and Captain Audubon had then reached his forty-fifth year.

Young Audubon, we may be sure, lost no time in exploring the resources of this fine estate, where every bird, tree and flower came to him as a new discovery. In following the Perkioming above the mill dam he found a cave, carved out of the rocks, as he thought, by nature's own hand, which was a favorite haunt of the unpretentious but friendly pewees, the first American birds to attract his serious attention. So delighted was the youthful naturalist that he decided to make the pewees' cave his study; thither accordingly he brought his books, pencils and paper, and there made his first studies of American bird life, in the spring of 1804, in the third year of the presidency of Thomas Jefferson. It was early in the season when Audubon chanced upon this quiet retreat; the buds were swelling and maples had already burst into bloom, but snow still lingered in patches through the woods, and the air was piercing chill. The pewees were not yet at home, but one of their nests, fashioned of mud and finest moss, was fixed above the vaulted entrance; their coming was not long delayed, and Audubon, marking the very night or day's dawn when the first pewee arrived, saw them beginning to restore their old home on the tenth of April.

Strange to say, almost at that very time another pioneer in American ornithology, Alexander Wilson, who will enter this history later, was teaching a rough country school at Gray's Ferry, Kingsessing, also on the Schuylkill, and not over twenty-five miles away. Though Audubon's early studies were very desultory, both naturalists began their observations at about the same time, for on June 1, 1803, Wilson wrote to a friend that many pursuits had engaged his attention since leaving Scotland in 1794, and that then he was "about to make a collection of all our finest birds."

---

[87] The lease, which was drawn up in English, April 10, 1789, reads in part as follows: "This indenture, made on the tenth Day of April in the Year of our Lord, One thousand Seven hundred & Eighty nine, Between John Audubon, of the Island of St. Domingo, Gentleman, now being in the City of Philadelphia, of the one party, and Augustine Prevost...." The lease included the messuages, grist mills, saw mills, plantation and tract of land, which is described, tools, implements, stock, and furniture of the mills and farm, and was drawn for one year; it was signed in the presence of Miers Fisher, agent and attorney for Jean Audubon.

In the inventory were included one windmill, one pair of scales, with weights of 56, 28 and 7 pounds, "skreen," four bolting cloths, two hoisting tubs, and one large screw and circle for raising the millstones. This lease was renewed in October, 1790, when Jean Audubon, who was then living at Nantes, agreed to keep the house in good repair from that time onward. It was the Prevost mortgage that Miers Fisher paid but forgot to cancel; it was finally cleared up by Dacosta in October, 1806.

Miers Fisher's Philadelphia residence, called "Ury," which Audubon often visited, was near Fox Chase, now in the Twenty-third Ward. See Witmer Stone, *Cassinia*, No. xvii (Philadelphia, 1913).

[88] For a photograph of this portrait of Lieutenant Audubon here reproduced, I am indebted to Miss Maria R. Audubon; the originals of both portraits are now in possession of Audubon's granddaughter, Mrs. Morris F. Tyler.

It must be set down to Audubon's credit that in the little cave on the banks of the Perkioming, in April, 1804, he made the first "banding" experiment on the young of an American wild bird. Little could he or anyone else then have thought that one hundred years later a Bird Banding Society would be formed in America to repeat his test on a much wider scale, in order to gather exact data upon the movements of individuals of all migratory species in every part of the continent. After a few trials, "I fixed," said he, "a light silver thread on the leg of each, loose enough not to hurt the part, but so fastened that no exertions of theirs could remove it." In the following spring he had the satisfaction of catching several pewees on their nests farther up the creek, and of "finding that two of them had a little ring on the leg," proving that the young of a migratory bird, steering by the "compass" which is carried in its brain, did sometimes return to its home region, if not to the actual cradle or home site.

Across the Philadelphia road, which today leads to the little railway station, and not more than a quarter of a mile from Audubon's farmhouse, stood another but more pretentious mansion of the Colonial era, called "Fatland Ford," pertaining to an extensive farm of that name which was noted for the fertility of its alluvial acres. A road from the present village of Audubon to the Schuylkill River and the ford runs through the "Fatlands of Egypt," as the most productive parts of this old farm were then called. From the house could be seen the camping grounds of the Revolutionary soldiers, and James Vaux, its owner and builder, is said to have entertained General Howe at breakfast and to have shown him the room which General Washington, his guest of the previous day, had left just in time to avoid an introduction.

Shortly before Audubon reached "Mill Grove," William Bakewell, an Englishman who had emigrated to New Haven in 1802, bought this farm, and with his wife and family took possession in the winter or spring of 1804.[89] Of the six Bakewell children, the two eldest, Lucy Green and Thomas Woodhouse, were but three years younger than the naturalist. The senior Bakewell, said Audubon, called at "Mill Grove" to pay his respects, but being then from home, and having brought with him a Frenchman's dislike for everything English, he failed to respond. In the autumn, however, when grouse had become plentiful in the woods, a chance meeting brought them together, and young Audubon, who was a great admirer of his neighbor's expert marksmanship and well trained dogs, duly apologized for his neglect and forthwith paid a visit to "Fatland Ford."

We shall let the naturalist tell in his own words of his first meeting with the young woman who afterwards became his wife:

> Well do I recollect the morning, and may it please God that I may never forget it, when for the first time I entered Mr. Bakewell's dwelling. It happened that he was absent from home, and I was shown into a parlor where only one young lady was snugly seated at her work by the fire. She rose on my entrance, offered me a seat, and assured me of the gratification her father would feel on his return, which, she added, would be in a few moments as she would despatch a servant for him. Other ruddy cheeks and bright eyes made their transient appearance but, like spirits gay, soon vanished from my sight; and there I sat, my gaze riveted, as it were, on the young girl before me, who, half working,

[89] See Vol. I.

half talking, essayed to make the time pleasant to me. Oh! may God bless her! It was she, my dear sons, who afterward became my beloved wife, and your mother.

When Mr. Bakewell returned, his daughter, Lucy, presided at the tea that was served, and Audubon received his first experience of hospitality in the English style, that was to be repeated in Britain at a later day on a more lavish scale. A hunting expedition was arranged and the men started out at once. Festivities of various sorts, and, later, skating parties, became the order of the day, and it was not long before hospitalities were exchanged, when Audubon, having secured, with the aid of his tenant's son, as many partridge as possible, had the whole Bakewell family to dinner under his roof at "Mill Grove."

Audubon's choice of a wife, thus quickly made, marked a turning-point in his career, and the curious fact remains that while he might have ransacked the country from Florida to Maine, as he afterwards repeatedly did in his search after birds, and woefully blundered, the woman who by her sterling qualities of mind and heart was the one to recognize and call forth the best that was in him, should have been placed by circumstances close by his door. Whatever the world has ever owed to Audubon is a debt due to Lucy Bakewell, for every leaf of oak that is plaited for his brow, another of lavender should be twined for hers.

During this gay but brief period of his life, Audubon has described himself as inordinately fond of dress, often cutting, as he said, an absurd figure by shooting in black satin breeches and silk stockings, and wearing the best shirts which the Philadelphia market could afford; he took pride, he adds, in riding the best horse that he could procure, and in having his guns and fishing tackle of the most expensive and ornate description. "Not a ball," he said, "a skating match, a house or riding party took place without me."

Figure 16. "Mill Grove" Farmhouse, West Front, Facing Perkioming Creek.

Figure 17. "Fatland Ford," The Childhood Home of Lucy Bakewell Audubon. This and the above after Photographs of August 16, 1914.

While freely acknowledging his follies at this time, he was able to say that he was addicted to no vices. His usual custom was to rise with the dawn, when his bird studies would begin, in the early hours which are best for this purpose. According to his own account, Audubon was extremely abstemious in his youth, for he declared that he had lived on fruits, vegetables and milk, with only an occasional indulgence in game and fish, and that he had not swallowed a single glass of wine or spirits until his wedding day. This was the more remarkable in a youth coming from a country which flowed with good wine, where school children are still served with watered wine for lunch, and where the cooks, as Goldsmith believed, could concoct seven different dishes out of a nettle-top, and who, if they had enough butcher's meat (a want that has since been abundantly supplied), would be the best purveyors in the world. Audubon attributed his iron constitution to this simple regimen, which had been followed, he said, from his earliest recollection, though he admitted that while in France it was extremely annoying to all about him; for this reason he would not dine out when his peculiar habits were likely to be the subject of unpleasant comment. To follow this account of himself:

> Pies, puddings, eggs, milk and cream, was all I cared for in the way of food, and many a time I have robbed my tenant's wife, Mrs. Thomas, of the cream intended to make butter for the Philadelphia market.... All this time I was as fair and rosy as a girl, though as strong, indeed stronger than most young men ... and why have I thought a thousand times, should I not have kept to that delicious mode of living, and why should not mankind in general be more abstemious than mankind is?[90]

William Gifford Bakewell, a younger brother of Lucy, has left this interesting record of a visit paid to "Mill Grove" in the summer of 1806:

[90] For this and the preceding quotation, see Maria R. Audubon, *Audubon and his Journals*, vol. i, pp. 18 and 27.

Audubon took me to his house where he and his companion, Rozier, resided, with Mrs. Thomas, for an attendant. On entering his room, I was astonished and delighted to find that it was turned into a museum. The walls were festooned with all kinds of birds' eggs, carefully blown out and strung on a thread. The chimney-piece was covered with stuffed squirrels, racoons, and opossums; and the shelves around were likewise crowded with specimens, among which were fishes, frogs, snakes, lizards, and other reptiles. Besides these stuffed varieties, many paintings were arrayed on the walls, chiefly of birds. He had great skill in stuffing and preserving animals of all sorts. He had also a trick in training dogs with great perfection, of which art his famous dog, Zephyr, was a wonderful example. He was an admirable marksman, an expert swimmer, a clever rider, possessed of great activity, prodigious strength, and was notable for the elegance of his figure and the beauty of his features, and he aided nature by a careful attendance to his dress. Besides other accomplishments he was musical, a good fencer, danced well, and had some acquaintance with legerdemain tricks, worked in hair, and could plait willow baskets.

Chapter 8

DACOSTA AND THE "MILL GROVE" MINE

If young Audubon was playing the *rôle* of a prodigal son at the "Mill Grove" farm, which in a certain sense was doubtless true, an episode soon occurred which put a check to his carefree existence. Not long after the naturalist had arrived, William Thomas, the tenant, called his attention to the lead-ore deposits, which he thought had been discovered by a Mr. Gilpin in 1791, and the news of this prospect was promptly communicated to the elder Audubon in France. Though the presence of this mineral at "Mill Grove" had been known, as we have seen, at a much earlier day, its rediscovery excited great interest, and may have been a factor of influence in the steps which were soon to be taken. It should be noticed, however, that before May, 1803, a young Frenchman from Nantes, bearing the Portuguese name of Francis Dacosta, had preceded young Audubon to "Mill Grove," and apparently had acquired at that time a certain interest in the farm.[91] Dacosta soon succeeded Miers Fisher as Jean Audubon's agent, and becoming enthusiastic over the lead mine, was anxious to exploit it. Acting also upon the senior Audubon's request, he assumed a sort of guardianship over the son.

Dacosta began to dig for ore in the following year. News of his enterprise spread rapidly, and this long neglected mine was heralded in the newspapers as "one of the first discoveries yet made in the United States."[92] On December 15, 1804, Dacosta purchased a one-half

---

[91] In Dacosta's final statement of his account, which was disputed, carried into court, and eventually settled by arbitration at Philadelphia, on August 1, 1807, these items occur: "Omitted, $300.00, paid by Francis Dacosta to Miers Fisher, on May 24, 1803;" and "Ditto $176.67, the proportion of Francis Dacosta in the rent of the first year, which has not been paid to him." (MSS. in possession of Mr. Welton A. Rosier.)
It seems probable that Dacosta was sent to this country by Lieutenant Audubon to act as his agent for the disposition of "Mill Grove," and to succeed Miers Fisher in the conduct of his business affairs. Interest in the neglected and forgotten mine may have diverted them from their original plans.

[92] The following notice, copied from *Relf's Gazette*, appeared in the *New York Herald* for Saturday, November 17, 1804:
"The lead mine discovered on Perkiomen creek, in Montgomery county, Pennsylvania, the property of Francis Dacosta, has been lately opened, and attended with great success. The vein proves to be a regular one, and of long continuance. Its course is N.N.E.; its direction is nearly perpendicular, and its thickness from one foot to 15 inches. Two tons of that beautiful ore were raised in a few hours, and one ton more at least was left in the bottom on the pit, which is yet but nine feet deep. From the situation of this mine, its nearness to navigation and market, its very commanding height, its richness in metal, and the large scale it forms on; it is thought by judges to be one of the first discoveries yet made in the U. S.
"From the analysis made of 100 parts, it contains:

undivided interest in "Mill Grove,"[93] giving, as we believe, a mortgage, and hoping to pay for his share out of the profits of the lead mine. Thereafter for about two years he continued to conduct the farm and develop the mine, upon the basis of a one-half interest, in addition to a small salary.[94] In case the mine proved a success, it was understood that young Audubon was to be taken into the business and thus obtain a means of self-support.

Dacosta was at first averse to forming a company, but the Quaker tenant, William Thomas, who caught the fever, and who was thought to possess more knowledge of the mine than he was ready to divulge, seems to have been taken conditionally into the partnership. Dacosta made full reports of his progress to the old sailor at Couëron, who came regularly to Nantes to send back to America his well considered answers and candid advice. Dacosta also called persistently for money, but as Lieutenant Audubon was unable to meet these demands, he applied to his friend François Rozier, a wealthy merchant at Nantes, to supply the needed capital. Rozier invested 16,000 francs, and to complicate matters took a mortgage upon one-half of the value of "Mill Grove," in which the earlier proprietor, John Augustin Prevost, as well as Francis Dacosta, was also interested. Jean Audubon, Dacosta and Rozier thus became partners in an enterprise which seems to have swallowed up all of the money which was advanced and never to have made any substantial returns.

The eventual failure of the lead mine must be attributed in part to the high cost of materials, as well as to the expense involved in uncovering the ore, a difficulty which all later exploiters seem to have found insuperable. Dacosta also discovered that the management of his youthful charge was quite as difficult as making a success of the mine. His grievances on this score were duly reported at Couëron, and if he was really trying to carry out the instructions which came from France, it was perhaps no wonder that he received the undisguised contempt of his rebellious pupil. How just the naturalist's charges against his hated tutor may have been, will be considered in the sequel, but Lieutenant Audubon's letters,[95] to be given presently at length, clearly show that in spite of the strained relations which later ensued, Dacosta continued to enjoy his confidence for some time after young Audubon's return to France in 1805. The more serious troubles that followed seem to have arisen from entanglements into which all were later drawn.

In the first two letters to be given, but the third and fourth of the series, Jean Audubon refers particularly to "Mill Grove" and the prospective mine, and to the proposed marriage of

Oxyd of lead	85
Oxyd of iron	1
Sulphuric acid	13
Water	1
	100

"The lead being coupelled, has proved to contain 2½ oz. fine silver to 100, which is nearly 3 dollars worth of that metal."

93 For the sum of 31,000 francs, or $6,200, a slight advance on the cost to Jean Audubon, when he had taken over the farm fifteen years before.

94 The following item appears in Dacosta's final account: "To compensation claimed by Francis Dacosta for making up half of his expenses, in managing the mining works, the mill-repairs, and taking up the formation of a company during two years of constant cares, troubles, and loss of time, at 300 dollars a year—$600.00." (From statement of disputed claim)

95 For copies of a part of the Audubon-Dacosta correspondence, which is perhaps half of what exists but all that it was possible to obtain, I am indebted to Monsieur Lavigne. The first letter, the present copy of which is incomplete, was evidently written in the winter of 1804-5. Lieutenant Audubon, who at this time was sixty-one years old, was living at Couëron, but came to Nantes to conduct his correspondence. All the letters were carefully transcribed in a separate copybook, and are here translated as literally as possible from the French.

his son to Lucy Bakewell, concerning which he was reluctant to give his consent for reasons which he specifies at length; his sanction was in fact withheld until the young man was on the road to self-support two years later.

Jean Audubon to Francis Dacosta

[Nantes, 1804-5]

I told you to sell to W. Thomas the portion on the other side ... but your letter of the 27th of September with that of Mr. Miers Fisher, who is not in favor of it, has made me change my mind in the meantime. If your plan succeeds, as I wish it may, this part of the farm would become almost indispensable for exploitation [of the mine]. Moreover, has not Mr. W. Thomas intentions, which we do not know? Might it not be possible that in this very same part he had made more valuable discoveries than those which he has shown us? In all these matters, however, I rely entirely on the wisdom of Mr. Miers Fisher and of yourself, and I thank you for your willingness to remain in charge of my affairs,[96] by accepting anew the power of attorney, which he sends me together with the indenture to be signed by my wife and by myself in presence of witnesses. But you ask that this should be done before the mayor of Nantes, while we have been living, since you departed, in the commune of Couëron; accordingly this will be taken before the mayor of that commune, and legalized by a prefect of the department. That, I believe, will fulfil the same obligations, for should it be necessary for my wife to come to Nantes in the weather that we are constantly having it might cause a delay that would be prejudicial to us. Remember, my dear Sir, I expect that if your plan succeeds, my son will find a place in the works, which will enable him to provide for himself, in order to spare me from expenses that I can, with difficulty, support. Your first letters have almost persuaded me that this so-called mine was of little or no account, but the arrangement that you have made with W. Thomas is so important that I do not doubt you made certain of the value of the object before deciding to grant him a recompense, which was to be only in the thing itself. In this work we should then be making a very great sacrifice, and it would be a loss. If, however, you propose to forestall the payment of the sums that you owe, I accept [the proposition] to be paid in Philadelphia; I will reflect upon it, and will look into it. If I can arrange matters for this [plan] with Mr. Dupuir, my next will be more explicit upon this subject. My son speaks to me about his marriage. If you would have the kindness to inform me about his intended, as well as about her parents, their manners, their conduct, their means, and why they are in that country, whether it was in consequence of misfortune that they left Europe, you will be doing me a signal service, and I beg you, moreover, to oppose this marriage until I may give my consent to it. Tell these good people that my son is not at all rich, and that I can give him nothing if he marries in this condition.

Jean Audubon to Francis Dacosta

Nantes, le 19 ventose, an 13 9 March, 1805

---

[96] That is, after having become a part owner of the "Mill Grove" property.

Mr. Dacosta, Philadelphia:

I have received at this very moment your duplicate of the twelfth of November, and your letter of December fifth, which is not so favorable for several reasons as the one preceding it, yet this impels us to hope that your last tunnel will not be a deserter, and that the oxides of iron which are present will not vanish upon further digging; this, at least, is my hope. You do well to make every effort to obtain associates. If this does not succeed, and if you should wish to work for our interests, I should always approve of everything that you do, since you have my confidence. In this case I believe ... that you should make the most urgent repairs, above all at the principal house, before going there to live. As to Mr. W. Thomas, you do well to keep him for yourself for every reason that you give me, and I believe that he will not be stubborn about withdrawing until he has, or has not, deserved his reward.

I am [vexed] Sir; one cannot be more vexed at the fact that you should have reason to complain about the conduct of my son, for the whole thing, when well considered, is due only to bad advice, and lack of experience; they have goaded his self-esteem, and perhaps he has been immature enough to boast in the house to which he goes, that this plantation should fall to him, to him alone. You have every means to destroy this presumption; it is known at Philadelphia that you have the same rights as I have, and that you are doing nothing but for our mutual advantage. I am writing to him on this subject, for he does not speak of it to me, and I am giving him the rebuke that his indiscretion deserves. Read this letter, and have the kindness to seal it before delivering it to him. You tell me that I can refer, in regard to his conduct, to the report that Mr. Miers Fisher has given of it in his long letter of the month of September; that, unhappily, I have not received, for Mr. Fisher tells me nothing about him, neither what is good nor bad. As to going to that country, this seems well nigh impossible; to recall my son is not easier; the reasons which made me send him out [there] still remain. Only an instant is needed to make him change from bad to good; his extreme youth and his petulance are his only faults, and if you have the goodness to give him the indispensable, he will soon feel the necessity of making friends with you, and he can be of great service if you use him for your own benefit.

It is necessary then, my dear Sir, that we endeavor, by gentleness, to reclaim him to his duty. If you are indulgent with him, it will be I who should be under every obligation to you. I hope that the enclosed letter will work a change with him. This is my only son, my heir, and I am old. When Mr. Miers Fisher shall have shown my letter to the would-be father-in-law, he will see that he is mistaken in his calculation upon the assumed marriage of his daughter, for if it should take place without my consent, all help on my part would cease from that instant; this, if you will have the kindness, is what you may say to the would-be father-in-law, that I do not wish my son to marry so young.

Your letters of the 28th of October and the 12th of November are in the country.[97] I cannot reply categorically upon their contents; I will examine them, and will tell you in my next what I think about them. Your family, which I have seen, is well. Our ladies thank you for your kind remembrance. I am....

When the preceding letter was written young Audubon was on his way to France, to protest, as he said, against Dacosta's treatment of him. At the date of the letter which follows, he was at Couëron, hunting birds with Dr. d'Orbigny.

---

[97] That is, at Couëron.

Jean Audubon to Francis Dacosta

Nantes, 14 June, 1805

To Mr. Dacosta, Philadelphia:

I have received, at this very moment, your letter of the 8th of April. I have replied to your preceding by duplicate. Like yourself I am greatly astonished that you should not have received the contracts which I forwarded to you at once. I have reserved copies of these papers, which I have literally copied.

If I had the least idea that they would not reach you, and that an accident had befallen the ship, I should forward them in duplicate, but as this boat, at the time of its departure, was long delayed by the embargoes as well as by bad weather, I am persuaded that this is the sole cause, and that they will have reached you since.

You are about to appeal to the supreme court to prove your ownership; is there a living being who can contest it? If our deeds, granted in France, have not their full force in that country, nothing can annul them for us who are French. You shall do in this matter what you like; the greatest objection is this, that it stops your operations; but who is to blame? It is due to distance, and not to any negligence.

You say that you will do nothing until you have these documents; if your intention is to work for our benefit, as you say in your preceding, a company still being disagreeable [to you], that ought not to stop you; you have every power, [and] time lost is irreparable. I am much annoyed at the delay that this Mr. Miers Fisher causes you; as you say, he is an honest man, but negligent, and this in consequence of his age, and absorption in his great business.

We now return to Mr. David Ross,[98] who in his letter tells a pack of lies. At the close of 1789 I presented myself at his house with the power of attorney of Mr. Formon,[99] when we settled the business of the "Count of Artois," and the "Annette."[100] There never has been, as he said, any dissolution of the partnership between Mr. Formon and myself. I settled the accounts at that time both with him and with Samuel Plaisance concerning these vessels, with the exception of a residue of three thousand francs which are due me from Mr. Edward, their associate, who died at London. When I asked him for his certificates, he gave me for excuse that they were at the iron factory above Richmond, and that he had given Mr. Formon a private obligation that he would be very glad to have an exchange for the certificates. This affair has rested there ever since, and according to his letter Mr. Formon has taken out seven thousand, four hundred dollars, which exceeds his share by 1,650 dollars. If the estate of Mr. Formon is not without resources, it is to his heirs that you must apply for this overdraft, and get from Mr. David Ross all that you can, for with such people one cannot rely upon getting anything except with iron hooks.

---

98 This name appears as "Rost" in all the letters.

99 Member of the firm of Audubon, Lacroix, Formon & Jacques, engaged in the Santo Domingo trade. In these letters the name usually appears as "Formont."

100 Vessels in which Jean Audubon was personally interested, and upon which he endeavored in vain to collect the money and interest due him. In a document in English, dated [Les Cayes] April 9, 1782, concerning the *Annette*, of which Jean Audubon was captain and part owner, and signed by him and David Ross & Company, it is stated this vessel was bound for Nantes with a cargo of tobacco, in the purchase and sale of which Captain Audubon was under orders of Mr. Ezekiel Edwards of Nantes.

The son-in-law of Mr. Formon doubtless will have found among his papers all that constitutes the legal basis of my portion; his certificates, his letter of attorney prove it, and this is a title, and I believe that I have proofs by accounts current. I salute you.

Jean Audubon to Francis Dacosta

Nantes, 22 June, 1805

To Mr. Dacosta:

I have just received your letter of April 23, and hasten to reply to it, in order to prove to you that not one of yours has been neglected, which could be readily seen by my copybook. I am not surprised that at this time you have not received your papers, because they cannot have left before the 10th or 15th of last March, having been held up by the embargoes and the bad weather, as you will see by the date of the letters which accompany them.

They were entrusted to the son-in-law of Mr. Paulin, and if the ship arrives safely as I trust it will, you have now received them.

What negligence on the part of Mr. Miers Fisher! In truth it is unpardonable, to let the mortgages stand after having paid them![101] Will you then, I pray, clear this up for the sake of our mutual peace of mind? You speak of repairs to the house,[102] it needs a complete cover; would it not be better for me to send some slate from here? This would perhaps be less expensive, and well nigh everlasting. Should you consider it advisable I will send you some at once.

I beg you not to neglect the affair of David Ross; if you can collect this sum, you will use it for our needs. I am annoyed that all these mishaps prevent you from working;[103] be well persuaded that it is no fault of mine, and that I am guilty of no negligence.

You speak of my going to that country; if such had been my intention I should have done it long ago. I am still troubled with an inflammation of the lungs; and one ought not to be ill in a foreign country, where he does not receive the care that he enjoys in his own home. You ask me to bring you money.... You know better than anyone else what was my [financial] position when I sold to you; by that alone you must know how difficult this would be for me. It is necessary to manage so that our object suffices us [or so that the mine pays its way], and if we cannot work on a grand scale, we must needs do the best with our affairs on a lower plane; for that I depend on you. I salute you.

P.S. When you shall have my papers from Mr. Miers Fisher, you will find a promissory note of Mr. Samuel Plaisance of Richmond, for the business of the widow Ross. If there were justice there this sum would be paid to me with the costs.

The foregoing letters show that Dacosta had been asked to oppose the proposed marriage of the younger Audubon to Lucy Bakewell until consent should be given; that he was calling for more money to exploit the lead mine and was urging Lieutenant Audubon to come to

[101] This was probably the mortgage which Jean Audubon gave to Prevost when "Mill Grove" was purchased in 1789, for in Dacosta's final account for 1806-1807 this item occurs under October 15, 1806: "To the recorder in Norristown for entering satisfaction of John Audubon mortgage to John Augustin Prevost ... $2.83."

[102] The principal house at "Mill Grove," which Dacosta was preparing to occupy.

[103] Owing to the delay in receiving his legal papers from France, Dacosta had threatened to carry his case to the courts, and had stopped work at the mine.

America; and that their relations were becoming strained, Dacosta, to prove his title to a one-half interest in the mine and farm, having threatened to take his case to the courts.

This mining experiment was spread over many years. Before turning to the sequel, let us glance at the picture which the naturalist has left of his unsympathetic tutor. "Dacosta," he said, "was intended to teach me mineralogy and mining engineering, but in fact" he "knew nothing of either; besides which he was a covetous wretch, who did all he could to ruin my father, and indeed swindled us both to a large amount. I had to go to France to expose him to my father to get rid of him, which I fortunately accomplished at sight of my kind parent. A greater scoundrel than Dacosta never probably existed, but peace be with his soul." In one respect only, said Audubon, did he receive any sympathy from his guardian: Dacosta commended his drawings of birds. "One morning," Audubon relates, "when I was drawing a figure of the *Ardea herodias* [the great blue heron], he assured me that the time might come when I should be a great American naturalist"; however curious it might appear, he adds, that praise "from the lips of such a man should affect me, I assure you that they had great weight with me and I felt a certain degree of pride in these words even then."

To follow Audubon's story further, not only did Dacosta take control of his finances, but he interfered with his personal liberty, first by objecting to his proposed marriage to Lucy Bakewell, and then by cutting off his stipend when he rebelled.[104] Audubon, being thoroughly aroused, determined to return to France and lay the case before his father in person. With this end in view he walked to Philadelphia, whither Dacosta had gone, to demand the money necessary to take him to Nantes. He was given, as he says, what purported to be a letter of credit to a Mr. Kauman, an agent and banker in New York. Returning with his letter to "Mill Grove," he then started on foot for New York, where he arrived on the evening of the third day. While there he stayed at the house of Mrs. Palmer,[105] "a lady of excellent qualities," who received him most kindly. Audubon called promptly upon Benjamin Bakewell, for whom he was the bearer of a letter from his brother, William Bakewell, of "Fatland Ford." Instead of an order for money, Kauman's letter, he said, contained only the advice that its bearer be "arrested and shipped to Canton." Perplexed and bewildered beyond endurance, Audubon said that for the first time he felt the call of murder in his blood, and his outraged feelings were not assuaged until his landlady, to whom he had opened his heart, and Mr. Bakewell, had come to his aid. Having secured from this gentleman the necessary funds, he bought a passage in the ship *Hope*, which was then about to sail direct for Nantes.

Thanks to an old cash account of William Bakewell, we can follow Audubon's movements at this time fairly closely. This record[106] extends from January 4, 1805, to April 9, 1810, during which time he advanced money to his future son-in-law and received credits due him from various sources. He did the same for the young partners when an association in business had been formed between Audubon and Rozier, and acted as their agent or attorney after the sale of their farm and their settlement in the West; as will be seen he aided Audubon very substantially later when money was needed at Louisville and for the more ambitious projects at Henderson, in which his son was also interested. This particular record shows that

---

[104] In the light of the preceding letters, Dacosta would appear in these respects to have been only attempting to carry out his instructions.

[105] Probably Sarah White Palmer, Benjamin Bakewell's sister-in-law, and widow of the Rev. John Palmer, who at one time was associated with Joseph Priestley in editing the *Theological Repository*, an organ of the Unitarians. Her son-in-law, Thomas W. Pears, was later a partner in Audubon's business ventures at Henderson, Kentucky. Her grave is in the Bakewell burying plot at "Fatland Ford."

[106] See Appendix I, Document No. 7.

he supplied Audubon with small sums of money on January 4 and 12, 1805, just before his departure from "Mill Grove," and that on the eighteenth of the same month he paid his brother, Benjamin Bakewell of New York, $150 on the young man's account. This was undoubtedly the passage money which Audubon had borrowed from his friend, and as the ship was then ready to sail, the date of his voyage on the *Hope* is very closely fixed.

After his vessel had passed Sandy Hook and was opposite New Bedford, the captain, in order, as he averred, to make necessary repairs, ran her into that port, where they passed a week. This was thought to be only a ruse on the captain's part to gain time, for, having recently married, he wanted a holiday on shore; accordingly he had ordered a few holes bored below the waterline in the bows of his ship. When they finally put to sea in earnest, they passed "through an immensity of dead fish floating on the surface of the water," a remark which now recalls stories of the famous tilefish, once thought to be extinct, which have been found floating dead in vast numbers in that part of the Atlantic. After nineteen days out the *Hope* entered the Loire and anchored at Paimbœuf, the lower harbor of Nantes; this was in February, and not far from the eighteenth of that month.

Chapter 9

AUDUBON'S LAST VISIT TO HIS HOME IN FRANCE

Reaching his home at Couëron in the spring of 1805, Audubon took his parents completely by surprise. He found his father, then in his sixty-first year, still "hale and hearty," and his "*chère maman* as fair and good as ever." It was a time of momentous events in France; Napoleon had placed the crown upon his head but a few months before; defeat and victory followed in rapid succession. But this did not prevent the young naturalist from spending a year in "the lap of comfort" at Nantes and in the quiet villa of "La Gerbetière," where as usual he hunted birds and collected objects of natural history of every sort.

At this time also Audubon formed a friendship with a young man after his own heart, Dr. Charles Marie d'Orbigny, who "with his young wife and infant-son" was then living near his home. "The doctor," he said, "was a good fisherman, a good hunter, and fond of all objects in nature. Together we searched the woods, the fields and the banks of the Loire, procuring every bird we could, and I made drawings of every one of them—very bad, to be sure, but still they were of assistance to me."[107]

Charles d'Orbigny, who was Audubon's most intimate early friend and in all probability his father in natural history, was always spoken of in terms of great affection. While at Paris in October, 1829, Audubon learned from the naturalist Lesson that D'Orbigny was then in charge of the museum at La Rochelle and that "his son, Charles, then twenty-one," whom "he had held in his arms many times," was in the city; on October 8 he wrote in his journal: "this morning I had great pleasure in meeting my godson, Charles d'Orbigny. Oh! what past times were brought to my mind."[108]

---

[107] Maria R. Audubon, *Audubon and his Journals*, vol. i, p. 39.

[108] Dr. d'Orbigny had three sons, all of whom were born in Couëron: Alcide Charles Victor in 1802, Gaston Edouard in 1805, and Charles in 1806; the youngest and eldest became distinguished naturalists. So far as known, Audubon was godfather only to the second, Gaston Edouard, who according to the records of the Catholic church at Couëron, "was born on the 3d day of the present [month], the issue of the legitimate marriage of Mr. Charles Marie d'Orbigny, doctor of medicine, and of Anna Pepart," was christened on August 20, 1805, in the presence of the godfather, John James Audubon, the godmother, Rosa Audubon, the father and mother, together with the "undersigned" (Extracted by Monsieur Lavigne). D'Orbigny appears as a witness to the powers of attorney which Jean Audubon and his wife issued jointly to their son and to Ferdinand Rozier at Couëron in 1805 and on November 20, 1806.

Figure 18. Early Unpublished Drawings of French Birds: Above, "Le Friquet Mâle De Buffon. The Sedge Sparrow. No. 13. Near Nantz, 1805. J. J. A."; Below, "Le Rossignol De Murailles—De Buffon—The Redstart. No. 50. Near Nantz, August, 1805. J. L. F. A."
Published by courtesy of Mr. Joseph Y. Jeanes.

In later life the elder D'Orbigny seems to have fallen on evil times. He appeared as a debtor to Lieutenant Audubon's estate, and the cordial relations that had long existed between the two families were broken; this is shown only too plainly by the following sharp letter[109] written by Gabriel du Puigaudeau and addressed to the doctor, on August 3, 1819, when the family had become reduced in means:

[109] For copies of this and the following letters, which are here translated from the French, I am indebted to Monsieur Lavigne.

Gabriel du Puigaudeau to Charles M. d'Orbigny

Your letter of the twenty-fifth of January reached me in due time. I am grieved to see that you are annoyed because I addressed you through the voice of the mayor of the town in which you live, since I had not the honor of knowing the mayor any more than the enmity which may exist between you; I was in duty bound to find out where you were; I heard it said that Esnaudes was your home and I wrote you more than a year ago; when I received no reply, the supposition was that I must have been misinformed. I wrote to the mayor of Esnaudes and he had the kindness to reply that you were practicing in his commune. I am writing to you under this cover, persuaded that my last will not have the same fate as my first, which surely had not reached you.

As to the claim that Madame Audubon has upon you, the different credits which you mention are assuredly more than enough to pay the amount, but with forfeitures; unfortunately there are many creditors who do nothing but this; Madame Audubon gets nothing, and finds herself in straightened circumstances, although her hands are full of notes. You say that your creditors can claim only thirty-five hundred francs. I have certain knowledge to the contrary, since already the mortgages on your house reach nearly three thousand francs, while Madame Audubon is your creditor in the sum of at least sixteen hundred francs. I wish in business to be frank, and to have others so with me. You say that you owe rather those who have supplied you with food; you are unwilling then to recall that the sums that the late Mr. Audubon lent you repeatedly were for the same purpose. You tell us to be patient, and who have been more patient than we for the past four years? You speak of reduction of interest; indeed it is impossible that you should have thought of this, or that we should be content with what you should be so good as to give us, and that when you deem it convenient, without our being able to file a protest. I leave you to reflect on what we must think of this matter, and I beg you to see in my manner of writing to you the interpretation that I have given to what you write yourself.

Madame Audubon does not think that she should exact at once the capital in addition to the interest, but she charges me to say to you that, having a right at least to the interest accrued, she begs you to have that money paid to her with the least possible delay.

The following letter concerning D'Orbigny's affairs was also written by Gabriel du Puigaudeau to J. Cornet of Esnaudes, on June 26, 1819:

Gabriel du Puigaudeau to J. Cornet

Your honored [letter] of the sixteenth was duly received. It is impossible to be more grateful to you than I am for the information that you have been kind enough to give me about Mlle. Bouffard[110] as well as about M. Delouche. I will use it to my profit. As to the question that you put to me concerning M. d'Orbigny, I have the honor to tell you that he has lived in the commune of Vue in this department, and was highly esteemed and regretted when he left to come here. He lived here fifteen years without anyone having cause to reproach him in any way. He has always been very well regarded and received by the best society here, and he carried from Vue the regrets of all. He left us to take part in a manufactory of soda, established at Noismoutiers, in the department of La Vendée.

---

[110] A daughter of Catharine Bouffard, regarding whom see Vol. I.

> I have had no news of him since. As to his pecuniary resources, I know him to have but one. His wife had a house, at Paimbœuf in this department, which was sold three years ago to satisfy the holders of mortgages. This is all that I can tell you about them; he owes my mother-in-law about fifteen hundred francs (money received at different times from my late father-in-law), for which we have his notes, but God only knows when we shall be paid.

As early as the autumn of 1805, if not before, plans were laid for getting young Audubon again safely out of France, for fear, no doubt, that the remorseless conscription officers of Napoleon would send him to the war if he remained. At that time Lieutenant Audubon and his wife issued jointly to their son and to Ferdinand Rozier a power of attorney for the conduct of their business affairs in America. Parts only of this punctilious document, which was written in French, have been preserved,[111] and these through the translation of a "notary public and sworn interpreter of foreign languages for the Commonwealth of Pennsylvania, resident in Philadelphia." The names of the grantors, who signed this letter on October 21, 1805, were attested under the signature and seal of the mayor of Couëron; this official upon the same day declared that, in conformity with the rigorous requirements of the laws of the State of Pennsylvania, since "no other act, not even a notarial instrument, can in any manner supply the same," he had examined Anne Moynet Audubon apart, when she admitted that she perfectly understood the nature of the act, which she had "signed, sealed, and delivered of her own free will and accord, without being compelled thereto by her husband, either by threats, or by any other means of compulsion whatsoever." The mayor's signature was authenticated three days later by the subprefect of Savenay, and the formality was finally closed by the attestation of his signature by the prefect, on the 27th of November.

It was during this last visit to his home in France that Audubon's sister, Rosa,[112] was married to Gabriel Loyen du Puigaudeau, who was not, however, as the naturalist has stated, either "the son of a fallen nobleman" or his father's "secretary." Du Puigaudeau came from a family of merchants in easy circumstances, and for a long time lived the life of a country gentleman of leisure—for a period at Port Launay, below Couëron, and later, after Lieutenant Audubon's death, at his own villa, "Les Tourterelles," in that commune, not far from "La Gerbetière." His father, though of a rich family, was not a "gentleman," that is, a member of the aristocracy, as the term was then used in France. Du Puigaudeau was without any settled business, but his revenues, upon which he depended, failed not long after the death of his father-in-law. He and young Audubon appear to have been good friends for many years, and after the latter's return to America they corresponded to as late as 1820, when for some reason their relations were broken.

In the spring of 1806 Lieutenant Audubon arranged a business partnership between his son and Ferdinand Rozier, to endure for nine years, and also secured passports for both to

---

[111] See Appendix I.

[112] The civil ceremony of Rosa Audubon's marriage was performed at the mayor's office in Couëron, on December 16, 1805 (*le 26 frimaire, an 14*), when the bride was in her eighteenth year; the contract had been drawn on the 12th day of that month (*le 22 frimaire, an 14*) by notary Martin Daviais, who was mayor of Couëron in the following year, and the religious ceremony was conducted by the Bishop of Nantes. "The following have assisted," so reads in translation the Couëron record, "at the marriage, aforesaid, on the side of the groom, M. André Loyen du Puigaudeau, his brother, and M. Honoré François Guiraud, his brother-in-law; by the side of the bride, her father, and M. Jean Audubon, her brother, [and these have] undersigned, together with the bridegroom." Audubon's signature reads "J. L. J. Audubon."

enable them to emigrate immediately to the United States. To the same hand can also be traced their "Articles of Association," which were drawn with the utmost care and designed to govern them in all their future business relations in the New World: these were signed by "Jean Audubon," and "Ferdinand Rozier," at Nantes, on March 23, 1806. Moreover, eight days before they embarked, a second and more elaborate letter of attorney was issued to them jointly by the Lieutenant, his wife, and, in this instance, the aged father of Ferdinand, under date of April 4, 1806.[113] According to the terms of this admirably executed paper the partners were entitled to conduct all the affairs of the grantors in reference to their property in the United States to the best of their judgment and ability; to carry on the "Mill Grove" farm, to the extent of their part ownership in the estate, or to dispose of this interest; "to exploit or cause to be exploited the mine recently discovered on the said farm, to consult in every important matter Mr. Miers Fisher, merchant of Philadelphia,—as a common friend and good counsellor, to keep all necessary books and registers, and at the end of each year, or sooner, to strike a balance of the receipts and expenses of the said farm and the exploitation of the mine, should there be reason for it."

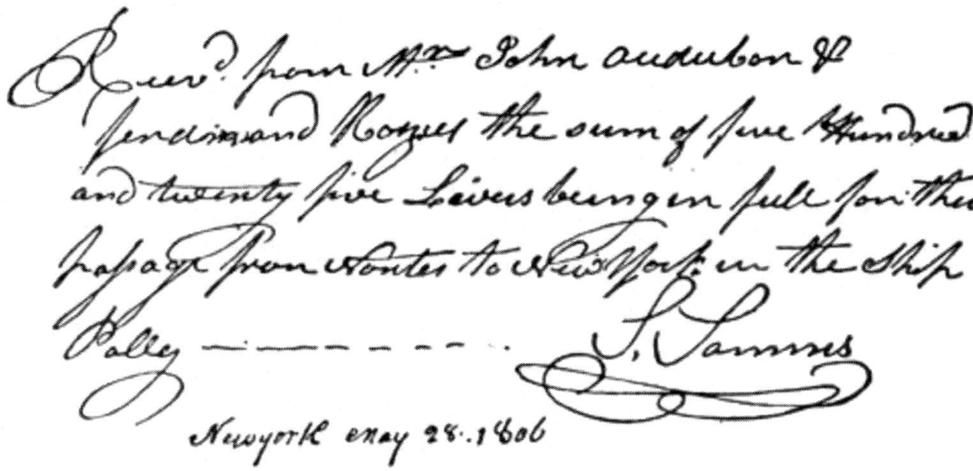
Rec'd from Mr John audubon & ferdinand Rozier the sum of five hundred and twenty five Livers being in full for their passage from Nantes to New York in the Ship Polly ------- S. Sammis

Newyork may 28..1806

Figure 19. Receipt Given by Captain S. Sammis of the "Polly" to Audubon and Ferdinand Rozier for Their Passage Money from Nantes to New York. From the Tom J. Rozier MSS.

To secure at this time the necessary passports for their young men no doubt taxed all the resources of the elder Audubon; Rozier's, said the naturalist, was written in Dutch, of which he did not understand a single word, while his own letter stated that he was born in New Orleans. These subterfuges worked so well that the inspection officer, after reading Audubon's paper, promptly offered him his congratulations, adding that he would be only too glad to leave his unhappy country under as favorable conditions. Audubon and Rozier sailed

[113] For the full text of these two documents, which are so interesting for our story, see Appendix I, Documents Nos. 9 and 10; and for translations, Documents Nos. 9a and 10a. For the privilege of examining and reproducing the first of these papers I am indebted to Mr. Charles A. Rozier, of St. Louis, and for the second, as well as the power of attorney of 1805, referred to earlier, to Mr. Tom J. Rozier, of Sainte Geneviève, Missouri. In the case of this second warrant it will be noticed that the grantors signed only the minute which was filed with the notaries, who, with the judge of the Court of the First Instance, affixed their names to the document itself. No better illustration could be given of the dignity which the French attach to the office of notary, to the honored incumbents of which their private affairs are unreservedly entrusted, than this elaborate judicial document.

from Nantes on Saturday, April 12, 1806, on the ship *Polly*, Captain Sammis, but they did not land in New York until Tuesday, May 28, after a perilous voyage of nearly eight weeks. A fortnight had been passed at sea when they sighted a suspicious looking vessel which immediately gave chase, fired several shots across their bows, and compelled the captain to heave to and submit to being boarded and searched. This proved to be an English privateer, named the *Rattlesnake*. She was rather considerate for a British cruiser of the period, for she merely impressed two of their best seamen and robbed them of their provisions, carrying off, said Audubon, all of their "pigs, sheep, coffee and wine,"[114] in spite of loud remonstrances of the captain and of an American Congressman who happened to be among the passengers. "The *Rattlesnake*," he continued, "kept us under her lee, and almost within pistol-shot, for a day and a night, ransacking the ship for money, of which we had a great store in the run under the ballast which was partially removed, but they did not go deep enough to reach the treasure. The gold belonging to Rozier and myself I put away under the ship's cable in the bow, where it remained until the privateers had departed."

Upon reaching a point thirty miles off Sandy Hook, they learned from a fishing smack that two British frigates lay off the harbor and were impressing American seamen, that, in short, they were even more unwelcome than pirates who sailed under letters of marque. The captain, thus forewarned of one danger, had the misfortune to run into another, for upon taking his vessel into Long Island Sound, she encountered a storm and was stranded in a gale; no great harm was experienced, however, for the vessel was finally floated off and reached New York on the following day. The passage money paid by Audubon and Rozier to Captain Sammis amounting to 525 livres, or $125,[115] was entered, according to their articles of agreement, as the first item of their "social expenses." After a brief visit with Benjamin Bakewell they hurried to "Mill Grove," and Audubon to the home of his sweetheart, Lucy.

[114] In the register of the Central Committee of Nantes it is noted, under date of October 4, 1793, that "owing to the friendly relations then existing between France and the citizens of the United States, and to the good feeling evinced by them in sending to us for food, four American ships are accordingly permitted to leave the port of Nantes, with cargoes of wine, sugar, and coffee, and also to take enough biscuit for the voyage."

[115] The receipt which the captain handed the young men, and which the methodical Rozier preserved, remains as a souvenir of this voyage (in the Tom J. Rozier MSS); it reads as follows:

Recvd. from Mr. John Audubon & ferdinand Rozier the sum of five Hundred and twenty five Livers being in full for their passage from Nantes to New York in the Ship Polly......... S. Sammis

[In Rozier's (?) handwriting] New York May 28, 1806

[Indorsed by Rozier on back] Payé le 11 avril 1806.

Chapter 10

"LA GERBETIÈRE" OF YESTERDAY AND TODAY

Before following further Audubon's history in America, we shall return for a more intimate view of the happy home which he had left behind him in France. This was at Couëron, a small commune in the arrondissement of Saint-Nazaire, on the right bank of the Loire, nine miles west of Nantes. Here, as we have noticed, his father had acquired a country place at about the outbreak of the Revolution. The old house still stands, though in decay, and is still known as "La Gerbetière," a name possibly referring to the wheat which is harvested from the surrounding fields as of yore. In the records of that district country places are always designated by their proper names, and it is a curious fact that while such names survive, they are seldom or never displayed on door or gate.

Figure 20. "La Gerbetière" Jean Audubon's Country Villa at Couëron, France, and the Naturalist's Boyhood Home. From a Photograph of August 18, 1913.

In a journal written before 1826, Audubon says: "My father's beautiful country seat, situated within sight of the Loire, about mid-distance between Nantes and the sea, I found quite delightful to my taste, notwithstanding the frightful cruelties I had witnessed in that vicinity not many years previously. The gardens, greenhouses, and all appertaining to it appeared to me of a superior cast." Though it was occupied for many years previously as a refuge from the turmoil or heat of the city, Lieutenant Audubon made "La Gerbetière" his permanent abode only when he retired from the navy in 1801, still maintaining, as we have seen, a foothold in Nantes.

Upon Audubon's first return from the United States in the spring of 1805, he said that his vessel entered the mouth of the Loire and anchored off Paimbœuf, the lower harbor of Nantes. "On sending my name to the principal officer of the customs," the narrative continues, "he came on board, and afterwards sent me to my father's villa, La Gerbetière, in his barge and with his own men." It is to be noticed, incidentally, that as the distance to be covered between the lower and upper harbors was twenty-five miles, or sixteen miles to Couëron, such journeys no doubt were made upon the arrival of incoming vessels for the regular business of the service.

It has been suggested, without proof, that Couëron represents the ancient town of Corbilo, mentioned by Strabo at the beginning of our era. Though unquestionably ancient, at the time of the Revolution it was a small and unimportant parish of poor but industrious farmers. It occupies rolling ground, but little raised above the Loire, to the east of Port Launay and nearly opposite Pellerin. As this commune was easily accessible by river-barge from Nantes, the revolutionists seem to have thought it worth watching, though Citizen Audubon found its people in a tranquil mood when he canvassed their district in behalf of the Central Committee in April, 1793. Couëron is still a farming community, but its population[116] has been considerably swelled in recent years by the development of a large industry for the treatment of lead; it is the shot tower and forest of chimneys of these great metallurgical works that arrest the eye of the traveler as he approaches Couëron by river at the present day. The town is also accessible by railroad, but the steamer journey from Nantes, which is made in less than an hour, is more attractive as well as more direct. In this section the Loire is flanked on either side by bottom lands, reduced in places to narrow strips, which are followed at intervals by elevations called, by courtesy, hills or buttes. To the west of Couëron, and especially at Pellerin, which stands high, these buttes come close to the river, which is eating them away.

My visit to Couëron, which was made on a warm midsummer's day in 1913, served to correct certain previous impressions, but I found the old Audubon homestead in its essential aspects but little changed, considering that over a century had rolled by since the naturalist's visit which we have just described. After leaving Nantes at the Gare de la Bourse by one of those quaint little trains which still do service in the less traveled parts of France, we traversed the broad Quai with requisite deliberation, passing shops, warehouses and factories in long array. A slight swerve from the river soon brought us to Chatenay, now a part of the city; it is still some distance from that point before the real countryside is reached, and scenes familiar to southern Brittany are in a measure reproduced. There were the old farmhouses of rough stone, dear to every painter's heart, mellowed by age and lichens, and surrounded by great ricks of straw, for the harvest had been gathered and the stubble fields were brown.

[116] The total population of Couëron, as given in the official directory for 1913, was 2,035, but the total working population is probably three times as great.

There also the farms were divided into small plats, marked by willows or ramparts of stone. On higher ground stood the windmills, characteristic of Brittany also,—stalwart towers of stone, with broad arms of latticed wood ever ready to take the sails.

The small station for Couëron lies in the commune of Sautron, and at this isolated point the traveler will sometimes find a country conveyance to take him to the village. While we were raising the dust from this old Couëron pike on the eighteenth day of August, swallows hawking with characteristic energy for their insect prey were the only birds we saw to remind us of the ornithologist, who as a youth had doubtless passed this way many times, over a hundred years before. The most direct approach to the old Audubon place from Sautron, as we afterwards learned, is by a path which diverges on the right and leads through stubble fields and cabbage patches, along hedgerows and stone walls. We, however, fared on to the town and soon began to pass shops and small modern houses. On the side of the village the traveler's eye is certain to be arrested by a great crucifix in stone,[117] which rises high above the street from a lofty pedestal, and is approached by tiers of stone steps. Nearly opposite stands the *secrétariat*, or official bureau of the commune, where a solitary clerk, who seemed to welcome my intrusion in a place where business was utterly stagnant, closed his office and with characteristic courtesy cheerfully showed me the way. This led directly westward to one side of the center of the town, and after passing down a street of old houses of the humblest description, we were again in the region of brown fields and old farmsteads.

Couëron village, which is marked by a modern church with an aggressive spire, extends along the river bank, but since its streets run parallel with it, the river itself is seen only at certain openings, occurring at irregular intervals. In going to "La Gerbetière" by the course I have described, the Loire was not visible at any point, and was not seen until we emerged from one of the village streets at the steamer's pier. My guide had said that from the rise at the next crossroads we should see the roof of the house which we had come to visit, and his prediction was verified when I recognized immediately its cupola raised above the gray stone walls which there bound every highway and field. The old villa is rather less than a mile from the village, but owing to the rolling nature of the country, it is completely hidden until at close approach it stands suddenly revealed. It lies in a fork of the road, securely inclosed by high, massive walls of stone, now hoary with age, while on the front it is further screened by a natural growth of bushes and trees. Immediately behind and to the west rises a prominent butte which cuts off the view to Port Launay on the river; this forms the one distinctive landmark of the district, as its two windmill towers are visible from all surrounding points. In Audubon's day the house commanded a wide view of the Loire, but the river is now so completely masked by foliage as to be visible only from the upper windows; apparently it once flowed nearer to the house but has been pushed away by the construction of modern dykes. The hilltop to which I have just referred, like the roof of the villa, commands a panorama of the whole region, including Nantes and all the surrounding communes.

"La Gerbetière" is now a small estate of less than fifty *ares*, or one and a half acres, of land. The buildings, which form a quadrangle with enclosed court, occupy a corner next the side street, and stand about 200 feet back from the main highway leading from Couëron to Port Launay. The extent of the original property cannot now be determined, but Lieutenant

[117] There is also the *grand calvaire*, which stands on an eminence in the village. This was erected in 1825 on the foundations of the chateau of the dukes of Brittany, the last of whom, Francis II, died at Couëron in 1488. His tomb is in the nave of the cathedral at Nantes; the *grand calvaire* was restored by two Couëron families in 1873, and is a very elaborate structure.

Audubon, who retired at the age of fifty-seven, was never a farmer on a large scale. The original house, which probably dates from early in the eighteenth century, has an easterly wing or L, continued into a long, low section through which the court is now entered from the road at the side; this was probably added by Jean Audubon, but the westerly end and wing are a more modern accretion, built for the accommodation of additional tenants, as many as three families having occupied the place in 1857.

"La Gerbetière" was entered from the main street by a small door which pierces the high enclosing wall, and leads the visitor into what was formerly an ornamental garden, the original design of which can still be traced. At the time of my visit, however, this entrance had long ceased to be an avenue of response. Encouraged by the sight of a peddler's cart, I walked up the side street and entered the court. Here the response was prompt and vigorous enough, and from the guardians of the place, one of which was chafing at his chain close to the doorway. I crossed rather gingerly to an open hallway, opposite the main entrance, and knocked repeatedly, noting here that rooms opened to this small entrance hall on either side, and that a steep stairway led to others above. At last, during a temporary lull in the barking of dogs, the "tok-tok" of sabots was heard on the stairs, and I handed up my card with one from the director of the Natural History Museum at Nantes. After various messages had been shouted back and forth, I was led through another passage to the tenant, who was talking with the peddler in the garden. Julien Lebréton, who was a farmer on a small scale, received me kindly and answered my questions to the best of his ability; it did not surprise me that he was both puzzled and suspicious, or that his first thought was of our coming to look over the place with a view to its purchase.

The decayed villa, which stands in the midst of scattered farmhouses of a humble order, reproduces a style characteristic of many parts of France. The original house, of two stories, was built of cream-colored limestone, similar to that for which many French towns are famous. It has a swelled slated roof with beveled gables. Surmounting the roof is a cupola which suggests a third story, carried out in harmony with the lower structure. A narrow balcony, resting upon a molding of stone and protected by an iron grill, without which no such house would be considered complete, runs the length of the second story, and is accessible from every room by glass doors. From the main entrance below one passes directly through to the court, about which are now grouped various stables and other low buildings, not all of which date from Audubon's day.

What was once a small formal garden is still marked by solid boundaries of cut limestone. This was evidently constructed by Jean Audubon, since it occupies the area in front of the original house and the easterly extension which is attributed to him. The remaining available land was devoted to fruit, vegetables, and possibly to the greenhouses which the naturalist mentioned. At one time an orangery occupied some part of the house or court. There are now no large trees on the property; the fruits are all of recent and inferior growth, while the garden I saw was planted to cabbage and running riot with weeds.

When Jean and Madame Audubon passed through the door leading from the main street, they entered upon a paved alley which ran parallel with the high wall, whence they could reach the house by any one of several walks or enter the fruit garden by another. If so inclined, they could turn to the right, ascend a flight of granite steps to a platform on a level with the top of the wall, and under a shady bower of vines and leafy shrubbery, look off on the racing waters of the Loire, scrutinize their visitors before admitting them, or observe such manifestations of life as lonely country roads of that period had to offer. As they passed up the central garden walk they could admire the beds of old-fashioned flowers, kept, we may be sure, in perfect order, for Jean was a very methodical man, and his wife, we believe, an excellent home maker. This walk led to a low terrace, flanked with a heavy wall, which ran the whole length of the house.

Figure 21. "La Gerbetière" and Couëron, as Seen from the Windmill Towers on the Ridge Overlooking Port Launay, on the Loire. "La Gerbetière," as Seen When Approached from Couëron Village by the Road to Port Launay. Port Launay on the Loire.

What little I saw of the interior of "La Gerbetière" was wholly devoid of interest, which agrees with the experience of another traveler who visited Couëron at a slightly earlier date;[118] at the time of his visit the place was unoccupied and forlorn, and the vegetation on the garden side so dense that it was utterly impossible to see any distance from the lower windows.

When "La Gerbetière" came into Jean Audubon's possession it was already venerable with age, and it was completely restored for him by an architect named Lavigne.[119] In an inventory drawn up shortly after Madame Audubon's death in October, 1821, the property of "La Gerbetière" is described by reproducing the account given in an early deed bearing date of November 11, 1769, which reads as follows:

> A house called La Gerbetière, situated near the port of Launay, consisting of a sitting room, drawing room, kitchen, upper chamber ... garret, and other quarters serving as a laundry, stable at the back, with pigeon loft above, court, parterre, vegetable garden to one side, an orangery with orange trees, in the middle of the house, the whole in front of a close surrounded by high walls except on the side of the setting sun, with land belonging to the heirs of M. de la Haye Moricaud, held mutually,[120] the whole bounded on all other sides by highways. Notice: The aforesaid house and parterre [stand] in an empty field, which serves as a fair-ground, and is partly planted with young trees in serial rows; held in common with the Marquis de la Musse, with another empty field containing about two journals of land....[121]

"La Gerbetière," never more than an unpretentious country house with an attractive garden, was idealized in the fervent imagination of Audubon when in after life he drew upon the memories of his youth in France; for it had meant to him escape from the city, which he detested, to the fields and river which he loved. Yet, in spite of the abuse which a long line of poor tenantry inevitably entails, with intervals of total neglect lasting for nearly a century, this decayed villa of pre-Revolutionary days still stands in marked contrast to its neighbors, and bears witness to a taste to which they were strangers. The greenhouses, the fruit and shade trees, if such it possessed, and all lesser adornments of the place have vanished long ago, but thanks to the durability of French stone and mortar, much about this old country seat is still well preserved. Whether Audubon ever saw his old Couëron home again after leaving it in 1806 is doubtful, though one of his sons visited the place, and the naturalist incidentally speaks of a pilgrimage to Les Sables d'Olonne which might have occurred in 1831 or a little later. In following the fortunes of the naturalist's family in France it will be necessary for us to return to La Gerbetière.[122]

[118] Mr. William Beer, who paid a visit to "La Gerbetière" with Dr. Louis Bureau in 1910, writes me that the woodwork was poor in quality, and that all the rooms had been altered in size and appearance.

[119] But not related to M. L. Lavigne, to whom I am indebted for extracts from the deed, a translation of which is given below, as well as for many other references.

[120] That is, the landlord to receive one-half the produce.

[121] A "journal" of land being as much as a man could cultivate in a day's labor.

[122] See Chapter 17.

Chapter 11

First Ventures in Business at New York, and Sequel to the "Mill Grove" Mine

When Audubon and Rozier reached "Mill Grove" at the beginning of the summer of 1806, they found the troublesome Dacosta installed as its master by virtue of his interest in the property and his former position as agent, to which they were now to succeed. No doubt they found difficulties in carrying out all the articles of agreement[123] in their business constitution, for they were to take possession and call Dacosta to account. They were also in duty bound to investigate the lead mine on the farm, and ascertain whether it promised any success, and if the expenses already incurred were warranted, before committing themselves to further development. One-half the product of the mine and farm was to be equally divided between them, and in order to visualize clearly their profit and loss, they agreed to keep a "special book for the purpose." "On one side," their third "Article" read, "will be entered the items of expense, day by day, and at the moment this is done, on the other side [shall also be entered] the sales and products of the farms, and of all that can result from this business, in such a way that the profit shall be always apparent by the addition of the items which compose the debit and the credit."

The house at "Mill Grove" was to be treated as an object separate from all business, "in order," so the "Articles" read, "that we may settle matters as completely as we desire." It was also agreed, in the fourth "Article," that they should "add to the expenses of this exploitation those necessary for life, and others of a mutual character, so long as it should suit them to live and dwell together." It was further stipulated that even if the mine proved a failure, they should remain six months on the farm, in order to gather useful information from the country, before embarking in any form of commerce, whether inland or maritime. The cost of their journey to America was to be entered as the first item of their "social expenses," and any

[123] For the privilege of examining Ferdinand Rozier's copy of their "Articles of Association" I am indebted to the kindness of Mr. Charles A. Rozier, of Saint Louis. This is written on three sides of hand-made, hand-ruled Government linen, small letter size, with printed revenue stamp (50 centimes) of the French Republic at top, and stamped with the seal of the Department of Registration and Stamps ("ADM. DES DOM. DE L'ENREG. ET DU TIMBRE REP. FRA.—*Administration des domaines de l'enregistrement et du timbre, République Française*"). The signature of "Jean Audubon" bears a close resemblance to that of the father, Lieutenant Jean Audubon, who was undoubtedly the author of the document. For the "Articles" in full, in French and English, see Appendix I, Documents Nos. 9 and 9a.

expenditure for travel in their mutual interests was to be considered under the same head. In case they should persuade any merchants in America to send goods to M. Rozier, Senior, at Nantes, he should be entitled to one-half the profits, while the partners should divide the other half between them. All other profits and losses resulting from their commercial transactions were to be shared equally. The partners resolved to maintain friendship and a mutual understanding, but "upon the least difficulty" each should choose one arbitrator, and the two thus chosen were authorized to select a third; the partners were bound to accept the decision thus reached without appealing to any court. In the case of the death of one of the associates, read the tenth "Article," the survivor should have sole charge of making a settlement of the business and should report to the proper heirs. The survivor, in such an event, would be entitled to a commission of ten per cent [in addition to his one-half interest], but in no case should the partnership be dissolved "until after nine years, counting from the day of the date of the present [instrument]." As will be seen, Audubon and Rozier were unable to fulfill all the conditions thus carefully laid down.

Young Audubon's dislike of Dacosta, the uncertainty of the mining project, and other difficulties of the situation soon decided the partners to cut short their stay at "Mill Grove." Both were equally interested in the lead mine, but after working several months without success in an attempt to form a mining company, they wisely decided to leave such experiments to the enthusiastic Dacosta and to seek an opening in trade, where the hazard would be no greater and their ignorance less profound. Following the advice of their Quaker friend, Miers Fisher, they decided to sell to Dacosta their remaining rights in "Mill Grove." As a preliminary it was necessary to divide the property which had been held in common by him and Lieutenant Audubon since 1804, and this division was effected by an agreement drawn up at Philadelphia on the fifth day of September, 1806.[124] Ten days later the remainder of "Mill Grove" was conveyed to Francis Dacosta, representing a number of capitalists whom he had managed to interest in the mine, of whom the astute Stephen Girard is said to have been one. The sale was subject to conditions,[125] dependent upon their success in mining lead, which, as will appear eventually, could not have been fulfilled. These various transactions are so clearly set forth by Ferdinand Rozier in writing to his father at Nantes that we shall reproduce his letter in full:[126]

---

[124] Among the elder Rozier's papers was part of an old letterbook belonging to his son; it is written in French, and labeled "Correspondence of Ferdinand Rozier." On one of the four sheets preserved this item occurs: "4 July, 1806, Philadelphia; record of an agreement with Mr. Dacosta proprietor of one half of the Mill Grove farm,— at least of the value of sale." The first entry is dated "19 février—1806, New York," which, if correct, would imply that Rozier spent two years instead of one in the United States when he visited this country in 1804 (or came a second time), and that he returned, with young Audubon, almost immediately after reaching France; the last record is "August, 1807, New York." (MS. in possession of Dr. Louis Bureau, Nantes.)

[125] According to the records of Montgomery County, as collated for Mr. W. H. Wetherill, the remaining half interest in "Mill Grove" was sold by J. J. Audubon (and Ferdinand Rozier) to Francis Dacosta & Company, for a consideration of $9,640.33. The business was conducted mainly by Rozier, acting under the advice of their friend, Miers Fisher.

[126] Translated from the French of Ferdinand's copy, in possession of Mr. Welton A. Rozier, to whom I am indebted for the privilege of reproducing it.

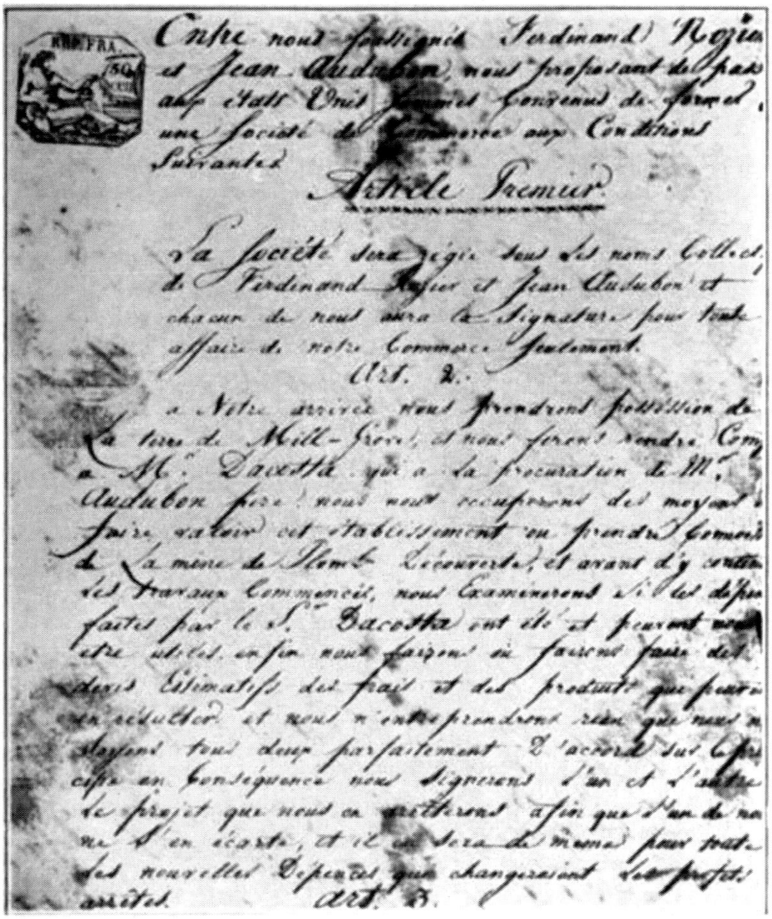

Entre nous soussignés Ferdinand Rozier et Jean Audubon, nous proposant de passer aux Etats Unis sommes convenus de former une société de Commerce aux Conditions suivantes

Article Premier

La société sera régie sous les noms collect. de Ferdinand Rozier et Jean Audubon et chacun de nous aura la Signature pour toute affaire de notre Commerce seulement.

Art. 2.

à Notre arrivée nous prendrons possession de la terre de Mill-Grove, et nous ferons rendre Compte à Mr. Dacosta qui a la procuration de Mr. Audubon père; nous nous occuperons des moyens de faire valoir cet établissement en prenant connaissance de la mine de Plomb découverte, et avant d'y continuer les travaux commencés, nous examinerons si les dépenses faites par le Sr. Dacosta ont été et peuvent être utiles, enfin nous ferons ou ferons faire des devis estimatifs des frais et des produits qui peuvent en résulter et nous n'entreprendrons rien que nous ne soyons tous deux parfaitement d'accord sur ce point. en Conséquence nous signerons l'un et l'autre le projet que nous en mettrons afin que l'un de nous ne s'en écarte, et il en sera de même pour toutes les nouvelles Dépenses qui changeraient les projets arrêtés.

Art. 3.

Figure 22. Beginning of the "Articles of Association" of John James Audubon and Ferdinand Rozier, Signed at Nantes, March 23, 1806.
After the Original Manuscript of Rozier's Copy, in Possession of Mr. Charles A. Rozier.

Ferdinand Rozier to Claude François Rozier

Philadelphia, 12 Sept., 1806

Duplicate. My very dear and venerable father:

Still in hope of cherished news from you, and replies to my letters of 31 May, 22 June, and 4 July, I have to tell you that we have since succeeded in closing all our business relations with Mr. Francis Dacosta, in the following manner: We are anxious that our method of procedure may be satisfactory to you; we have followed the advice of Mr. Miers Fisher, and have had his approval in all that we have done. What should set you at rest is that as regards your investment, you will find that I have made quite a neat profit. Here is a copy of the agreement.

"It is agreed between Mr. Dacosta and Mr. J. Audubon that the farm of "Mill Grove," which they now hold in common, shall be divided between them as follows:

"1. Mr. Dacosta shall have the lot of 113 and a half acres, situated on the N.E. side of Perkioming creek, with all the buildings, mines, et cet., and in general all that it contains.

"2. Mr. Audubon shall have the lot of 171 acres, situated on the other side of the creek.

"3. Mr. Dacosta shall pay to Mr. Audubon for the difference [in value] of the lot of 113½ acres, and of that which it contains:

"1. The sum of eight hundred dollars, payable with interest, in three years from this day;

"2. The sum of four thousand dollars, upon the first products of the lead mine.

"4. The contract made with Mr. Thomas shall remain to the charge of the two parties.

"Note. Mr. Duponceau is begged to draw up the necessary deeds to put this agreement into execution, which [deeds] we undertake mutually to exchange at the first requisition."

"[Executed] at Philadelphia, this 5th of Sept, 1806."

[Signed] "Fcis Dacosta"

"Ferdinand Rozier"

"J. Audubon"

The futile attempt that we have made to form a company [to work this mine], which is a condition [of success], the slight resources at our command, as well as our lack of knowledge in work of this kind, all have determined us to abandon our rights for the offer of four thousand dollars[127] upon the first products that shall come from the mine. The expense that must be incurred in [working] it will be very heavy; to this must be added the uncertainty of success. The mine may promise much at the beginning, and after that yield nothing. In short an enterprise of this kind can be properly conducted only by a capitalist or by a company. We have regarded this mine as a lottery which can make the fortune of the promoter, or lead him into great losses. As to the agreement with Mr. Wm. Thomas, we do not consider it as very serious; since it is quite uncertain whether he will be paid in whole or in part, as he has not kept his agreements. This is Mr. Dacosta's opinion. As to our half we are decided not to let it go under eight thousand dollars, which is its value as estimated by several farmers. So you see, my dear papa, that our half [as worth] 8,000 dollars, at least, the sum of eight hundred dollars by mortgage, with interest, and that of four thousand dollars upon the first products from the mine, will cover easily the interest on the purchase of sixteen thousand francs.

Since expenses are at least double what they would be in France, owing to the cost of products of every sort, we are determined to go into trade, to cover our expenses, and to choose for ourselves some kind of serious work that can lead us to an honorable establishment. You should be at ease about the manner we shall adopt for our operations, as we wish only to go slowly, and especially [to be] guided by the advice of the respectable persons whose acquaintance we are so fortunate as to enjoy, and who beyond a doubt will aid us along this thorny path.

"By our letter of the 4th July we have sent the account current of Mr. Dacosta, by which Mr. Audubon is charged with 315 dollars and 5 cents; we have begged you to send the documentary evidence which may put us in a position to prove that Mr. Audubon ought not to pay Mr. Dacosta's private expenses, as the matter is to be decided here by arbitrators. We beg Mr. Audubon to use the utmost speed in sending his documents. It is our ardent [hope] also that you have received our first [letter] of May 31, with that of Mr.

[127] "*Gourdes*," that is, piasters or Spanish dollars.

Bakewell, the merchant in New York, with a remittance of 3,000 and a few francs for the purchase of divers objects. I assure you that we are in the greatest anxiety [as to] what is the state of your health,[128] as well as that of the family, and to learn if you have received our letters. The nephew of Mr. Bakewell writes us that his uncle in New York has despatched several vessels consigned to you, for which I congratulate you sincerely. We have also received your letter of the 30th of June, but I cannot reply to it, since the boat is leaving this evening for Amsterdam, but you can count upon my conforming to its contents.

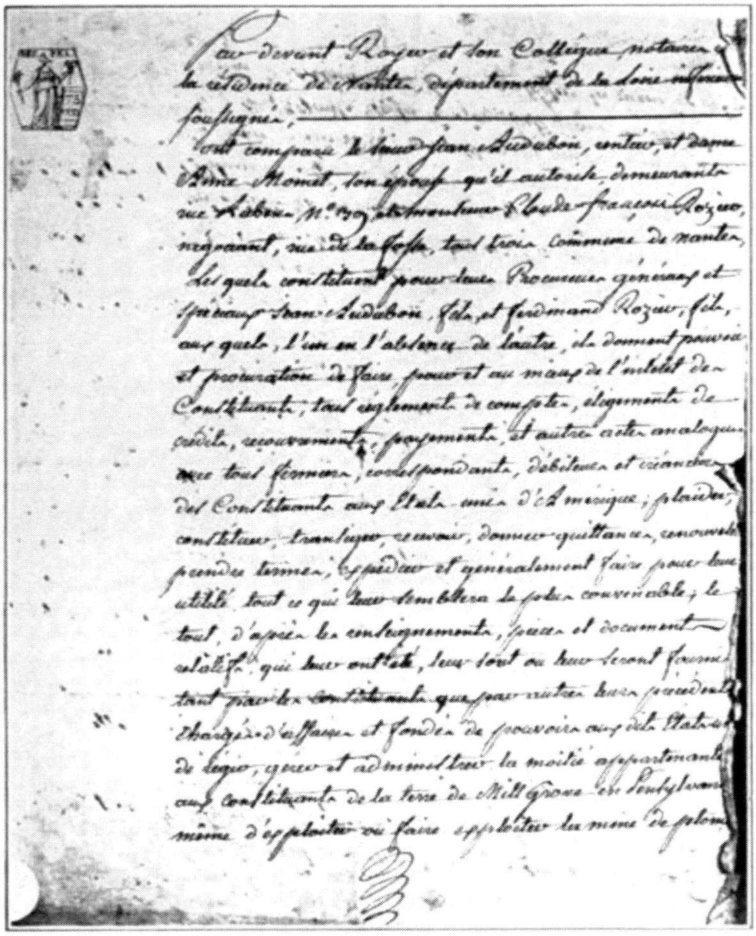

Figure 23. First Page of the Power of Attorney Granted by Jean Audubon, Anne Moynet Audubon and Claude François Rozier to John James Audubon and Ferdinand Rozier, Nantes, April 4, 1806. After the Original Manuscript of Ferdinand Rozier's Copy, in Possession of Mr. Tom J. Rozier.

Your personal letter grieved me particularly by your last expressions, and I should wish that you would have done me more justice; I can have made mistakes, but for ... the idea alone has made me shudder. I am delighted that all the family is enjoying perfect

[128] Claude François Rozier, at this time an aged man, died at Nantes on September 7, 1807; he had two sons and six daughters, of whom Ferdinand was the second son and the fifth child; his wife, Renée Angelique Colas, died at Nantes, February 9, 1824.

health. Embrace dear Mama for me; my kind regards to my brother and sisters; do not forget to remember me to all the family, and to our friend, Mr. Audubon, the father, and his family. Finally, my dear Papa, be assured that I shall forget nothing to increase our intimacy. You give me the means of supporting it with labor. Believe in my sincere and enduring attachment.

Your respectful son,
Ferdinand Rozier.

We are eager to hear of the receipt of our letters, and we beg you to address them to Mr. Bakewell of New York.

The inbred caution, sound sense, and sterling integrity which this letter displays would be a good foundation for any career, and we are not surprised to find that in after life Ferdinand Rozier became a keen and successful trader on the western frontier.

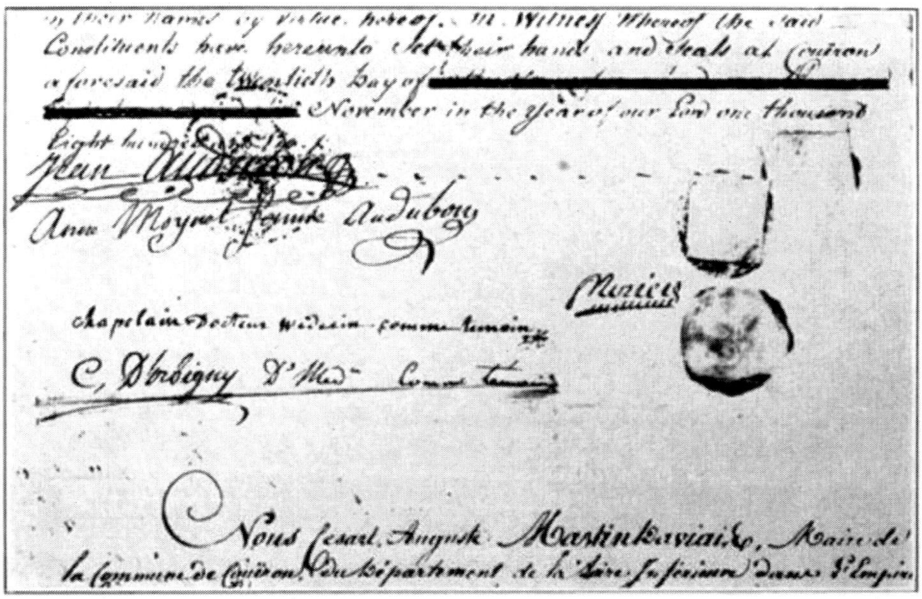

Figure 24. Signatures of Jean Audubon and Anne Moynet Audubon, and Dr. Chapelain and Dr. Charles D'orbigny, as Witnesses, to the Power of Attorney Granted to John James Audubon and Ferdinand Rozier, Couëron, November 20, 1806. After the Original Manuscript in Possession of Miss Maria R. Audubon.

The division and sale of "Mill Grove" probably ended the joint interests of the elder Audubon and Rozier, for in November, 1806, a new power of attorney[129] was given to the

[129] This was issued, so the letter reads, to "their son, John Audubon, and Ferdinand Rozier, both of the said city of Philadelphia, Gentlemen," by "John Audubon, late of the city of Philadelphia, in the commonwealth of Pennsylvania, now residing in the commune of Couëron, near the city of Nantes in France, Gentleman, and Anne Moynette, his wife," to apply to all lands and other property belonging to them in the United States, with the power to "raise or borrow money on the whole or any part or parts of the said lands, tenements, or hereditaments, to secure the repayment of said monies by bond, warrant of attorney, to contest judgment of the mortgage of the said lands, tenements, or hereditaments, or any part or parts thereof...." Written in French and English; signed by Jean Audubon, Anne Moynet, his wife, by Doctors Chapelain and C. d'Orbigny as

young men by Lieutenant Audubon and his wife; as later events will prove, however, their rights in the property were not completely surrendered with its transfer to Dacosta and his mining company in the autumn of this year. The partners were now free to "choose some kind of serious work," and Ferdinand, who was then twenty-nine, was anxious to make a beginning at once. Since he was not as yet proficient in the English tongue, Rozier engaged as a clerk in the French importing house of Laurence Huron, of Philadelphia, while Audubon, following the advice of his future father-in-law, entered the office of the latter's brother, Benjamin Bakewell, in New York.

In the autumn of 1806 Benjamin Bakewell was conducting a successful wholesale importing business at 175 Pearl Street. He then owned several vessels, and his correspondents were scattered over England, France, the West Indies and the Southern States. With him were associated at this time a number of young men, including his nephew, Thomas W. Bakewell, Thomas Pears, a nephew of his wife, Thomas Bakewell, his son, as well as John James Audubon. The hospitable family to which young Audubon was now admitted on terms of intimacy, in accordance with the custom of the day, lived in the rear of the counting-house during the winter months but in summer migrated to the country, the Bakewells going five miles out on the Bloomingdale Road. Benjamin Bakewell had come to this country in 1794, in the same year as the famous chemist, Joseph Priestley, whose friendship he enjoyed and whose religious teachings had drawn both him and his brother, William, from rigid Calvinism to the greater tolerance of the Unitarian belief. At twenty-four he was an independent mercer in Cornhill, London, and was well acquainted in France, where he had spent considerable time during the Revolution, which had destroyed his trade. One of his patrons at this time was Claude François Rozier of Nantes, and inasmuch as the correspondence with him had to be conducted in French, and may possibly in this instance have been due to young Audubon's initiative, it was naturally intrusted to him.

Seven letters of the naturalist, dating from January 10, 1807, to July 19 of that year, by good fortune have been preserved, and they throw into full light another shaded corner of his interesting life. From the contents of these letters,[130] as well as from other facts, we know that Audubon remained in Bakewell's office for nearly a year, from the autumn of 1806 to the summer of 1807. Bakewell's house imported linens, lace, gloves, wines, firearms and any kind of merchandise that promised a ready and remunerative sale in New York; in return they forwarded coffees, sugars and other commodities to Rozier, receiving from him also prices current and introductions to other merchants in France. Another correspondent was the Huron

---

witnesses, by the mayor of Couëron, the prefect of the arrondissement and the prefect of the department; countersigned on December 4, 1806, by W. D. Patterson, of the "Commercial Agency of the United States at Nantes." For the favor of examining this paper, I am indebted to the kindness of Miss Maria R. Audubon.

130 For the privilege of examining these letters I am indebted to the courtesy of Dr. Louis Bureau, Director of the Museum of Natural History and Professor in the School of Medicine at Nantes, maternal great-grandson of François, and grandnephew of Ferdinand Rozier. The letters were found in an old trunk that once belonged to his grandfather, François Denis Rozier. Five were written in French (Nos. 1, 2, 4, 6 and 7), and addressed from New York to François Rozier at Nantes; one (No. 3) in English and another (No. 5) in French were sent in care of Rozier, to his father, John Audubon, Esq., Nantes, with the direction to be delivered as soon as possible; all are on unruled foolscap, wafer-sealed, and each also bears an outside seal in wax, stamped with Bakewell's initial (B). It is not possible to say whether Lieutenant Audubon ever received these letters of his son; if received, it is not very obvious why they should have been left in the old merchant's hands, unless his ill health at the time, and subsequent death were the cause. I am further indebted to Mr. William Beer, for the perusal of his copies, which have been followed to a large extent.

Since all of these early letters throw an interesting light upon the times as well as upon Audubon's personal history, we shall give them in full, rendering the French into English as literally as practicable.

firm in Philadelphia, so it is probable that Ferdinand owed his employment there to Benjamin Bakewell.

While Audubon expressed himself at this time as freely in English as in French, in the former language the tendencies of his French tongue and the influence of his Quaker friends were strangely blended. He never bothered with accents, and took as many liberties with the spelling of French as of English. Some of these lapses are purely phonetic, while others are more original, as "schacket" for "packet," "fither" for "Fisher"; two variations of Rozier's name and of Nantes occur in the same letter. It should be remembered, however, that at the beginning of the nineteenth century bad or random spelling was a very venial offense, which gentlemen of quality, or even scholars, could commit with impunity. In this respect Audubon's early essays in English would probably compare favorably with Gibbon's youthful French.

John James Audubon to Claude François Rosier

[Letter No. 1, addressed]
M. Fr. Rozier,
Merchant-Nantes.

New York, 10 January, 1807.

Dear Sir:

We have had the pleasure of receiving by the Penelope your consignment of 20 pieces of linen cloth, for which we send our thanks. As soon as we have sold them, we shall take great pleasure in making our return.

I am truly sorry that you had not received any letters from us when you wrote, and I am also very disconsolate at having no news from my good father. You did us a most acceptable service in making us acquainted with your friends in different parts of France, and in offering to send us such goods as you shall deem suitable. Upon the same proposals I sent you orders several months ago, and did I dare, I should tell you that all articles having much show and little value are the very things that are à la mode, and these in one hundred per cent, [and] I assure you that we should be very happy to receive some small consignments. As soon as we shall have realized our funds, we will make our orders, in accordance with our means. Mr. Bakewell has made a great profit on the consignment that you made him shortly after our arrival. We should be flattered by another like it. Have the kindness to write us often, and to send us prices current as far as possible. I hope that you will have had our letters concerning a plan of business with Mr. Huron. If you will have the kindness to see him,[131] he can communicate to you his ideas on the subject. His plan, I believe, will be advantageous both to you and to us.

Your son is just about to come from Philadelphia, to live in New York until there is some news; but we will write you more at length by Capt. Sammis, who brought us to this country. I even venture to hope that you will send back some merchandise for us. Have the kindness to forward us invoices, with the goods consigned to us, in order to avoid the penalty and the expense of having them taken to a public warehouse, [a proceeding] which is often a great disadvantage on account of the fees. Consign always to Mr. Benjamin Bakewell, who treats us, so far as possible, as good friends.

[131] This Philadelphia merchant was evidently in France and intending to visit Nantes at this time.

Present my respects to your family, and believe me ever
Your faithful servant,
J. J. Audubon.

John James Audubon to Claude François Rozier

[Letter No. 2, addressed]
Monsieur Fr. Rozier,
Negociant,
Nantes.
Loire Inferieure.

New York, April 24, 1807.

My dear Sir:

I am profiting by a good opportunity for Bordeaux to apprise you of the receipt of a duplicate of the orders that you gave us several months ago. You will also know that the wines, consigned to Mr. L. Huron, have arrived in this city and the insurance has been saved. Your son has gone to the spot [the dock in Philadelphia], and by one of his letters advised me that the 60 cases of wine are sold. He tells me that you can count on a net profit of nearly 20 p. c. If it turns out very good, the remainder will not fail to find a purchaser. Mr. Le Ray has arrived and has brought with him a small box of lace for Mr. Benjamin Bakewell here; it ought to arrive in a few days from Philadelphia. Mr. B. B. appeared satisfied with the sale of his squared timber; he is anxious only to see the returns; he is unhappy that the commerce of your town with this country cannot be regularly conducted except by Bordeaux, whence we have vessels every month. As our friend, Ferdinand, will write you from Philadelphia concerning Mr. Huron, I shall not enlarge about him. In several of your letters you intimate that if we decide upon establishing a retail shop, you can keep us constantly employed; our ideas upon this subject are in perfect accord, and it would be indeed a pleasure if we could start under the auspices and good advice of Mr. Bakewell here; objects well chosen, favorably bought, and shipped with care, are always sure of meeting a good sale. I venture to hope that the ship La Jeanne, Capt. Sammis, will have arrived in your port, and that the Indigoes shipped by Mr. Bakewell will reach there in time for the sale of this merchandise, of which I have some fears, in view of the sum they have cost him.

We thank you for the prices current that you have sent us. In one of my last, directed by way of Bordeaux, I begged you to call on Mr. Fleury Emery for a box of seeds, from Martinique and from this country, for you and for my father. This was aboard the ship, the Virginia, Capt. Roberts, from this section. We hope shortly to send you some merchandise, and possibly Mr. Bakewell will profit by an opportunity that we shall have in a few days for your port. A little more than three weeks ago I was at Mill Grove, and I rented it for a year, being unable to do better for the present. Your son, now in Philadelphia, is trying to settle the accounts of my father with Mr. Dacotta [Dacosta], who does not easily forget the role of chicaner. Present, I pray you, my respects and compliments to your good family and wife, and believe in me as your devoted and constant servant,

J. J. Audubon.

Have the kindness to deliver the enclosed to my good father.

The following quaint and charming letter, which young Audubon enclosed with the preceding and under separate seal, but which his "good father" may not have received, will be transcribed in full, without the change of a letter or mark. Lieutenant Audubon, who was then in his sixty-third year, was living, as we have seen, at Couëron, the small river town nine miles west of Nantes, the center of the mails for the Loire Inférieure, and came frequently to that city to conduct his business correspondence.

John James Audubon to Jean Audubon

[Letter No. 3, enclosed with No. 2, addressed]
John Audubon, Esq.,
Nantes.
pr Bourdeaux

New York April 24th 1807

My dear Father

I send thee by a good opportunity, but going to Bordeaux I deed send about a month ago a small Box containing some very curious seeds & some useful ones the whole was directed to Mr. Fleury Emery it was given here to the Care of Capt.. Roberts of the Virginia I do hope they are now in thy possession thou have been so often disappointed that it always pains me to think that they have been Miscarried: thou shalt found some of the Best Whatter Missions and Girmonds Called here St. Domingo Schachet[132] as in a few days I shall have again a good opportunity for Nantz I will send thee a Duplicate of the same Seeds, I have seen in the News Paper that a ship called the Betzey had been in Nantz do make some Enquiries for it there are on board of her Many Birds and a collection of seeds from America for thee The Caps.. McDougal; pray when thou answer to this be kind enough to mantion these little things. I hope that the Jane Cap.. Sammis as reached your Port and given thee some Turtle fit to be eaten in soupe. Mr. L. Huron deed few days ago. Received some Wines on a/c of M. Rozier and hits they prove goods[133] and will bring a good profit. Mr. F. Rozier the son speaks of going to France sometime this summer he is now near Mr. Huron at Philadelphia and will try while he is there to settle the Business between M. Dacotta and thee M.. Rozier had shosen M.. Huron for arbitrator but I would not agree to it until M.. Miers fither[134] was to have part in it. I am now waiting for an answer. I am allways in Mr. Benjamin Bakewell's store where I work as much as I can and passes my days happy; about thee weeks ago I went to Mill Grove for a/c of the latter and had the pleasure of seeing there my Biloved Lucy who constantly loves me and makes me perfectly happy. I shall wait for thy Consent and the one of my good Mamma to Marry her. could thou but see her and thou wouldst I am sure be pleased of the prudency of my choice; M..B. Bakewell is allways willing to oblige me and will do many things for me: do not participate the Ideas of M. Rozier Going to France to his father it would perhaps Injure us for a while. I wish thou would wrights to me ofnor and

[132] "Of the St. Domingo packet."
[133] "Mr. L. Huron did, a few days ago, receive some wines on a/c of M. Rozier, and hopes they prove good," etc.
[134] Miers Fisher, for many years Jean Audubon's trusted agent and attorney in America.

longuely think by thy self how pleasing it is to read a friend's letter. Give my love to all my friends and thine and kiss mamma, Rosa and Brother Pigaudeau[135] for me I hope they continue to be all happy, do remember to send me thy portrait in miniature dressed as an officer[136] it will cost thee little and will please me much. Some of thy hair and ask my sister for the Music she does not want. I wish to receive some letter from M.. Dorbigny[137] whom I have often wrighten and send some curiosities. he is yet to answer to my first.

When thou seeist Mr Rozier pray him and try to engage him to send us some-goods then we feel very inclined to set up in a retail store which would do, us a great deal of good. I will send him a letter by this opportunity—Good by farwell good father believe me for life thy most sincere friend be well be happy thy son. J.J. Audubon

J'espere que tu poura lire—adieu—adieu.

John James Audubon to Claude François Rozier

[Letter No. 4, addressed]
Mr. Fccis Rozier,
Merct, Nantes—*Ocean.*

New York, May 6th, 1807.

Dear Sir:

I wrote you recently by a ship going to Bordeaux; the letters were carefully intrusted, and I hope that they were received. I notified you of the arrival of the wines to the address of Mr. Huron of Philadelphia, and told you that part of the cases were sold. Your son informed me this morning that wine of so good quality ought never to be exported in cask, and that the profit would have been greater if the whole had been in case. Mr. Benjamin Bakewell has received the bill of lading of Mess Gereche brothers, and the gloves and the lace are at present on the road from Philadelphia to this place; perhaps we shall have them tomorrow; I am afraid that they may be dear. In several of your letters to Ferdinand you speak of a retail store, and my friend begs me tell you that nothing could suit us better than that you should have the kindness to send us enough [goods] to set up a shop at once on a good footing. As soon as advised, we shall order you to stock it with merchandise of your choice. You should have already received the bill of sale of a bale of linen cloth. You can judge that I have learned to shave Messrs the Americans, since I have been with Mr. B. B. In conscience, however, [the goods] have been sold at one third above their value. Should you decide upon sending another [shipment], do not count upon so good a sale. You must know, however, that I am always disposed to do everything for your interests, and that I shall always seek to merit your approbation. Should you decide to make [us] a consignment for a retail shop, have the kindness to follow, point by point, the following bill:

---

[135] Gabriel Loyen du Puigaudeau, his brother-in-law.

[136] That is a miniature or an old portrait of his father in the uniform of a lieutenant-commander, which with its companion, representing Mme. Jean Audubon, his stepmother, then hung in the house of "La Gerbetière" at Couëron. The original portraits, which are reproduced, measure 23½ by 18½ inches, and were painted probably between 1801 and 1806; they were inventoried in documents bearing date of November 14, 15 and 17, 1821, shortly after Mme. Jean Audubon's death. They were restored in Paris about ten years ago for Monsieur Lavigne, to whom I am indebted for the photographs and this information.

[137] Audubon's intimate friend.

60	doz. morocco leather powder flasks—green or gray, copper mounted, like those that you sell at the shop for 25 sols [soldos].
60	doz. d. d. of leather, mahogany color, at the same price.
100	boxes d.
100	music boxes,[138] in prices from 10 to 18 francs, good pieces and gay music.
100	boxes of seal-wafers, containing 1 gross each, assorted in color [but] more of the red than any other.
10	gross of small boxes of seal-wafers.
3	boxes of pastels, good, well assorted, and chosen by the sons of M. Belloc; more would not return us anything.

If you could procure us good books in English at Paris, M. Bakewell assures me [that we would realize] a great profit on them, and upon the other articles as given above, if well chosen. We hope to sell Mill Grove, and we will credit you with a great part of the profit in colonial merchandise. It is with impatience that I await some news of the indigo of Mr. B. Bakewell. Have the kindness, I pray you, to forward the enclosed letter to my father as soon as possible, and will you take from the ship *Ocean*, the carrier of this letter, a little box [sent] to your address for him, and will you send this to him also? Present my respects to your ladies; accept mine and those of the Bakewell family. Ferdinand is well. I salute you, and I am your devoted friend, Audubon.

Herewith the bill of lading of the box. The captain did not wish to make any charge, and has been perfectly polite.

John James Audubon to Jean Audubon

[Letter No. 5, inclosed with No. 4, in French, and addressed]
Mr. Fccis Rozier
Mercht
Nantes
pour Mr. Audubon *père*
aussitôt que possible

My dear Friend:

Thou wilt find herewith a bill of lading of a small box containing nineteen species of seeds, a bottle of reptiles for Mr. Derbigny [D'Orbigny], and some dried plants also for the latter. I will write thee of Mr. Kauman, by the ship *Mentor*, which is to leave a little while after this one. Adieu, my good friend! The box will be addressed to Mr. Audubon, *Md*,[139] Nantes, with "American seeds" written above; besides two Bs, like this which follows[140]

B.B

The Capn. promises me to take care of it, and of my letters also. If thou findest in my letter anything which displeases thee, remember that I am thy son. Adieu! Farewell, my good friend! Thine for life. J. J. Audubon. New York, *May 6, 1807*.

---

[138] "*Serinettes*," the old time music boxes, or bird-organs, of Swiss origin, that were very popular in America down to the time of the Civil War, or even later. They were manufactured at St. Croix as late as 1880; instruments of similar type, with dancing figures, have been adapted to the penny-in-the-slot machines common in Switzerland to-day.

[139] *Marchand*, or retail merchant.

[140] Initials of the head of his firm, Benjamin Bakewell.

Do not forget, I pray thee, to send me for the good Mrs. Bakewell the complete works of Mr. Genlis[141]by the first opportunity, and for me an exact copy of the departments of France like that which I made, and which is in thy cabinet. I wish thee to copy them for my brother-in-law.[142]

John James Audubon to Claude Francois Rosier

[Letter No. 6, addressed]
Monsieur Fccis Rozier,
Merchant
Nantes
p. Brig *Mentor*

New York, May 30th, 1807.

Mr. Francis Rozier,
Merchant, Nantes.
Dear Sir:

By my last, sent on board the ship *Ocean*, Capt. Bunken, I apprised you of the arrival of the gloves and lace, shipped by your order at Rochelle for the account of my good friend, Benj. Bakewell. I can now inform you of their sale, which is also advantageous, although the principal part was fine and of very great price. The gloves in prices of 23# 28# D, are what is needed for this market here, and especially if they are of any other color than yellow or bottle green they are less apt to soil; further they conceal defects more, and find in consequence more purchasers. The laces were better, although there was a heavy duty. You should know that here the extravagance of the women equals or rather quite balances the circumspection of the men, so that all articles for women should be beautiful, that is to say, conspicuous. I await with a kind of pleasure the arrival of Cap. Sammis, for although I am convinced that the indigoes will meet with no success at Nantes, their return here will compensate us. I am sorry that I did not order from you some little pistols and the guns which would serve perfectly. Believe nothing as to Mr. Bakewell, and be well assured that he is our friend. Have then less fear: I hope shortly to consign, that is to say, Mr. B. B. will consign for us, coffee and sugar from Martinique to your address. Your son is still at Philadelphia with Mr. Huron. They have sold the wines quite well.

But in truth I have been astonished that Mr. Huron did not make you an immediate return. I thank you sincerely for the little package that you said had been prepared for us. Be sure that Mr. B. B. will aid us to a sufficient degree, and always in a way that anything which you send us will be promptly returned in merchandise assigned to you. The land, which we cannot sell without a great disadvantage, keeps us very short of cash, and prevents us for the moment from dealing on as large a scale as we should desire; but with your kindness in sending us the materials for starting a grand retail shop with different articles, it will aid us very much. As you well say, it is a little unfortunate that there is no longer a boat from your port here.

[141] The reference was to Mme. Stephanie-Felicité de Genlis (1746-1830), teacher of the children of the Duke of Orleans, Philippe-Egalité, and authoress of many works on education, once popular, but now known only to the antiquary and the ragman.

[142] Meaning possibly his prospective brother-in-law, Thomas W. Bakewell, a fellow clerk in the office.

I write to my father by the same opportunity. Will you, I pray, get it to him as soon as possible, and I beg you to go aboard for the live birds for him and for you.

Present my respects to your good family, and believe me for ever, Your faithful friend and Servant, Audubon.

I should be very happy if you would send me a good box of pastels, chosen by Mr. Belloc, the younger, at 2 c 3 Louis.[143]

John James Audubon to Claude François Rozier

[Letter No. 7, addressed]
Monsieur Fr. Rozier,
Negociant,
Nantes.
Loire Inférieure.

New York, July 19, 1807.

Dear Sir:

Mr. Benjamin Bakewell as well as myself have received your letters by the *Comet*, which had a passage of 42 days. We have at present in the warehouse a great part of the merchandise of the latter [vessel], and in good condition; Mr. B. B. appears to be satisfied; he is about to send some teas that you have ordered from him. It has grieved me much to see him send a boat to Nantes, and not consigned to you, but his reasons were, I believe, so sound that I did not dare remonstrate. The agents of the house of Rossel and Boudet paid him the ⅔ of the invoice, or a draft upon London for an equivalent sum, that neither Ferdinand nor I were authorized to do; the latter is at Philadelphia. In a short time we are leaving for a voyage upon the Ohio, the details of which you will learn [from him], or from my father, and which I believe will be very advantageous to us. We hope to sell Mill Grove this autumn, which we shall do, however, only at a profit. We received this morning a letter from Mr. Fleury Emery, who urges Mr. B. B. to give him some shipments, but regarding this I do not know his intentions. I have also received a letter to-day from our friend, Fd, who is quite well, and longs to be doing something.

Mr. Emery advises me of the receipt of a little box of seeds for my father and you. I think that your gardens are now embellished with foreign trees.

Mr. B. B. is loading tea for you, a thing that gives me much pleasure. I am sending you a letter from Ferdinand that I received yesterday. Presenting you as well as your whole amiable family with humble respects, I continue to be your faithful servant, Audubon.

My regards, I pray to you, to my cousin, the younger.

Audubon's loyalty to his kind-hearted employer is evident in every one of these amiable letters, yet it is plain that they were written upon his own initiative, and a merchant of today might seriously object to such a candid exposition of his dealings as young Audubon's friendly epistles occasionally revealed.

The numerous references which these letters contain regarding the disposition of the "Mill Grove" farm may well puzzle the reader who has followed the story to this point; we

[143] One Louis was equal to twenty francs, or four dollars.

must therefore attempt to unravel the tangled threads of this intricate affair. In the spring of 1807 Audubon, who was then anxious to start a "retail shop," complained that the land, which could not be sold to advantage, kept them short of capital and prevented them from dealing on so large a scale as they could wish. On the 24th of April he wrote that three weeks before he had gone to "Mill Grove" and closed an agreement for renting the property (evidently referring to the farm as distinct from the mine) for a year, being unable to do better, and that Ferdinand was then in Philadelphia trying to settle his father's accounts with Dacosta, who did not readily forget his trickster's *rôle*. In Audubon's letter of the same day, inclosed in the same packet with the request that it be delivered to his father, there is a similar reference, with the note that Ferdinand, who had charge of the settlement, had chosen Mr. Huron as arbitrator, but that he would not agree unless honest Miers Fisher had a part in it. Finally, as late as the 19th of July of that year he wrote to Rozier, the elder, that they were hoping to sell "Mill Grove" in the autumn, but would do so only at a good profit; yet at this time the property had been out of their possession, technically at least, for nearly a year.

Still more curious is this statement in Audubon's autobiography,[144] relating to the year 1813; "I bought a wild horse, and on its back travelled over Tennessee and a portion of Georgia, and so round till I finally reached Philadelphia, and then to your grandfather's at Fatland Ford. He had sold my plantation of Mill Grove to Samuel Wetherill, of Philadelphia, for a good round sum, and with this I returned through Kentucky and at last reached Henderson once more."

When "Mill Grove" was conditionally sold to Dacosta and his mining company in September, 1806, he gave a mortgage and bond to Miers Fisher, who again became Lieutenant Audubon's agent. Many months elapsed before the necessary legal papers could arrive from France, and meanwhile Dacosta's yearly accounts were contested, and gave no end of trouble.[145]

After operating the lead mine for five years, Dacosta's company failed, and "Mill Grove" again passed into other hands; it was finally sold to Samuel Wetherill in 1813.[146] If our inferences are correct, the mortgages by which the Audubon and Rozier interests were

---

[144] Maria R. Audubon, *Audubon and his Journals*, vol. i, p. 32.

[145] Especially his account current, from June 1, 1806, to July 25, 1807, with the "Mill Grove" farm, and "John Audubon of Nantz," drawn up and signed at Philadelphia on the latter date. Dacosta then claimed a balance due him of $950.64 above the returns from farm and mine, of which he was entitled to one-half; this sum included his salary and numerous minor expenditures. When his account was contested and taken out of court for settlement, it was cut by the arbitrators to $530.

The following is a "copy of the Award given by John Laval & Laurence Huron appointed referees by Francis Dacosta and John Audubon the elder by a rule of reference in the Common Pleas of this county to have their differences in accounts settled:"

"We the within named referees, having heard the parties and examined their respective accounts & vouchers, do award that there is due by the defendant, John Audubon the elder, to the plaintiff, Francis Dacosta, the sum of five hundred and thirty dollars, which we find to be the full balance of all current accounts between them, and we award that the said ballance be paid by the said John Audubon the elder to the said Francis Dacosta by defalking the same from the account of the condition of the Bond of Eight Hundred Dollars—mentioned in the within rule of reference conformably to the agreement endorsed on the said Bond."

"Witness our hands Philadelphia 1st August, 1807."

"Signed—John Laval."

"Laurence Huron."

(Copy of original MS., in possession of Mr. Welton H. Rozier.)

[146] In 1811 "Mill Grove" was conveyed by Francis Dacosta & Company, to Frederick Beates, who in 1813 sold it to Samuel Wetherill, Jr., for $7,000, the property having shrunk to less than one-half the value placed upon it in 1806. For the enterprises of the Wetherills.

protected were repeatedly transferred, and the first considerable amount of ready money that had appeared in the entire series of transactions was furnished by Mr. Wetherill. It is doubtful if Jean Audubon ever received any returns from his American farm after the advent of Dacosta in 1803. The ultimate failure of the lead mine was assuredly not the fault of this exploiter, but his dubious methods of accounting and probable failure to keep his contracts no doubt led the naturalist to denounce him as a swindler.

It may be recalled that in their "Articles of Association" Audubon and Rozier had agreed that the house at "Mill Grove" should be "an object separate from all business, in order that we may control this property as long as we desire," but the conditional sale to Dacosta apparently included the farmhouse as well as the land.

Many of Audubon's references to "Mill Grove" were apparently wide of the mark, but viewed in the light which we have endeavored to shed upon this involved affair, they would be in harmony with the essential truth; in writing to the elder Rozier, who became a partner in the enterprise, there was no motive which could have led him to depart from it.[147]

We will now return to the story of Audubon's life in New York. While he was supposed to be learning the exporting business with Benjamin Bakewell, his heart was in the woods and fields, and every hour that could be snatched from the counting-room found him in the pursuit of birds or drawing their portraits. He used the pencil and black crayon point combined with pastels, and while much of his artistic work at this time was hastily done, he was capable of producing excellent likenesses. A very delicate drawing of the Wood Thrush, signed with his initials, and dated at "Mill Grove, Pennsylvania, 14 *aout*, 1806," is numbered 209, showing that his collection of American birds was already extensive, even if it did not include many that were well known. In the winter of 1806-7, while in New York, Audubon paid most attention to the waterfowl, frequently visiting the shore and the markets for his subjects. The sketches which he then made were all in full size, and, as an evidence of the rapidity with which he worked, it may be noticed that he would often complete two or more large drawings of ducks on the same day. New York at this time was a city of about 75,000 people; Audubon said that by walking briskly he could pass from one end to the other in a few minutes.

In the foregoing letters we have seen young Audubon sending seeds and live birds to his father and to Francois Rozier, and reptiles and dried plants to Charles d'Orbigny, and ordering for his own use the best drawing materials from France. While at New York he had the good

[147] Since we have been obliged to enter rather minutely into the history of "Mill Grove," in order to trace the relations of the Audubons to it in an important period of the naturalist's career, the reader may be interested in the anticlimax which its famous mines reached at a later day. The Ecton Consolidated Mining Company had been in operation at "Mill Grove" for a considerable period, when, in 1848, the Perkioming Association was formed and ten thousand dollars was at once invested in machinery. In 1851 these two companies were combined under the name of the Perkioming Consolidated Mining Company, which issued 50,000 shares of stock, at six dollars each, thus representing a capital of $300,000. A mining settlement quickly sprang up on Audubon's old farm, where numerous buildings of stone, a general store, and miners' houses were to be seen. In the first annual statement issued by this company, the buildings were said to represent an outlay of $15,000, while $140,000 had been expended on machinery, both above and below ground. A Cornish expert, who was summoned from England, was paid $1,414 for a verbose report, the substance of which, it was said, was expressed in conveying the information, already known, that the "mineral mined is copper ore" (copper pyrite occurring in association with lead). This company closed its business in 1851, by assessing its stockholders one dollar a share, thus bringing the total loss in this final effort to $350,000, nearly one-third of which had been drawn from Philadelphia. After one, or two, further unsuccessful attempts had been made, all the substantial buildings of the mining works became a quarry, from which stone was sold by the perch, the ruins of the old engine house alone remaining to this day as a witness of the follies of the generations that are gone. (This account is based upon reports which have appeared in the press of Philadelphia or in other Pennsylvania newspapers.)

fortune to become a friend and *protégé* of the most distinguished naturalist of the metropolis, Dr. Samuel L. Mitchell,[148] eminent in many walks of life, and at that time a member of the United States Senate. Audubon prepared many birds and mammals for Dr. Mitchell's collections, and the friendship thus early formed proved of much service to him later. He was probably working for Dr. Mitchell when, as the story goes, some of his neighbors lodged a complaint with the municipal authorities on account of the strong odors that habitually issued from his workroom, and a constable was sent to investigate.

Audubon remained in New York as late as August 22, 1807, for on that day he made a drawing of the "Sprig-tail Duck," but without doubt he had come to feel the incongruity of his position in a business to which his heart was a stranger. As an instance of his preoccupation at this time, he confesses to have once forwarded but forgot to seal a letter containing $8,000. If Benjamin Bakewell failed to make a business man out of Audubon, it was not from lack of kindness, and probably no one else would have been more successful. As it happened, Audubon did not leave his employer any too soon, for at the close of 1807 Benjamin Bakewell's exporting business was ruined by the Embargo Act, through which President Jefferson had hoped to bring Great Britain and France to terms by cutting off their American trade, and for a year or more his estate was in the hands of creditors for settlement.

The naturalist has left a characteristic sketch of himself at this time: "I measured," said he, "five feet, ten and one half inches, was of fair mien, and quite a handsome figure; large, dark, and rather sunken eyes, light-colored eyebrows, aquiline nose and a fine set of teeth; hair, fine texture and luxuriant, divided and passing down behind each ear in luxuriant ringlets as far as the shoulders." The habit of wearing his hair long, thus early acquired and later favored by his wandering mode of life, appears to have lasted more than twenty years.

---

[148] Samuel Latham Mitchell (1764-1831), physician, naturalist, politician and voluminous writer on many subjects. In 1797 he founded, in association with Dr. Edward Miller and Dr. Elihu H. Smith, the *New York Medical Repository*, and was its chief editor. He began also, at the University of New York, one of the earliest collections in natural history, and in 1817 appealed to the Historical Society of his city for the foundation of a Zoölogical Museum; in the same year he organized the Lyceum of Natural History, and was its first president, Joseph Le Conte serving as corresponding secretary, and John Torrey as one of its curators. On April 9, the following subjects were assigned to different members for investigation, "Ichthyology or fishes, Plaxology or crustaceous animals, Apalology or mollusca, and Geology or the earth" being reserved for the president; Samuel Constantine Rafinesque took charge of "Helmintology or worms, Polypoligy or polyps, Atmology or Meteorology, Hydrology or waters, and Taxodomy or classification;" John Torrey, who became a distinguished botanist, was more modest, and assumed charge only of "Entomology or insects;" while to John Le Conte were given "Mastodology or mammalia, Erpetology or reptiles, and Glossology or nomenclature." See the *American Monthly Magazine and Critical Review* (New York) for August, 1817, p. 272.

Chapter 12

EARLY DRAWINGS IN FRANCE AND AMERICA

Audubon began to draw birds and other animals when a child, and, like most children, was ready to believe that his crude sketches were finished pictures if only they possessed some sort of a head, a tail, and sticks in place of legs. But, unlike the majority of youth, he went direct to nature for his subjects, and his "family of cripples" failed to satisfy him long. He gradually developed a high ideal, and at an early age felt stirring within him the impulse and the power to express it. On stated anniversaries his masterpieces, he tells us, were burned, in spite of the praise and flattery they had evoked; he would then exert all his powers to do better, and this commendable practice was kept up for years.

In this respect the child was father of the man, for on the 5th of March, 1822, when Audubon was living in New Orleans, too poor to buy even a blank-book for a journal, he thus wrote of his work during the previous months: "Every moment I had to spare I drew birds for my ornithology, in which my Lucy and myself alone have faith. February was spent in drawing birds strenuously, and I thought I had improved by applying coats of water-color under the pastels, thereby preventing the appearance of the paper, that in some instances marred my best productions. I discovered also many imperfections in my earlier drawings, and formed the resolution to redraw the whole of them." Seldom satisfied with the results attained, he kept up this laborious process of revision and selection by which he approached more closely to his ideal, the truth of living nature, for more than forty years, until, in fact, the last plates of his *Birds of America* came from the press in England in 1838. An examination of the originals of those plates today[149] proves that many of their defects were inevitably caused by the makeshifts to which he was sometimes forced by lack of time.

Audubon has credited his father with the only judicious criticism which he ever received at the youthful stage of his art. "He was so kind to me," said the son, "that to have listened lightly to his words would have been highly ungrateful. I listened less to others and more to him, and his words became my law." When he was about seventeen years old, or probably not far from the year 1802,[150] he was sent to Paris to study drawing under Jacques Louis David, the acknowledged leader of French art during the period of the Revolution. This popular artist, who had uttered fierce invectives against "the last five despots of France," became

---

[149] See Vol. I.

[150] Cuvier stated in his report on Audubon's *Birds*, delivered at the Academy of Sciences, Paris, September 22, 1828, that the author had been twenty-five years before a pupil in the school of David. This would place the date in 1803, but earlier than the autumn of that year, when Audubon started for America.

nevertheless court painter under Napoleon; like many another Conventional regicide, he was destined to end his career as an exile from France, and died in Brussels in 1825.

Figure 25. Early Unpublished Drawings of French Birds: Above, European Crow with Head of Rook, "Le Corbeau Ou Corneille Noire De Buffon, English Crow, En Compagne Autour De Nantes, Grole, Petite Grole, No. 155;" Detail, "Bec De La Fragonne Ou Freux Bufton—Bec of the Jaig Daw No. 166;" Below, White Wagtail, "La Lavandiere De Buffon. Wag-Tail, Whatter Wag-Tail, White Whatter Wag Tail, Common Dish Washer, the 22 of Décember, 1805. Near Nantz. No. 65."
Published by Courtesy of Mr Joseph Y. Jeanes.

Audubon has said but little of this Paris experience, but he remarked: "At the age of seventeen when I returned from France, whither I had gone to receive the rudiments of my education, my drawings had assumed a form. David had guided my hand in tracing objects of large size."[151] An interesting sidelight is thrown upon this incident by the fact that, not many years before, David had been warmly welcomed in the city of Nantes, when it is not unlikely

[151] *Ornithological Biography*, vol. i, p. viii.

that the naturalist's father was one of the throng of citizens who made his acquaintance. The occasion to which I refer was so noteworthy in the annals of Audubon's paternal city as to make a digression at this point of our narrative inevitable. In March, 1790, Daniel de Kervegan, a wealthy merchant who was then serving his second term as mayor, had aroused so much enthusiasm by his public spirit and sterling character that the citizens had voted the sum of 300 livres, or about $60, for his portrait, to be executed in oils and placed in one of their public buildings. The commission was offered to David, who accepted it, and with such enthusiasm did he set to work, that upon reaching Nantes he asked the privilege of paying his respects to the Municipal Assembly, which was in session. Upon being admitted to the Chamber, on the 24th of March, he expressed these sentiments:

> If ever my art has brought me any gratification, or any success, never before have I had better excuse for boastfulness.
>
> I have made it a duty to respond to the worthy invitations, inspired by patriotism and gratitude, that hallow this most timely and most astounding revolution.
>
> It is your work, gentlemen, and the respect which you render to the chief of your administration which speaks in praise of your sentiments and virtues and which will transmit their memory, along with your glory, to posterity.[152]

David worked on this portrait for about a month, and on April 23, before his departure for Paris, he asked the privilege of again addressing the Assembly. Not only was the request granted, but he was publicly thanked for the trouble he had taken in coming to their city, and a committee was appointed to express the sentiments of esteem with which he had inspired the whole community. We may add that David seems to have taken this canvas to his studio in Paris, where it was subsequently lost or destroyed in the period of turbulence that followed.

David's radical speeches from the tribune, added to his popularity as an artist, no doubt brought him pupils in plenty from every quarter of republican France. Young Audubon was probably admitted to the most elementary class, for he received no instruction in the use of oils but was directed to study the rudiments of drawing from the cast. As he had hoped to perfect himself in the art of depicting animals, he was disappointed. "Eyes and noses belonging to giants," he said, "and heads of horses, represented in ancient sculpture, were my models." He also spoke of drawing "heads and figures in different colored chalks," and of "tolerable figures" obtained by use of the manikin, but adds: "These, although fit subjects for men intent on pursuing the higher branches of the art, were immediately laid aside by me"; yet he "returned to the woods of the New World with fresh ardor,"[153] and there began a series of drawings which were later published.

While this is virtually all that has been recorded of this incident in Audubon's career, a number of interesting facts might be added which throw light upon the surroundings of his life at Paris while under the tuition of this master. At that time David was enjoying the privilege, accorded to eminent artists from an early day, of living with his family and of having his studios in special quarters set apart for the purpose in the palace of the Louvre; this

---

[152] F. T. Verger, *Archives curieuses de la ville de Nantes et des départements de l'ouest* (Nantes, 1837-41); for further references to David in this chapter I am mainly indebted to Georges Cain, *Le Long des Rues* (Paris, 1812), and Charles Saunier, *Louis David* (Paris, no date).

[153] The implication as to time, which is repeated above, contradicts an earlier statement, which is probably more nearly correct, for when Audubon returned to America in 1806 he was twenty-one.

was continued until all the artist tenants were turned out by one of Napoleon's peremptory orders in 1806. David's principal studio was at the corner of the Quai de Louvre and the square, facing the church of St. Germain l'Auxerrois, at a point occupied in the present structure by the grand staircase leading to the Egyptian Gallery. It was here that his more advanced pupils studied; the appearance of its interior, with his pupils at work, as well as the view from one of its windows, by means of which its exact position can be determined, may be seen today in the interesting painting by Matthew Cochereau. This small picture, first exhibited in the salon of 1814, now hangs in the Louvre in company with some of the finest of David's works, and immediately beneath his huge canvas representing the coronation of Napoleon. Over his principal room David had also a private studio, and at one time he had another on the Quai, opposite the Institute of France, while his numerous pupils occupied a series of rooms, one above another, not remote from the first. Access to these apartments was gained from the street by means of a spiral stairway, the opening of which may still be seen in the Egyptian Hall.

It is common to speak of this gifted man as if he alone had stifled all the art of the eighteenth century in France, as if he were the molder of his age and not a part of it. Too often has he been judged on the basis of a few, unfortunately conspicuous, theatrical pieces, while his excellent portraits, of which there are many, entitle him to the gratitude of posterity. Buchanan remarked that the mannerism of David could "still be traced in certain pedantries discernible in Audubon's style of drawing," which is a fancy without any basis in fact. If it could be shown that drawing from the casts of antique statues could develop mannerisms in the careful delineation of birds and mammals, it would still appear that Audubon's style was really formed at a later period.

This brief Paris episode, which at most could have lasted but a few months, represented all the formal instruction which Audubon ever received in drawing, although he enjoyed some private tuition at a much later day. As to the sciences now embraced in biology, that is, zoölogy and botany, which would have been most useful to him, the score was blank; even books on any of these subjects were rare in America at the beginning of the nineteenth century.

When Audubon first came to the United States, he brought with him all his drawings of French birds, and a few pieces which may belong to this early period have been described.[154] Done in a combination of crayon and water color, they represent a European Magpie, a Coot and a Green Woodpecker, the latter especially, which bore the number "96," showing evidence of care and skill. The year passed at "Mill Grove" was not particularly fruitful, but during the Couëron visit which followed in 1805 and 1806, Audubon said that he made drawings of "about two hundred species of birds," all of which he brought to America and gave to his Lucy. After finally reaching this country in the latter year, these studies were continued, with an alacrity that seldom failed, until 1822, when he began to revise much of his earlier work, substituting water colors more completely for pastels, pencil and crayon point.

In writing to Bachman in 1836, Audubon thus referred to the work of his apprenticeship: "Some of my early drawings of European birds are still in our possession, but many have been given away, and the greatest number were destroyed, not by the rats that gnawed my

[154] See R. W. Shufeldt, in *The Auk* and the *Audubonian Magazine*.

collection of the 'Birds of America,' but by the great fire."[155] When the naturalist was in Philadelphia in 1824, in search of a publisher and sadly in need of funds, he made the acquaintance of Edward Harris,[156] who looked at the drawings he had for sale and said at once that he would take them all and at Audubon's own prices. Upon his leaving that city, this generous friend, we are told, pressed a $100 bill in his hand, saying: "Mr. Audubon, accept this from me; men like you ought not to want for money." "I could only express my gratitude," continues the naturalist, "by insisting on his receiving the drawings of all my French birds." The worthy Harris cherished this large series of Audubon's early studies and added to it many specimens of his later work. The entire collection remained in his family unbroken and unimpaired until 1892.[157]

This beautiful and unique collection, which represents *The Birds of America* in the bud, illustrates the development of Audubon's art from about 1800 or a little later to 1821,[158] and clearly shows that the fuller mastery which he attained after the latter date was manifested in no small degree at a much earlier period. His drawings of the Wood Thrush (1806), the Whippoorwill and Kingfisher (1810), the Carolina Parrot (1811), and the Nighthawk (1812), though detached and less ambitious as pictures, for truth of line and delicacy of finish would compare favorably with the best of his later work. After 1820 his ability had so far outstripped his ambition that there was needed only the stimulus of a powerful motive and a well defined plan to bring his powers into full fruition at once. A little later, when he began to revise, enrich and standardize all of his previous work, he used the brush and water colors more freely than ever before. Hundreds of his earlier studies were cast aside; many, to be sure, were hastily drawn in pastel, crayon and pencil, and had not time failed him at the end, nothing of his earlier American period would have remained in the final product.

Nearly all of these rejected drawings bear serial numbers, which from the lack of sequence now observed, show that they were subject to constant change and that their total number must have been great. All bear the scientific and common names in French or English or both, and many are signed with the artist's initials or name; besides giving the place and date, in some cases the weights and measurements of his subjects are added, with detailed sketches of foot, bill, or eggs.[159]

A large crayon sketch of a groundhog, in excellent drawing, is labeled "Marmotte de sauvage, No. 159, le 6 juin, 1805." The Redstart, executed in August of the same year, is a good example of Audubon's more delicate early work; it shows also the attention which he

---

155 Referring to the fire of 1835, in New York.

156 See Chapter 21.

157 When it passed into the equally worthy hands of Mr. Joseph Y. Jeanes, of Philadelphia. Mr. Jeanes purchased from the estate of Mr. Edward Harris, 2d, directly or indirectly, and at different times, about 110 of these early originals; others were dispersed, four of early date being in the Museum of Harvard University. Mr. Jeanes also possesses a large section of the Audubon-Harris correspondence, which extended over nearly a quarter of a century, and of which little has been published; to his kindness I am indebted for the privilege of reproducing some of the drawings, as well as numerous extracts from the letters, in the present work.

158 Audubon said that some of the originals of *The Birds of America* were "made as long ago as 1805," which may well have been the case, but the earliest date which has been preserved on the drawings is that of July 1, 1808, for "Rathbone's Warbler," later recognized as an immature form of the Summer Warbler. The Carbonated Warbler was drawn May 7, 1811. Seven bear the date of 1812, namely: Yellow-rumped Warbler, April 22; Le petit Caporal, April 23; Wood Pewee, April 28; Blackburnian and Bay-breasted Warblers, May 12; Chestnut-sided Warbler, May 17; and Cuvier's Wren, June 8.

159 For a list of Audubon's early dated drawings see Appendix II. Through the courtesy of Mr. Jeanes, I am able to reproduce a fuller series of Audubon's early drawings of French and American birds than has hitherto been published, and have chosen the subjects to illustrate the development of his style.

was then beginning to pay to accessories, his bird being perched on a spray of ripening blackberries. The Wagtail, on the other hand, was a rough crayon sketch, dashed off on December 22 of the same year. A pencil and crayon drawing of the Mountain Titmouse, which is a European bird, was probably made from a captive, and at sea, since it bears the date of January 22, 1805, when Audubon was, I believe, aboard the *Hope*.[160] The latest of these French pieces, designated "No. 94. Woodpecker, le 8 mars, 1806. près Nantes; 12 to the tail," was executed, about a month before the naturalist finally left France with Rozier to settle permanently in the United States. The excellence of such a drawing as that of the Wood Thrush (1806) is in marked contrast to the more ambitious "Fish Hawk or Osprey, A. Willson, Perkioming Creek, 1809," in which the bird holds a white sucker in its talons but is less happily rendered. Nine large pastels of waterfowl and two smaller pieces, representing a Robin and Brown Thrush, in the same style, are good examples of Audubon's cruder efforts of that time; they were merely hurried sketches or practice work, with no attempt to finish with all the perfection of detail of which he was then capable.

In a full-size pastel of the Black Surf or Velvet Duck, drawn on December 28, 1806, and signed "J. J. L. Audubon," the note is added: "the only specimen of the kind I have ever seen." He became well acquainted with the Velvet Ducks, now better known as the White-winged Scoters, and in his account of the species says: "As we approached the shores of Labrador, we found the waters covered with dense flocks of these birds, and yet they continued to arrive there from the St. Lawrence for several days in succession. We were all astonished at their numbers which were such that we could not help imagining that all the Velvet Ducks in the world were passing before us."[161]

Several of these drawings are credited to "The Falls of the Ohio," as the rapids of this river at Louisville were then generally called; a number to "Red Banks," the old name of Henderson, Kentucky; while five were done in Pennsylvania, probably when Audubon was at the home of his father-in-law, William Bakewell, in the spring of 1812. An excellent drawing of the Chuck Wills Widow was probably made on the Red River,[162] in Arkansas, when Audubon was exploring that country and slowly making his way to New Orleans in June, 1821, though it should be noticed that a steamboat on which he sometimes traveled was called the *Red River*.

Audubon began in the usual way, by representing his birds in profile, and often on a simple perch, but gradually introduced accessories which eventually became such an important part of his plan that, after 1822, his plates took on more the character of balanced pictures, literally teeming with the characteristic fruits and flowers of America, as well as with insects and animals of every sort, suggestive of the food and surroundings of his subjects, not to speak of American landscapes drawn from many parts of the country.

Dissatisfied with the older methods of drawing birds in the stereotyped attitudes of most stuffed specimens, Audubon made many experiments at "Mill Grove" before hitting upon what he called his "method" of using wires to pierce and hold the body of the bird in any attitude which he desired to represent. His device, which was simple only for one who possessed the requisite knowledge and skill, was publicly exhibited at a meeting of the Wernerian Society at Edinburgh on December 16, 1826.

---

160 See Vol. I.

161 *Ornithological Biography*, vol. i, p. 354.

162 See Appendix II.

Figure 26. Early Unpublished Drawing of the Groundhog: "Marmotte De Savage, Le 6 Juin, 1805, No. 159." Published by Courtesy of Mr. Joseph Y. Jeanes.

Figure 27. Water-Color Drawing of a "Young Racoon of This Year, September 10, 1841." Published by Courtesy of the American Museum of Natural History, New York.

A recently killed bird was fixed in the position desired by means of wires, and placed against a background ruled with division lines in squares to correspond with similar lines on Audubon's paper. The parts, measured if necessary with compasses, were then drawn in, and every part was rendered in due proportion. As to the difficulty of thus securing natural attitudes, aside from any question of draughtsmanship, we have only to recall the bungling work of most taxidermists; there are careful students of animal life who are able to reanimate their subjects, even when reduced to dried and mounted skins, but such ability is not easy to acquire or impart. Method is always subordinate to power, and Audubon at his best, when not hampered by lack of time, was able to represent the living, moving bird in a hundred attitudes never attempted before, which surprised the world of his day by the remarkable skill, freshness and fidelity they displayed.

Some have complained that Audubon, in striving for effect, too often exaggerated the action of his subjects; his birds, like the Frenchman he was, gesticulate too much, while Wilson's were more cautious or sedate, as became a canny Scot. The complaint may be well founded, but the explanation is too trivial for serious consideration. Wilson, like his predecessors, regardless of nationality, merely followed custom, which led by the path of least resistance. Barraband and all the best French artists before him in depicting bird and animal life had done the same, and in their hands the perch, were the subject a bird, became stereotyped to the last degree, as if inserted with a rubber stamp. Audubon followed the same course until he became imbued with the desire of endowing his animals with all the moving energy of which they were capable, whether in seizing their prey, feeding their young, or fighting their enemies. It is well known that many an animal, though ordinarily cautious or even timid, can be roused to vigorous action under the spur of emotion, as when its young are suddenly threatened, and be it warbler, bluebird, or cuckoo, may become a contortionist at a moment's notice. Very few of the 1,065 life-size drawings of birds which appear in his large plates could be truly described as fantastic or unnatural.

Audubon's problem was rendered more difficult by the fact that all of his animals were drawn to the size of life, and because his desire and style compelled him to represent the utmost detail, even to the barbs of a feather or the individual hairs of a mammal. When a landscape was to be included it was not an easy task to harmonize life-sized objects in the foreground with receding objects, and here he sometimes failed. Some of his least happy compositions, however, were the result of haste, as an examination of the originals of his *Birds of America* has clearly shown; when hard pressed for time he would resort to the scissors and paste, in order to combine the parts of several distinct drawings into one plate, and often leave the backgrounds to be supplied entirely by the engraver. One of the few grotesque results of such methods is seen in plate 141, wherein are represented the Goshawk and the Stanley Hawk; the latter, which was originally designed for different surroundings, has quite lost its center of gravity on an islet amid stream. An early reviewer thought that the artist must surely have intended this for a caricature, as in the case of one of Hogarth's famous prints, in which a man on a distant hill is lighting his pipe at a candle held out of a window in the foreground.

The action of Audubon's subjects was sometimes exaggerated; his birds on the wing were occasionally ill drawn, and other defects might be mentioned. But we must admire his boldness for attempting so many difficult positions, and admit that, when all is considered, he succeeded to admiration, and set a new standard for the illustration of works on natural history.

Chapter 13

AUDUBON'S MARRIAGE AND SETTLEMENT IN THE WEST

In the summer of 1807 Audubon and Rozier had decided to try their fortunes in the West, which then meant the Ohio Valley and the wilds of Kentucky, and had fixed upon Louisville as a promising point for pioneer trade. On August 1 they purchased a considerable stock of goods through the commission house of their friend, Benjamin Bakewell, and three days later gave their note, payable in eight months, for over $3,600.[163] Then, or a little later, they had dealings also with Messrs. Robert Kinder & Company, of New York, as well as the French importing house of Laurence Huron, with which Ferdinand had been recently associated in Philadelphia; apparently also they sent goods to François Rozier at Nantes, and from him received imports through the Bakewell firm, but, as we shall see, all foreign trade was soon cut off. When their plans were complete and their goods had started for the frontier, they set out themselves for Louisville on the last day of August, 1807.

Ferdinand Rozier kept a record[164] of this journey, the formidable nature of which will be best appreciated by reading his matter-of-fact narrative composed from notes daily jotted down. In these easy-going times, when oceans and continents are crossed with ever increasing ease and speed, this simple chronicle of early travel in America is worth preserving, if only for its historical contrasts.

> On the thirty-first day of August, 1807, in company with Audubon, I left Mill Grove for Louisville, Kentucky, where we anticipated engaging in the mercantile business.

---

[163] See Appendix I, Document No. 11.

[164] This diary was first brought to my attention by Mr. Ruthven Deane, and for permission to reproduce it I am indebted to the kindness of a great-grandson of Ferdinand, Mr. Welton A. Rozier, of Saint Louis. Mr. Rozier writes that the original French notes have been mislaid or lost, but that they were closely followed in this translation, whenever complete. Though numerous verbal changes have been made in the present draft, these have not altered the meaning in any respect. Ferdinand Rozier's narrative begins as follows:
"I left Nantes, France, in company with John James Audubon, on Saturday, the 12th day of April, 1806, bound for the city of New York, U. S. A., on an American ship named the *Polly*, commanded by Captain Sammis, and arrived at New York on Tuesday, the 27th day of May. While on the voyage across the ocean our vessel was stopped, overhauled, searched, and robbed by an English privateer, named the *Rattlesnake*, which detained us a day and a night.
"We remained in New York City for a few days, and then removed to Mill Grove, on Pickering [Perkioming] Creek, in Pennsylvania, a tract of land owned by our fathers, and at that time thought to contain valuable minerals."

Leaving Philadelphia by stage we traveled to Lancaster, Pennsylvania, a distance of sixty-one miles, where we arrived at four o'clock in the afternoon; we dined, and proceeded to Big Chickers, distant nine miles farther, where we spent the night. The roads from Philadelphia to Lancaster were in excellent condition, and at about every two miles we found good taverns. The only remarkable thing we noticed in agriculture was hemp, there being little else of interest. The city of Lancaster was attractive, but the short duration of our stay prevented us from having more than a casual view of it. The tavern where we slept was not very good; from our chambers, however, we could discern a new bridge, which had two immense arches spanning the river.

At eight o'clock in the morning we left Lancaster for Elizabethtown, distant nine miles. The roads were miserable, and we suffered a severe jolting and shaking up. Arriving there, we procured two additional horses, which made six all told, and went on to Middletown, where we breakfasted at a tavern named the "Eagle"; the village was small, with few houses, and nothing of interest.

Journeying on to Harrisburg without mishap, over roads somewhat improved, we finally arrived, and discovered a very beautiful river called the Susquehanna. The city of Harrisburg itself appeared very attractive to us, and its situation is beautiful; proceeding, we were first compelled to cross the river, which was accomplished by means of a large flatboat propelled by a sweep of generous proportions. The captain, who proved a most voluble person, informed us that the river abounded in fish, and then related marvelous tales of the remarkable catches that had been made; many of his stories, however, were of such glaring improbability that we were forced to doubt his veracity.

Carlisle, sixteen miles distant, was reached in due course, and there we changed horses at a tavern called the "John Mason." This city, though small, presented a fine appearance, having a market place, two large churches, many brick buildings, a large academy, and several attractive taverns. Continuing, we finally came to Walnut Bottoms, where we engaged chambers at a very imposing tavern; this proved far superior to any we had hitherto visited; it was clean and inviting; its appointments were good, and its service excellent. On our journey we were impressed by a tree of great size, that resembled an oak, but upon inquiry learned that it was called Hackberry,[165] and produced a fruit similar in size to a cherry. On the north and south of us were high mountains which presented an imposing appearance; the foliage was heavy and luxuriant; the soil of the foot-hills appeared fertile, but the crops were inferior.

We were awakened early in the morning so as to begin our journey in good season, and having had a heavy storm during the night we expected to find the roads very bad, but to our delight they were none the worse for the rain. Journeying most of the way through woods, we came to Shipensburg and breakfasted; this village had only one long street, and presented an appearance far from pleasing. A lady with her sock [knitting work] proved a great talker and asked us many questions. This village was intersected by a creek, called the Middlespring. We next came to Chambersburg, ten miles away, and there rested and purchased tickets for continuing our journey. That village lies in a valley, and is composed of two squares containing a post office, an academy, a factory, market place and tavern.

When the stage was at last made ready for its journey we took our places in it, but no sooner was the village left behind than we encountered very rough roads, which for a

[165] In the rich bottom-lands of the Ohio River basin the hackberry or sugarberry (*Celtis occidentals*) sometimes exceeds one hundred feet in height, and has a diameter of from four to five feet.

time caused great discomfort; our feelings were expressed by all the passengers, but at length we reached a tavern named "Cable Roussed," where our horses were changed. We next stopped at the "John Campbell" tavern, and saw many drunkards about; then at "Peter White's," almost at the foot of the mountains, where we were each treated to a glass of excellent fresh milk. Still going on and approaching the mountains, the roads became so excessively rough that Audubon and myself decided to proceed on foot. Though this was a three-mile climb, we managed to cover it in three and a half hours. So bad in truth was the road that it seemed well nigh impossible for any vehicle to ascend the mountain; the stage did go up, however, and reached the summit soon after us. On the heights of the mountain was a small tavern where refreshments were served, and while partaking of a light lunch there we were waited on by a couple named Currie, and James, their hired man. While we were refreshing ourselves, our host told harrowing tales of wild-animal hunting in the mountains, and assured us that there were many beasts in the surrounding woods. Leaving the summit in the stage, we continued for some distance, but the jolting, rolling and swaying was so frightful that we decided to descend on foot. The three miles down the mountain was covered quickly, but we were utterly worn out with fatigue when we reached McConnelsburg; this village lies in a valley, has few houses and but little of interest; we made forty miles during the day. Leaving early on the next morning, after traveling thirty-two miles, over better roads, we spent the night at the tavern of B. Mastin.

Having breakfasted at an early hour, we were again on our way by sunrise, and after driving two miles came to the Juniata River, which was crossed in a leaky flatboat. Eight miles beyond this point we saw a very fine and stately mansion which was said to belong to a Mrs. Haily. Finally after a hard and tiresome day we arrived at Bedford. The Juniata River flows along Bedford in a narrow bed, between high mountain walls; the village is situated in the valley, and boasts many fine stores and residences. We were told that about fourteen miles farther on there were mineral springs, the waters of which possessed great curative properties, and that many people visited them each season; time, however, did not permit us to visit this resort.

Six horses were hitched to our stage when we departed the next morning. The mountain roads ascended more gradually, and were less rough; the weather being exceptionally fine, forty miles were easily made before reaching our destination at a village called Somerset, which contained a courthouse that marked it at once as the county seat. At four o'clock of the morning following we were again on our way, and left Somerset in a heavy fog, which at that early hour sharply accentuated the chill in the air. At the end of the day we found ourselves at Laurel Hill, where we passed the night at the tavern of John Arranats.

Again at four in the morning we resumed our journey, and after crossing Laurel Creek once more encountered rough roads, but soon reached a tavern called the "Jacob Hoff," where we breakfasted. Still pushing forward, at noon we came to the small house of a family called Margennefs, and procured a meager lunch. At a short distance from this place a change of horses was made, and after driving all the afternoon we entered the attractive village of Greensburg, where we spent the night. Rising reluctantly at peep of day, we continued on our course and made ten miles before breakfasting at a tavern, the "Stewart Auberge" by name. After leaving this point we came to Turtle Creek, when the road descended so abruptly that it was decided to dismount and walk, but the heat was sultry and oppressive, and we suffered greatly. At last, however, the city of Pittsburgh was reached, and there we found good and commodious lodgings at the Jefferson Hotel,

> conducted by Mr. Galland, a most genial and agreeable host. We remained in Pittsburgh several days, and became acquainted with many of its citizens, among whom were several countrymen of ours who were engaged in business and were very congenial and hospitable. The city does not present a pleasing appearance; it has been increasing in size with astounding rapidity,[166] and possesses a remarkable commerce; the Ohio River there is most beautiful.
>
> The remainder of our journey was by way of the Ohio, and we made it entirely in an open flatboat, a cumbersome unwieldy craft, managed by hand, and in this particular instance very badly. One who has never had this experience can little understand the terrible monotony, hardships and deprivations encountered on a long journey such as we endured. We were unprotected from the elements, and our beds consisted of bare pine boards, upon which we slept as best we could, enveloped in our great coats.
>
> There were times without number when our boat would run upon hidden sand bars to become grounded, and we were then often obliged to get into the cold water and assist in the work of extricating her. At other times, unprotected as we were, the rains drenched us to the skin, and our clothing was so saturated that it took many hours to dry. At night when it was clear, we continued our course down the river, but, in bad weather, or when very cloudy and dark, we were obliged to tie up to the shore, frequently to the bank of some wild, uninhabited island, and wait there for daylight; then we would resume our slow, tedious and seemingly never ending journey. Added to these hardships, our boat was commanded by a most disagreeable and ungentlemanly captain, named Harris; his language, and demeanor marked him as a person of low birth and bad character.
>
> Among some of the places which were passed *en route*, I remember the following: Wheeling, Marietta, Market Slough, famous for the conspiracy of Colonel Burr, Belleville, Litards Falls, Point Pleasant, Manchester, Maysville, Cincinnati, and finally our journey's end, Louisville.

At Louisville the partners were attracted by the country and its prospects, as well as by the hospitable character of the people. Their choice, as they then thought, had been well made, and they decided to make it their future home. "We marked Louisville," said Audubon, "as a spot designed by nature to become a place of great importance, and had we been as wise as we now are, I might never have published *The Birds of America*; for a few hundred dollars laid out, at that period, in lands or town lots near Louisville, would, if left to grow over with grass to a date ten years past [this being 1835], have become an immense fortune, but young heads are on young shoulders; it was not to be, and who cares."[167]

Rozier did not say when either they or their goods reached the pioneer settlement, but from an item in their account current with the Bakewell house,[168] it is evident that they opened a retail shop in Louisville at once, for on September 29 they were charged with $57 for an order of powder horns and shotbags. In the same record there is a more interesting entry under date of December 31, 1807: "advanced per [sailing packet] *Jane*, for indigo and expences ... $1,516.43," ordered evidently through Mr. Bakewell, presumably for export to France. This incident Audubon must have had in mind when in after life he wrote: "The mercantile business did not suit me. The very first venture which I undertook was in indigo; it

---

[166] The population of the second city of Pennsylvania in 1800 was 1,565; in 1840, 4,768; and in 1910, after the annexation of Allegheny, 533,905.

[167] Maria R. Audubon, *Audubon and his Journals*, vol. i, p. 28.

[168] See Appendix I, Document No. 11.

cost me several hundred pounds, the whole of which was lost." It may be recalled that in his letter of April 24 of this year, Audubon wrote Francois Rozier[169] that the Bakewell house had sent him a consignment of indigo by the same ship, Captain Sammis, and hoped for its favorable sale in France. No doubt the venture succeeded so well that the young traders were induced to repeat the experiment. As it happened, however, on December 22, a week before this entry for the indigo was made, the famous Embargo Act of President Jefferson had taken effect, with the result of cutting off all exports to England and France and at the same time of paralyzing American trade. The Bakewell house, as we have already noticed, like so many others, immediately went down, and the partners found that their tobacco and other western produce found so little sale in New York that by April 7, 1808, they were obliged to call for an extension of their notes.

Notwithstanding the gloomy outlook for trade, Audubon had no fears for the future. As early as March, 1808, he left Rozier in Kentucky and returned to Pennsylvania. No time was lost in making known his plans to Lucy Bakewell and her family, and having received their approval, the lovers prepared for the adventurous journey that was to celebrate their wedding. Audubon was married to Miss Bakewell, at "Fatland Ford," on Friday, April 8, 1808, by the Reverend Doctor Latimer, an Episcopal clergyman of Philadelphia, and on the next morning started with his bride for the frontier. This event must be regarded as the most auspicious in his career, for in all probability the world would never have heard of Audubon had it not been for the spur to his ambition and the balance wheel to his character which came through his admirable wife.

The first stage of their honeymoon involved the long ride of over 250 miles to Pittsburgh, the hazards and discomforts of which we have learned from Rozier's description; it was marked in this instance by an accident, for in crossing the Alleghany mountains their coach was upset and Mrs. Audubon did not escape without severe bruises. At Pittsburgh the Audubons met a number of young emigrants bound westward like themselves, and in their company they prepared to float down the beautiful Ohio in a flatboat or ark. Their entire journey, which, owing to the windings of the river, could not have been much less than a thousand miles, was made in twelve days, and without further mishap.

The wild and varied beauty of the Ohio of that day had great attractions for the naturalist, who often regretted that no facile writer had left a true and vivid picture of it for the benefit of posterity, for he foresaw with great concern the inevitable changes which advancing civilization would quickly produce along its delightful banks. Audubon traversed this mighty highway countless times in after life, and some of his musings have lost none of their interest with the flight of time, for he had witnessed the advance of the white man and the retreat of the red, along with the great herds of deer, elk and buffalo that once found peaceful pasturage on its banks. Speaking of a later but hardly less romantic journey,[170] he said:

> As night came, sinking into darkness the broader portions of the river, our minds became affected by strong emotions, and wandered far beyond the present moments. The tinkling of bells told us that the cattle which bore them were gently roving from valley to valley in search of food, or returning to their distant homes. The hooting of the Great Owl, or the muffled noise of its wings as it sailed smoothly over the stream, were matters

[169] See Chapter 11.

[170] When Audubon was returning with his wife and infant son from Pennsylvania to Kentucky in the autumn of 1810; see "The Ohio," *Ornithological Biography*, vol. i, p. 29.

of interest to us; so was the sound of the boatman's horn, as it came winding more and more softly from afar. When daylight returned, many songsters burst forth with echoing notes, more and more mellow to the listening ear. Here and there the lonely cabin of a squatter struck the eye, giving note of commencing civilization. The crossing of the stream by a deer foretold how soon the hills would be covered by snow.

Many sluggish flatboats we overtook and passed; some laden with produce from the different head-waters of the small rivers that pour their tributary streams into the Ohio; others, of less dimensions, crowded with emigrants from distant parts, in search of a new home.

The margins of the shores and of the river were at this season amply supplied with game. A Wild Turkey, a Grouse, or a Blue-winged Teal, could be procured in a few moments; and we fared well, for, whenever we pleased, we landed, struck up a fire and provided, as we were, with the necessary utensils, procured a good repast.

Louisville at this time was a small trading and agricultural center of barely a thousand people.[171] Though the early promises of business there were not fulfilled, Audubon and his wife at once entered upon a happy period, for they made many friends in a new country settled by whole-hearted, well-to-do planters; the men were fond of good horses and of hunting, and the naturalist, who was also a merchant, was welcomed among them as a kindred spirit. But, said Audubon, "birds were birds then as now, and my thoughts were ever and anon turning towards them as the objects of my greatest delight. I shot, I drew, I looked on nature only; my days were happy beyond human conception, and beyond that I really cared not.... I seldom passed a day without drawing a bird, or noting something respecting its habits, Rozier meantime attending the counter."

To revert again to the business affairs of the Audubon-Rozier firm at Louisville, an interesting record has been preserved in a letter[172] written by Thomas Bakewell, a former fellow-clerk of the naturalist in the senior Bakewell's counting-house in New York; this was included with the statement of account, referred to above.

Thomas Bakewell to Audubon & Rozier

[At bottom of account sheet] New York, December. 13th. 1808

Messrs. J. Audubon & F. Rozier
Louisville
Gentn.

I have now the pleasure to hand you your account current with my Father's Estate according to your desire as expressed in your letter to Mess Robt. Kinder & Co. under date the 21st. of Novr. last. I cannot tell what error you allude to of $93. I suppose it is the amount of commission returned $93.94/100 which you will perceive is duly at your Cr. in the a/c. I am sorry to say that the tobacco is still unsold & that there is no prospect of selling it so as to cover the balance of your a/c. Messrs R. Kinder & Co. request me to say that they wish the yarn mentioned in their letter of the [word omitted] to be made of

[171] In 1800 the population of Louisville was 600, and in 1810 it had risen to 1,350; see Charles Cist, *Cincinnati in 1841* (Cincinnati, 1841).

[172] For this and the letter of Thomas Bakewell's uncle, William Bakewell, which follows later, I am indebted to Mr. Tom J. Rozier, and for accompanying "Account Current" of Audubon & Rozier.

water rotted Hemp & that they will write you or next post with their account against you as requested by you— I remain Gentn with Your mo. obt. Servt. Thos. Bakewell for the assignees of my Father's estate—

Give my love to Mrs. A. my aunt a recd. hers last night—S. & is much as usual—she remains very sick yet.

T B

[Superscribed] Messrs. Audubon & Rozier
Merchants
Louisville
Kentucky

Audubon fraternized with the sporting men of his district, who gladly sent him every rare bird that fell to their guns. At Shippingport also, then an independent center below the falls or rapids, he found a sympathetic spirit in Doctor W. C. Galt, a local botanist, as well as in Nicholas Berthoud, who had become his wife's brother-in-law, and who was a friend on whom he could always rely. The spirit of hospitality so manifest in all these new friends won the heart of Audubon and of his attractive wife, to whom the door of a neighbor's house was sure to open whenever business or adventure called her husband away. "We lived," said Audubon, "two years at Louisville, where we enjoyed many of the best pleasures which this life can afford; and whenever we have since passed that way, we have found the kindness of our former friends unimpaired." It was while they were living at Gnathway's hotel of the "Indian Queen," in Louisville, that Victor Gifford Audubon, who was destined to become his father's right hand in the publication of his most important works, was born on June 12, 1809.

When Audubon had reached his twenty-fourth year, nature, his fond nurse from infancy, was calling to him more loudly than ever before, but to most of his contemporaries his devotion to natural history could have seemed little else than sheer madness, or, at best, an utter waste of time. By the year 1810 his portfolios were swelling with upwards of two hundred pictures of American birds, produced, to be sure, without any plan, and far inferior to the best of his later work, but still done to the size of life, in the natural colors, and far excelling in fidelity and charm anything that had been attempted before. At this time, however, the young traders needed money for more practical affairs, and Audubon's father-in-law, William Bakewell of "Fatland Ford," consented to sell a portion of this estate, amounting to 170 acres, in order that his daughter, Lucy, might immediately realize her interest in it. From this sale nearly $8,000 was obtained; the money was deposited with Messrs. Robert Kinder & Company of New York, a firm with which Audubon and Rozier had dealt from the opening of their business at Louisville. This is clearly shown by the following interesting letter:[173]

William Bakewell to Audubon & Rozier

Fatland Ford *10 Apl 1810*

Messs Audubon & Rozier
Gentn

[173] See Vol. I.

I have at last settled the whole business with Mr Josh Williams I have allowed him for the two thirds in cash 3 per cent & have emitted to Messrs Kinder's 7838.50 on your account.—The quantity was surveyed to 170 acres at 47.5 per acre 7998.50, from which was deducted 160 dols for discount

As I have had a great deal of trouble & anxiety in this business & had to find assistants in surveying with several days attendance, dinners &c for the whole party several journeys to Norris Town and also to Philada with the carriage to convey the money—postages &c.—I charge you 1½ per cent on the purchase money which I hope you will think not unreasonable as I believe it is under the charge of the land brokers in Philada & they have no trouble in the business compared to what I have had—I feel as if a great burthen was taken off my back now it is all finished. Out of this you will please to present Lucy with 38 dols which was the price the mare sold for—I expected one of you Gentn would have come to the Eastward before now it is I expect Mr Roziers turn this Spring

I had one forged note returned at the Bank out of the money of Mr Williams & one dollar a counterfiet, but I had stipulated that he should take any faulty ones back. He paid about a third of the money in specie so that I was obliged to take the carriage with it. I took it to the Pennsylvania Bank & got an order on the Manhattan Bank in N York & have Mr Kinder's receipt for the order

They have got a considerable quantity of ore out of the mine[174] some lead & some copper but I do not hear of any being yet sold.

Present the kind regards of our family circle to my daughter, Mr Audubon, & my Grandson[175] who I hope are well. I remain Gentn,

Yours truly, Wm Bakewell

PS. Mr Kinder is of opinion that there ought to be a renunciation by Lucy of any claim of dower upon this estate to make the title good this may be sent on when you are coming this way

[Addressed] Messs Audubon & Rozier
Merchs
Louisville
Kentucky—
[Endorsed] Recd. May 5th. 1810

Lucy Green Bakewell, Audubon's wife, was three years younger than her husband, having been born at Burton-on-Trent, England, in 1788. Her family were descended from John Bakewell of "Castle Donnington," in Leicestershire; Robert Bakewell, the geologist, who came to the naturalist's defense many years later, and who lived until 1843, was a nephew of her grandfather, Joseph Bakewell of Derby. Left an orphan at an early age, Lucy's father, William Bakewell, was brought up by an uncle, Thomas Woodhouse, a rich bachelor of Crith, Derbyshire, who eventually left him a fortune.

When William Bakewell succeeded to his uncle's estate and manor, he lived the life of a country gentleman, devoting himself mainly to shooting and to the study of chemistry and natural philosophy, while he enjoyed the friendship of such men as Joseph Priestley and Erasmus Darwin. His advocacy of Priestley's republican and liberal religious doctrines is said

174 The lead mine at "Mill Grove," which with the remaining Audubon and Rozier interests in the farm had been taken over by Dacosta's company in September, 1806. The failure of Dacosta followed in about a year after the date of this letter.

175 Victor Gifford Audubon, who was then nine months old.

to have cost him the honorary office of justice of the peace in his community and to have determined his emigration to America. His first visit to America was made in the summer of 1798, when, with his brother Benjamin,[176] he started an establishment for brewing English ale at New Haven; through his chemical knowledge and skill he is said to have reproduced to perfection the famous Burton ales. William Bakewell brought his family to the United States in 1802, and when a disastrous fire destroyed his business at New Haven, he took up the large farm of "Fatland Ford" in 1804, as already related. In that retired spot he devoted much time to his library and laboratory, while living a life of easy independence. If abrupt in manners and inclined to severity in discipline, he was generous, kind-hearted and an ardent republican. Mrs. Audubon's mother, who felt keenly the separation from her own people, died in September, 1804, a few months after reaching "Fatland Ford," and in the following year William Bakewell was married to Rebecca Smith. This lady seems to have taken a strong dislike to Audubon, for when her death was announced in 1821,[177] he referred to her as "my constant enemy ... God forgive her faults."

At this time Audubon studied nature for the pure love of it, without the faintest expectation that his labors in natural history would ever be of any service to the world. But in the year 1810 occurred an event, of seemingly small moment at the time, which nevertheless left a distinct mark upon his career, as will be now related.

---

[176] See Vol. I.

[177] William Bakewell died at Philadelphia on March 6, of the same year, after suffering from the effects of a sunstroke, and was, eventually, buried at "Fatland Ford;" in 1822 his farm, originally of 800 acres, passed into the hands of Dr. William Wetherill. W. G. Bakewell, *Bakewell-Page-Campbell.*

Chapter 14

A MEETING OF RIVALS, AND A SKETCH OF ANOTHER PIONEER

On January 30, 1810, a man of rather coarse features, with a head of sandy hair, and possessed of manners that could be winning or aggressive according to his mood, might have been seen leaving Philadelphia afoot, for he had planned to keep his expenses down to a dollar a day and traveling by coach or on horseback suited neither his purse nor the objects of his mission. His clothing was coarse; his luggage, with the exception of a fowling-piece and two red-backed volumes of quarto size, was of the lightest description. But, could we have peered between the covers of those books, our curiosity would have been whetted, for they were filled with colored plates of American birds, the first-fruits of their bearer's untrained eye and hand; the text, moreover, was printed in a style which would have done honor to any country.

This man was Alexander Wilson, who, like Audubon, was a pioneer in the study of the birds of his adopted land, but who was twenty years his predecessor in point of publication. The books which he then carried were part of the first edition of his now famous *American Ornithology*, the second volume of which had appeared in Philadelphia at the beginning of that year. Though not destined to be completed until after his death, this work was to become one of the scientific and literary treasures of the nation, but it is not likely that one in ten thousand had then ever heard of him, whether as poet or as ornithologist, or cared anything about his work or his mission.

Wilson at that moment was starting on his last long journey through the West and South, in search of new birds. He also carried in his pocket a subscription list, and therefore belonged to that class of visitor which is seldom welcomed with rapture. At Lancaster, Pennsylvania, Wilson's first important stopping-place, and at that time the capital of the State, Governor Snyder put down his name for $120, the price of the completed work. This seemed a good omen, but, at Hanover, in the same state, an incident occurred which might have discouraged a less determined man; the interview has become historical, and we shall give Wilson's own relation of it:[178]

---

[178] In a letter to Alexander Lawson, written from Pittsburgh, on February 22, 1810; see Elliott Coues, "Private Letters of Wilson, Ord, and Bonaparte," *Penn Monthly*, vol. x, pp. 443-455 (Philadelphia, 1879).

Having a letter from Dr. Muhlenburgh to a Clergyman in Hanover, I passed on through a well cultivated country, chiefly inhabited by Germans, to that place, where a certain Judge *Hustetter* took upon himself to say, that such a book as mine ought not to be encouraged; as it was not within the reach of the commonalty; and therefore inconsistent with our Republican institutions! By the same mode of reasoning, which I did not dispute, I undertook to prove him a greater culprit than myself, in erecting a large elegant three story Brick house, so much more beyond the reach of the *Commonalty* as he called them, and therefore grossly contrary to our Republican institutions. I harangued this Solomon of the Bench more seriously afterwards, pointing out to him the great influence of Science on a young rising Nation like ours, till he began to show such symptoms of *intellect*, as to seem ashamed of what he had said.

At Pittsburgh Wilson met Audubon's old employer and relative by marriage, Benjamin Bakewell. The picture which he then drew[179] of that growing hive of industry will be read with interest:

On arriving at the town, which stands on a low flat, and looks like a collection of Blacksmith shops, Glass houses, Breweries, Forges, and Furnaces, the Monongahela opened to the view on the left running along the bottom of a range of hills so high that the sun at this season sets to the town of Pittsburgh at a little past four. This range continues along the Ohio as far as the view reaches. The ice had just begun to give way in Monongahela, and came down in vast bodies for the three following days. It has now begun in the Alleghany, and at the moment I write it is one white Mass of rushing ice. The country beyond the Ohio to the west appears a mountainous and hilly region. The Monongahela is lined with Arks, usually called Kentucky Boats, waiting for the rising of the river, & the absence of ice, to descend. A perspective view of the town of Pittsburgh at this season, with the numerous arks and covered keel boats preparing to descend the Ohio, the grandeur of its hills, and the interesting circumstance of its three great rivers—the pillars of smoke rising from its Furnaces Glass works &c. would make a noble picture. I began a very diligent search, in the place the day after my arrival for subscribers and continued it for four days. I succeeded beyond expectation having got 19 names of the most wealthy and respectable part of the inhabitants. The industry of the town is remarkable; everybody you see is busy; & as a proof of the prosperity of the place an eminent lawyer told me that there has not been one suit instituted against a mercht. of the town these three years! The Glass Houses, of which there are 3, have more demands for Glass than they are able to answer. Mr. Bakewell the proprietor of the best, shewed ... yesterday a Chandelier of his manufacture highly ornamented, ... for which he received 300 dollars. It would ornament the ... in Philada. and is perfectly transparent.

Eight days after he had reached Pittsburgh, Wilson bravely launched a little skiff, which he christened the *Ornithologist*, and began an arduous and perilous journey to Cincinnati, Louisville and New Orleans, a distance of two thousand miles. "In this lonesome manner," he wrote, "with full leisure for observation and reflection, exposed to hardships all day, and hard berths all night, I persevered from the 24th of February to Sunday evening, March 17th, when I moored my skiff safely in Bear Grass Creek, at the rapids of the Ohio, after a voyage of seven hundred and twenty miles."

[179] See Elliott Coues, *loc. cit.*

Cincinnati, then a town of five hundred houses, was reached on the ninth of March; while there Wilson made the acquaintance of Dr. Daniel Drake, who was later Audubon's friend, and examined a collection of Indian relics which had been taken from a freshly opened mound. He left Cincinnati convinced that its well-to-do class must be a very thoughtful people, so many of them, when approached for a subscription to his work, having replied that they would "think about it." Upon nearing Louisville at nightfall he became alarmed lest he should be drawn into the suction of the Falls, as no lights could be seen on the banks: cautiously coasting along the shore, where he encountered many logs and sawyers, at last he entered the Creek and secured his skiff to a Kentucky boat; then, "loading myself with my baggage," he wrote, "I groped my way through a swamp up to the town."[180] When Wilson had seen the Falls by daylight, he felt that his fears of the night before had been groundless, and declared that he should have no hesitation in navigating them single-handed.

It will be interesting to follow Wilson's journey a little further, before returning to the Louisville visit. After passing a few days in Audubon's town, he struck out into the heart of Kentucky, calling at Shelbyville, Frankfort and Lexington, and eventually reaching Nashville, Tennessee. Not far from the latter place he met a landlord of admirable discrimination, Isaac Walton by name, who showed himself worthy of his illustrious ancestor by declaring that Wilson was evidently traveling for the good of the world, and added: "I cannot, and will not charge you anything. Whenever you come this way, call and stay with me; you shall be welcome."

At Nashville Wilson wrote to Miss Sarah Miller, the lady to whom he was engaged but whom he did not live to marry: "Nine hundred miles distant from you sits Wilson, the hunter of birds' nests and sparrows, just preparing to enter on a wilderness of 780 miles—most of it in the territory of Indians—*alone* but in good spirits, and expecting to have every pocket crammed with skins of new and extraordinary birds before he reach the City of New Orleans." Continuing on his course in search of new birds and subscribers, Wilson arrived at Natchez on May 18, and, passing through Louisiana, on the sixth day of June he entered New Orleans, where his spirits were immediately raised by the accession of sixty new names to his list. After six months of continuous effort, traveling now in a small boat, now on the back of a horse, but frequently on foot, drenched by torrents of rain or scorched by the unaccustomed heat, often compelled to drink the poisonous water of cane brakes in Mississippi (to which must be attributed an attack of malarial fever, which he was able with difficulty to throw off, but from which, in all probability, he never fully recovered), he returned to New York by sea, and on September 2, 1810, was again in Philadelphia.

On this journey Wilson was a pioneer in much of the territory which Audubon had hardly begun to explore, but which later became the scene of his wanderings and adventures for many a year. At Louisville the two naturalists met, but they did not become good friends; though devoted to the same objects, differences in temperament might in any event have kept them apart. Unfortunately, the feelings of jealousy which were then aroused, or which were stirred up at a later day, were fostered by some of Wilson's injudicious friends to such an extent that from the moment Audubon's work became known, and long before he had published a line, they became as thorns in his path, and they continued to vex him for thirty years. It is not easy to reach a fair judgment in this matter now, and it would be impossible to

---

[180] Letter to Alexander Lawson, dated at Lexington, April 4, 1810; see Grosart, *Poems and Literary Prose of Alexander Wilson*, vol. i, p. 189.

do so without a better understanding of the man who suddenly appeared upon Audubon's horizon at Louisville in 1810 and then vanished. Because of the peculiar relations which existed between these two pioneers, we must follow the history of the elder man a little more closely.

Alexander Wilson was the son of a weaver at Paisley, Scotland, where he was born in 1766; he was thus Audubon's senior by nineteen years. His father, who was esteemed for his honesty and intelligence, had tasted prosperity, but irremediable poverty fell to his lot in later life. Alexander, the younger son, was motherless at ten, and the stepmother that soon appeared seems to have shown him scant sympathy, or, at all events, never won his affection. Alexander Wilson's youth unhappily coincided with an era of bad feeling in his native land; the times were hard in bonny Scotland, education was stagnant, and the public morals were debased. Wilson was a child of his times; like thousands of other youths, he was bound to suffer from the conditions of his early environment, but unlike many thousands of his day, he was possessed of talents and ambition which bitter adversity tended to sharpen and could never repress.

At thirteen young Wilson was taken from school and apprenticed to a weaver, William Duncan, his brother-in-law, and for three years he was no stranger to hard work and the birchen rod. For nearly three years more, as master weaver, he knew little beyond the grind and grime of the factory and the society of factory hands. At eighteen, however, his rebellious spirit struck, and for ten years he appeared in the *rôle* of itinerant peddler, poet and orator, and as socialist to the extent of championing the oppressed weaver class. At one time Wilson came into correspondence with Robert Burns and later made his acquaintance. His best dialect poem, "Watty and Meg, or The Taming of a Shrew," published anonymously as a penny chap-book in 1782, was his one popular success in the character of poet; according to report it was attributed to Burns, who admitted that he would have been glad to have written the verses, which sold so freely that a hundred thousand copies were disposed of in a few weeks.[181] In the disputes between capital and labor which arose at Paisley, Wilson took an active part. In connection with them he published a number of lampoons in verse, for which he was convicted of libel and was compelled to burn his satires at the town cross. In one instance, which occurred in February, 1793, a petty tyrant whom he had riddled exacted the fine,[182] and because of his inability to pay Wilson was sent to jail, where he languished for over three months.

Under the pressure of such persecutions, hard times, and possibly from disappointment in an affair of the heart, Wilson decided to emigrate. Practically driven out in rags from the country which one day was to raise a monument to his memory, at the age of twenty-eight he sailed from Belfast with his nephew, William Duncan, for the Eldorado of the New World. Wilson slept on deck throughout the entire voyage of fifty-three days, and landed at New Castle, Delaware, with the clothes on his back and an old fowling-piece as his only possessions. This was on July 14, 1794, nine years before John James Audubon left Nantes. Taking train "number 11," in the parlance of knights of the road, the two immigrants first walked to Wilmington in search of employment, and finding none there, went on twenty-nine miles farther to Philadelphia.

[181] See Grosart, *Poems and Literary Prose of Alexander Wilson*, vol. i, p. xxiv.

[182] For "The Shark, or Lang Mills Detected," a satire directed against William Sharp, a manufacturer of Paisley; Wilson was fined £12 13s. 6d.

The story is told that while they made their way through the woods of Delaware, Wilson shot a Red-headed Woodpecker and met with the Cardinal Grosbeak; as he often referred to the pleasure which the sight of these beautiful birds had given him, the incident, if it really occurred, may have played a part in the inspiration, which later came to Wilson, of becoming the historian of American bird life.

After eight hard years of shifting about, during which Wilson tried day-labor, weaving, peddling and school teaching, working long hours at miserable pay, he finally settled as a country school teacher near New York. On the twelfth of July, 1801, he wrote to a fellow teacher and friend, Charles Orr, who was then living at Philadelphia: "I live six miles from Newark and twelve miles from New York, in a settlement of canting, preaching, praying, and snivelling ignorant Presbyterians. They pay their minister 250 pounds for preaching twice a week, and their teacher 40 dollars a quarter for the most spirit-sinking, laborious work—6, I may say 12 times weekly." To the same friend, in 1802, he confided: "My disposition is to love those who love me with all the warmth of enthusiasm, but to feel with the keenest sensibility the smallest appearance of neglect or contempt from those I regard."

In 1802, at the age of thirty-six, Wilson decided to take up a school at Gray's Ferry, on the Schuylkill River, in Kingsessing Township, then a small settlement four miles from Philadelphia. A year later, in 1803, John James Audubon was sent to America to learn English and enter trade, and, as chance would have it, settled on the banks of the same river, not many miles from Wilson's old schoolhouse. In one respect the older man was the more fortunate, for, as will be seen, he found close by his door an excellent naturalist who played the part of mentor.

On February 14, 1802, while at Philadelphia, Wilson wrote to Orr:

> On the 25th. of this month I remove to the schoolhouse beyond Gray's Ferry to succeed the present teacher there. I shall recommence that painful profession once more with the same gloomy, sullen resignation that a prisoner re-enters his dungeon or a malefactor mounts the scaffold; fate urges him, necessity me. The agreement between us is to make the school equal to 100 dollars per quarter, but not more than 50 are to be admitted. The present pedagogue is a noisy, outrageous fat old captain of a ship, who has taught these ten years in different places. You may hear him bawling 300 yards off. The boys seem to pay as little regard to him as ducks to the rumbling of a stream under them. I shall have many difficulties to overcome in establishing my own rules and authority.

At Gray's Ferry, where he was then settled, Wilson again wrote in July: "Leave that cursed town at least one day. It is the most striking emblem of purgatory, at least to me, that exists. No poor soul is happier to escape from Bridewell than I am to smell the fresh air and gaze over the green fields after a day or two's residence in Philadelphia...."

George Ord, Wilson's staunch friend, literary executor, biographer, and editor of the last two volumes of the *American Ornithology*, thus characterized him: "He was of the *genus irritabile*, and was obstinate in opinion." He would acknowledge error when discovered by himself, "but he could not endure to be told of his mistakes. Hence his associates had to be sparing of criticism, through fear of forfeiting his friendship. With almost all his friends he had occasionally, arising from a collision of opinion, some slight misunderstanding, which was soon passed over, leaving no disagreeable impression. But an act of disrespect he could ill brook, and a wilful injury he would seldom forgive."

In 1801, while teaching and studying German at Milestown, Pennsylvania, Wilson had another unfortunate love affair, in this instance with a woman already married. To this he alluded in letters written in the summer of that year to his friend Orr, with whom he later quarreled. On August 7, 1801, he wrote: "The world is lost forever to me and I to the world. No time nor distance can ever banish her image from my mind. It is forever present with me, and my heart is broken with the most melancholy reflections."

Figure 28. Alex Wilson. After an Engraving by W H. Lizars, from a Painting by James Craw.

At Gray's Ferry, however, Wilson soon found in the estimable William Bartram, then in his sixty-first year, the sympathetic adviser, kind teacher, and judicious friend that he most needed, for though Wilson took the initiative in his ornithological plans, it was the kindly Bartram who eventually extended a helping hand. Both Bartram and Lawson, the engraver, urged him to devote his leisure to drawing, as a foil to his melancholic tendencies. Wilson did not hesitate long, for on June 1, 1803, he confided to a friend in Scotland that he had begun to make a "collection of our finest birds." Early in 1804 his purpose was clearly fixed, and on March 12 of that year he wrote to Alexander Lawson: "I am most earnestly bent on pursuing my plan of making a collection of all the birds in this part of North America.... I have been so long accustomed to the building of airy castles and brain windmills, that it has become one of my earthly comforts, a sort of rough bone, that amuses me when sated with the dull drudgery of life." A little later in the same month we find him appealing to Bartram for exact names, when he writes:

> I send for your amusement a few attempts at some of our indigenous birds, hoping that your good nature will excuse their deficiencies, while you point them out to me.... They were chiefly coloured by candle-light. I have now got my collection of native birds considerably enlarged, and shall endeavor, if possible, to obtain all the smaller ones this summer. Be pleased to mark on the drawings, with a pencil, the names of each bird, as, except three or four, I do not know them.

Figure 29. Will. Bartram. Reproduced from "Cassinia" for 1906.

Wilson, practically self-taught in everything, with no experience or training in drawing from nature, thus began at the age of thirty-eight to make his drawings of birds, before he knew the names of his subjects, and twenty years before Audubon's talents were known to any but members of his own family and a few intimate friends. The only aid in drawing which Wilson ever received appears to have come from the hints which Lawson supplied. Nevertheless, the best of Alexander Wilson's original drawings represent a degree of excellence and honest workmanship of which he had no need to be ashamed, and in many instances he owed far less to his engraver, Alexander Lawson, than did his great rival to Robert Havell.

In 1880 Dr. Elliott Coues examined a large collection of original Wilson and Audubon drawings and manuscripts, "owned and kept with the greed of a genuine bibliomaniac" by Joseph M. Wade, then editor of *Familiar Science and Fancier's Journal.* If not Wilson's portfolio itself, its contents, at least, said Dr. Coues, were then in Mr. Wade's possession, and this series of Wilson's drawings included, he thought, more than half of the originals of his famous plates. To quote Dr. Coues:[183]

> In handling these drawings and paintings, of all degrees of completeness, one of sensibility could but experience some emotions he would not care to formulate in words.... I was fairly oppressed with the sad story of poverty, even destitution, which these wan sheets of coarse paper told. Some of Wilson's originals are on the fly-leaves of old books, showing binder's marks along one edge. One of the best portraits, that of the Duck Hawk, is on two pieces of paper pasted together. The man was actually too poor to buy paper! Some of the drawings are on both sides of the paper; some show a full picture on one side, and part of a mutilated finished painting on the other. Some show the rubbing process by which they were transferred. They are in all stages of completeness, from the rudest outlines to the finished painting.

I know full well that in 1804, when Wilson had fairly begun his work on birds, he was poor enough, but I hesitate to believe upon such evidence that he was too poor to buy decent drawing materials. Wilson doubtless practiced economy in these matters as in everything else, through his ingrained habit of Scotch thrift, and he was probably quite as well-to-do then as five years before, when out of his slender earnings he was able to lay money aside.[184] Later, to be sure, his modest savings were quite consumed by his *Ornithology*, and then William Bartram came to his aid, even giving him a home in his own house. It is also wide of the mark to conclude from his fugitive letters or from his drawings, as this critic has done, that Wilson was possessed of genius only, and "had nothing else, not even talent and ability." Wilson certainly had a talent for writing and cultivated it with marked success; even his verse was not all of a "despicable mediocrity." In the art of drawing, however, his natural gifts were of a very modest sort, and what he achieved was the result of the most painstaking effort. Of course he was not a finished scholar, as graduates from the school of adversity seldom are, but he had a passion for knowledge and the determination to excel. His genius was not fully displayed until a powerful motive, the ambition to make known the birds of his adopted land, had possessed his spirit and taxed his powers to their utmost capacity.

Shortly after he had settled at Gray's Ferry, Wilson's susceptible nature was touched by another romance, which was again unfortunate for the poet and dreamer, but was probably the making of the ornithologist. Bartram's Botanic Gardens, on the outskirts of Philadelphia, had long been famous for their large and choice collection of native plants, gathered by the indefatigable zeal of their worthy founder, John Bartram, Quaker philosopher, traveler, botanist, agriculturalist and nurseryman; but the fairest flower in the whole collection at that time is said to have been Miss Anne Bartram, daughter of John the younger, niece of William,

---

183 See Bibliography, No. 43.

184 At Lancaster, Pennsylvania, on October 3, 1799, Alexander Wilson sent George Simpson, Esq., a State Treasurer's check in favor of Joseph Brown for $475, to be entered to the credit of Mr. Brown as one installment on 38 shares of scrip in the new loan at eight per cent, in the names of Thomas Eyes, 14 shares; Alexander Wilson, 14 shares; and Kenneth Sewell, 10 shares.

who then superintended the "Kingsess Gardens," granddaughter of the founder, and heiress to the estate. To this Quaker maid Wilson addressed a number of his poems, and he interested her in the drawing of birds; on March 29, 1804, he wrote to her uncle: "I send a small scroll of drawing papers for Miss Nancy. She will oblige me by accepting it." This little incident would show that Wilson was no stranger to the use of good drawing materials, however frugal his habits in this respect may have been. The young lady is said to have been not indifferent to her poet lover, and some of her family were friendly; the father, however, had no notion of bestowing his daughter's hand upon a poor schoolmaster, and for the third time Wilson's dreams of domestic bliss were shattered.

Such experiences no doubt tended to chasten the sensitive spirit of this real genius, whose whole life seemed to have been a continuous and losing struggle, while he felt within him an inspiration and a power that had failed to find adequate expression in labor at the loom, in verse, or in the hated vocation of teaching rough country schools at starvation wages. Though depressed by his misadventures in love, Wilson does not seem to have been embittered, and by way of diversion, he set out in the autumn of 1804, on a long walking tour from Philadelphia to Niagara Falls and back; in the following winter the experiences of this journey were embodied in a descriptive poem of 2,018 lines which he called "The Foresters," an effort which would have been less prosaic if frankly expressed in prose. Wilson's friendship for the Bartrams continued under the changed conditions, and he was invited to make his home under their hospitable roof. He was now free to devote himself heart and soul to birds and to birds alone.

Wilson etched the first two plates of his *American Ornithology* before he had obtained an engraver or a publisher. In April, 1806, he resigned his school at Gray's Ferry to accept an editorial position on a *New American Cyclopædia*,[185] then in course of preparation, at a salary of $900 a year. Samuel F. Bradford, the publisher of this work, soon became interested in Wilson's projected *American Ornithology* and agreed to publish it. It became the ambition of both author and publisher to produce the work in a superior style, and to make it as perfect and complete an American product as possible. Only the pigments used in coloring some of the plates were imported from Europe.[186]

Wilson issued in April, 1807, an elaborate prospectus of his proposed *Ornithology*, in which he stated that the completed work would comprise ten volumes, to cost $120, and that it would be illustrated by plates, engraved and colored by hand, after the manner of a carefully prepared sample which was issued with the printed announcement. In September, 1808, as already intimated, the first volume of the *American Ornithology*[187] appeared in an

[185] This was the American edition of Abraham Rees' revision of Ephraim Chambers' *Cyclopædia*, which had appeared in London in 1728; it was published at Philadelphia, in forty-one quarto volumes of text and six volumes of plates, by Samuel F. Bradford and the Messrs. Murray, Fairman & Company, 1810-1824.

[186] "The types," said Charles Robert Leslie, "which were very beautiful, were cast in America, and though at that time paper was largely imported, he [Mr. Bradford] determined that the paper should be of American manufacture; and I remember that Ames, the paper maker, carried his patriotism so far that he declared that he would use only American rags in making it." (*Autobiographical Recollections*, Boston, 1860.)

[187] The *American Ornithology: or, the Natural History of the Birds of the United States*: Illustrated with Plates Engraved and Colored from Original Drawings taken from Nature, by Alexander Wilson, was published in nine imperial quarto volumes by Messrs. Bradford and Inskeep, at Philadelphia, 1808-1814. Each volume contained nine plates and from 100 to 167 pages of text, exclusive of prefatory and other matter. The eighth volume, which was nearly ready for press at the time of the author's death, was edited by George Ord, Wilson's friend and executor; the final volume, which was wholly by Ord, and which was issued in the same year, contained a life of Wilson. After the appearance of the initial volume, the edition was extended to 500 copies

edition of 200 copies. Wilson immediately started on a canvassing tour of New England, in the course of which he visited the principal towns and colleges, going east to Portland, Maine, and as far north as Dartmouth College, in New Hampshire, where President John Wheelock and the professors received him with marked attention. On this journey Wilson did not average one subscriber a day, and he was forced to conclude that he had "been mistaken in publishing a work too good for the country"; "it is a fault," he said, "not likely to be repeated, and will pretty severely correct itself." Daniel D. Tompkins, Governor of New York, coolly said to him: "I would not give one hundred dollars for all the birds you intend to describe," not even if "I had them alive"; but a future Governor of that State, De Witt Clinton, the friend of science and scientific men, gave him the substantial encouragement he craved. When his second volume was ready for issue, Wilson wrote to Bartram: "This undertaking has involved me in difficulties and expenses which I never dreamt of, and I have never yet received one cent from it. I am, therefore, a volunteer in the cause of Natural History impelled by nobler views than those of money."

In the autumn of 1808 Wilson made a long and arduous tour of the South, in the course of which he visited every important town along the southern Atlantic seaboard, and though it cost him dear, he obtained 250 subscribers; it was then that his publishers decided to extend the original edition of his work to 500 copies. His longer and more perilous journey of 1810, when his meeting with Audubon occurred, has already been described. In 1812, after the sixth volume of the *Ornithology* had appeared, he again resumed his travels in the East and went as far north as Burlington, on Lake Champlain; at Haverhill, New Hampshire, he was summarily arrested and thrown into jail, the people of the town, utterly unable to comprehend the nature of his pursuits, suspecting that in his real capacity he was acting as a spy in the employ of the Canadian Government. The seventh and last volume of the *Ornithology* which Wilson lived to complete made its appearance in the spring of 1813. He had then been obliged to relinquish his work on the *Cyclopædia*, and was reduced to the pittance derived from the coloring of his own plates.

Alexander Wilson died at Philadelphia, after a brief illness, on August 23, 1813. A story was current that his end was saddened, if not hastened, by the dishonesty of his publishers, but I cannot vouch for it. Audubon may have had this report in mind when he wrote his name in the hotel register at Niagara Falls[188] on August 24, 1824; and added that he would never die, like Wilson, "under the lash of a bookseller." Even as late as 1879 Miss Malvina Lawson, daughter of Wilson's friend and engraver, left no doubt as to her belief when she wrote: "and to his other trials was added the fact that killed him,—the dishonesty of his publisher."[189]

---

and the first volume was entirely reset. Ord's life of Wilson was expanded for a three-volume edition of the *Ornithology*, and from oversheets of this work was produced as a separate volume in 1828.

Wilson's published lists of subscribers show 449 names, calling for 458 copies, more than half of which were taken by residents of Pennsylvania, New York and Louisiana; 70 were subscribed for in Philadelphia, chiefly by business men, artists, and "those in the middle class of society;" New Orleans in seventeen days gave him 60 subscribers; Europe supplied 15, among whom were William Roscoe, later a patron of Audubon, and Benjamin West, the artist. Wilson figured and described 278 species of American birds (within the limits of the United States), of which 56 were supposed to be new, and the total number, given by Wilson and Ord, is said to be 320. Twenty-three species were erroneously supposed to be identical with their European counterparts, yet all of Wilson's birds except the "Small-headed Flycatcher," referred to at the end of this chapter, have been identified. Considering the time and the difficulties under which he labored, his mistakes were remarkably few.

[188] See Vol. I.

[189] See a letter to Professor S. S. Haldeman, dated February 6, 1879, in *Penn Monthly*, vol. x (Philadelphia, 1879).

When we consider that Wilson's entire working period on the *Ornithology* was not over ten years, and that at the age of forty-seven he was called to lay down his pen and brush forever; that he produced in this brief space a work of great originality and charm, which did inestimable service in promoting the cause of natural history in both America and England, and which is likely to be read and prized for centuries to come, the achievement of this man is little short of marvelous. Knowing also the disabilities under which he labored, we are more than ready to temper our judgment with sympathy, and to overlook any faults which his character may have displayed. These indeed, we believe, were for the most part of a very trifling nature; those who knew Wilson best have all testified to his kindness of heart, his liberality, and his high sense of honor.

We must now return to the meeting of our two pioneers, which has been the bone of so much acrimonious contention. On his long journey to the Middle West and South, Wilson reached Louisville on a Saturday evening, the seventh of March, 1810, and put up at the tavern of the "Indian Queen," where, as it happened, Audubon was then living with his family; after spending five days in and about the town, he again set out on foot for Frankfort, on the morning of Friday, the twenty-third. Audubon has given the following account in the "Episode" of "Louisville in Kentucky":[190]

> One fair morning, I was surprised by the sudden entrance into our counting-room [at Louisville] of Mr. Alexander Wilson, the celebrated author of the "American Ornithology," of whose existence I had never until that moment been apprised. This happened in March, 1810. How well do I remember him, as he then walked up to me! His long, rather hooked nose, the keenness of his eyes, and his prominent cheek-bones, stamped his countenance with a peculiar character. His dress, too, was of a kind not usually seen in that part of the country; a short coat, trousers, and a waistcoat of grey cloth. His stature was not above the middle size. He had two volumes under his arm, and as he approached the table at which I was working, I thought I discovered something like astonishment in his countenance. He, however, immediately proceeded to disclose the object of his visit, which was to procure subscriptions for his work. He opened his books, explained the nature of his occupations, and requested my patronage.
>
> I felt surprised and gratified at the sight of the volumes, turned over a few of the plates, and had already taken a pen to write my name in his favour when my partner rather abruptly said to me in French, "My dear Audubon, what induces you to subscribe to this work? Your drawings are certainly far better, and again you must know as much of the habits of American birds as this gentleman." Whether Mr. Wilson understood French or not, or if the suddenness with which I paused, disappointed him, I cannot tell; but I clearly perceived that he was not pleased. Vanity and the encomiums of my friend prevented me from subscribing. Mr. Wilson asked me if I had many drawings of birds. I rose, took down a large portfolio, laid it on the table, and shewed him, as I would show you, kind reader, or any other person fond of such subjects, the whole of the contents, with the same patience with which he had shewn me his own engravings.
>
> His surprise appeared great, as he told me he never had the most distant idea that any other individual than himself had been engaged in forming such a collection. He asked me if it was my intention to publish, and when I answered in the negative, his surprise seemed to increase. And, truly, such was not my intention; for, until long after, when I

[190] *Ornithological Biography*, vol. i, p. 437.

met the Prince of Musignano in Philadelphia, I had not the least idea of presenting the fruits of my labours to the world. Mr. Wilson now examined my drawings with care, asked if I should have any objections to lending him a few during his stay, to which I replied that I had none: he then bade me good morning, not, however, until I had made an arrangement to explore the woods in the vicinity along with him, and had promised to procure for him some birds, of which I had drawings in my collection, but which he had never seen.

It happened that he lodged in the same house with us, but his retired habits, I thought, exhibited either a strong feeling of discontent, or a decided melancholy. The Scotch airs which he played sweetly on his flute made me melancholy too, and I felt for him. I presented him to my wife and friends, and seeing that he was all enthusiasm, exerted myself as much as was in my power, to procure for him the specimens which he wanted. We hunted together, and obtained birds which he had never before seen; but, reader, I did not subscribe to his work, for, even at that time, my collection was greater than his. Thinking that perhaps he might be pleased to publish the results of my researches, I offered them to him, merely on condition that what I had drawn, or might afterwards draw and send to him, should be mentioned in his work, as coming from my pencil. I at the same time offered to open a correspondence with him, which I thought might prove beneficial to us both. He made no reply to either proposal, and before many days had elapsed left Louisville, on his way to New Orleans, little knowing how much his talents were appreciated in our little town, at least by myself and my friends.

Some time elapsed, during which I never heard of him, or of his work. At length, having occasion to go to Philadelphia, I, immediately after my arrival there, inquired for him and paid him a visit. He was then drawing a White-headed Eagle. He received me with civility, and took me to the Exhibition Rooms of Rembrandt Peale, the artist, who had then portrayed Napoleon crossing the Alps. Mr. Wilson spoke not of birds or drawings. Feeling, as I was forced to do, that my company was not agreeable, I parted from him; and after that I never saw him again. But judge of my astonishment sometime after, when on reading the thirty-ninth page of the ninth volume of American Ornithology, I found in it the following paragraph:—

"March 23, 1810.—I bade adieu to Louisville, to which place I had four letters of recommendation, and was taught to expect much of everything there; but neither received one act of civility from those to whom I was recommended, one subscriber, nor one new bird; though I delivered my letters, ransacked the woods repeatedly, and visited all the characters likely to subscribe. Science or literature has not one friend in this place."

What actually happened at this meeting of the two naturalists will never be certainly known, beyond what can be gathered from their rather widely divergent accounts. It should be noticed, however, that the paragraph which Audubon quoted was extracted from Wilson's private diary; it was no doubt written on the spur of the moment, possibly to humor his own mood, and certainly with no thought of its later publication. It was inserted by George Ord in the biographical sketch of his friend appended to the ninth volume of the *American Ornithology*, which appeared in 1814, the year after Wilson's death. Audubon was not concerned, either directly or by implication, except in the last sentence, for it is evident that he was not one of those to whom Wilson had carried letters of introduction. Thus the matter stood until 1828, when Audubon's *Birds of America* were being engraved in England. In all probability the incident would never have been noticed by Audubon, had not Ord seen fit to revive it when his life of Wilson[191] was issued as a separate volume in that year. In this edition of the biography Ord inserted fuller extracts from Wilson's journal, with the evident purpose of placing the rival of his friend in an unenviable light.

Wilson's diary, which apparently was never seen by any of Audubon's friends, is now known to us only through such extracts as Ord and Waterton, his bitter enemies, have seen fit to make public; the original has probably been destroyed, for it cannot be traced later than 1840, when it was still in the hands of George Ord.[192] Charles Waterton gave similar extracts from this famous journal in one of his philippics against Audubon in 1834, when he said that it was the testimony of this record that defeated Audubon's friends in their initial attempt to bring him into the Academy of Natural Sciences at Philadelphia. Wilson's narrative of his adventures at Louisville in 1810, as given by Ord and Waterton, is as follows:[193]

> *March 17.* Take my baggage and grope my way to Louisville—put up at the Indian Queen tavern, and gladly sit down to rest myself.
>
> *March 18.* Rise quite refreshed. Find a number of land-speculators here.[194]
>
> *March 19.* Rambling round the town with my gun. Examined Mr.——'s drawings in crayons—very good. Saw two new birds he had, both Motacillæ.
>
> *March 20.* Set out this afternoon with the gun—killed nothing new. [People in taverns here devour their meals. Many shopkeepers board in taverns—also boatmen, land-speculaters, merchants &c.] *No naturalist to keep me company.*
>
> *March 21.* Went out shooting this afternoon with Mr. A. Saw a number of Sandhill Cranes. Pigeons numerous.
>
> *March 22.*
>
> *March 23.* Packed up my things which I left in the care of a merchant here, to be sent on to Lexington; and having parted with great regret, with my paroquet, to the gentleman of the tavern, I bade adieu to Louisville, to which place I had four letters of recommendation, and was taught to expect much of everything there, but neither received one act of civility from those to whom I was recommended, one subscriber, *nor one new bird*; though I delivered my letters, ransacked the woods repeatedly, and visited all the characters likely to subscribe. *Science or literature* has *not one friend in this place.* [Everyone is so intent on making money, that they can talk of nothing else; and they absolutely devour their meals, that they may return sooner to their business. Their manners correspond with their features.]

In this fuller record we learn that Wilson spent five days in Louisville; he examined Audubon's drawings on Monday, March 19, hunted alone on the 20th, went out shooting with Audubon on the 21st, and finally left Louisville on the morning of the 23d; no record was admitted by Ord for Sunday, the 18th, or for the 22d, a Thursday. Wilson noticed the

---

[191] *Sketch of the Life of Alexander Wilson, Author of the American Ornithology*, by George Ord, F. L. S. &c. pp. i-cxcix, Philadelphia, 1828; taken from vol. i of an octavo edition of Wilson, edited by Ord, and issued by Harrison Hall, in three volumes, at Philadelphia in 1828-29, with folio atlas of plates reproduced from the original work; *supra*.

[192] See Ord's charge of plagiarism against Audubon in the *Proceedings of the American Philosophical Society*, vol. i (1840). So far as could be ascertained in the summer of 1915, Wilson's diary of 1810 was not in the possession of any library or scientific society in Philadelphia, nor was it in the large collection of books which was given by Ord to the College of Physicians and Surgeons of that city at the time of his death in 1866.

[193] The bracketed lines are from Waterton, who once stated that he had examined the original.

[194] This sentence is quoted from Burns' biographical sketch of Wilson, but tenses are changed to correspond with other entries.

drawings of two new *Motacillæ*, or Warblers, in Audubon's collection, and it would have been only natural that he should have felt a strong desire to copy them, yet not a word was said about the loan of drawings to which Audubon refers; Wilson merely stated that from those to whom he was recommended he had received not "one act of civility,—one subscriber, nor one new bird." Audubon was evidently regarded as one of the "many shopkeepers" who boarded "in taverns," and not as a "naturalist," for Wilson said that he had none to keep him company, and it is rather significant that Audubon's name is not once mentioned in his *Ornithology*.

Twenty-nine years after Wilson's visit to Louisville, when Audubon came to publish the fifth and last volume of his *Ornithological Biography*, he maintained that Wilson had copied his drawing of a certain bird, called the Small-headed Flycatcher,[195] without any acknowledgment. To quote Audubon's words:

> When Alexander Wilson visited me at Louisville, he found in my already large collection of drawings, a figure of the present species, which being at that time unknown to him he copied and afterwards published in his great work, but without acknowledging the privilege that had thus been granted to him. I have more than once regretted this, not by any means so much on my own account as for the sake of one to whom we are so deeply indebted for the elucidation of our ornithology.

This troublesome bird was first described by Wilson in 1812, when he rightly pronounced it "very rare," and said that the specimen from which his drawing was made had been shot in an orchard, presumably near Philadelphia, on the twenty-fourth day of April, and that several had been obtained also in New Jersey. His friend Ord, who came to his defense in 1840, confirmed this statement by declaring to the American Philosophical Society of Philadelphia that he had been with Wilson on the day in question and had examined the specimen. Lawson also affirmed that in engraving the plate he had worked directly from the bird which Wilson had given him.

What has become of this mysterious phantom that has been a wandering and disturbing voice among ornithologists for over a century? It has given rise to no end of conflicting and sharp discussions between the partisans of the two naturalists chiefly concerned, the only thing certain being that if this supposititious species ever existed, it has forsaken its old haunts, if not the earth itself, and has never returned. No doubt it was simply a case of mistaken identity, and both Wilson and Audubon were wrong, each having had in hand and mind an immature representative of one of our numerous Warblers, which are now so much better known.[196] If Wilson copied Audubon's drawing of the bird, he must have replaced it with one of his own, for the figures of the two naturalists are very unlike. Certainly Audubon should not have made so serious a charge without offering more substantial evidence in proof;

[195] *Musicapa minuta*, which appears in Figure 5, Plate 50, of volume vi of Wilson's *American Ornithology* (pp. 62-63 of the text), and in Figure 2, Plate ccccxxxiv, of Audubon's *Birds of America* (*Ornithological Biography*, vol. v, pp. 291-3).

[196] Nevertheless so careful and discerning a naturalist as Thomas Nuttall confidently asserted that his friend, Mr. M. C. Pickering, had "obtained a specimen several years ago near Salem (Massachusetts)"; see *A Manual of the Ornithology of the United States and Canada* (Cambridge, 1832). Dr. Elliott Coues at one time thought that it might have been the Pine-creeping Warbler, and Professor Baird identified it as the female or young of the Hooded Warbler.

perhaps what he had intended to convey was that Wilson had obtained from him his first knowledge of the bird, and he was nettled to find that he had been studiously ignored.[197]

Among the originals of Audubon's *Birds of America* in possession of the Historical Society of New York, there is an early drawing of a Warbler which bears in pencil, in the naturalist's hand, the following note: "This bird was copied by Mr. Willson at Louisville."[198] The misspelling of Wilson's name, which was common with Audubon as late as 1820, would indicate that the note was not added after that time, but if Wilson copied this drawing, there is no evidence that he ever used it.

Figure 30. The "Twin" Mississippi Kites of Wilson (Left) and Audubon (Right), the Similarity of Which Inspired Ord's Charges of Misappropriation against Audubon.

Ord made another charge in which Audubon does not appear to such good advantage; though it refers to a later day, it is best to consider it now. This critic thought that a complaint of misappropriation came with ill grace from one who had been guilty of it himself, and maintained that Audubon had copied Wilson's figures of the female Red-wing Blackbird (*The Birds of America*, Plate LXVII), and had also stolen his drawing of the Mississippi Kite (Plate CXVII). Ord was probably mistaken in regard to the blackbird, but without a doubt the lower bird in the Kite plate was taken from Wilson (*American Ornithology*, Plate 25), though the copyist has reversed the outlines, left out one of the toes, added minor details, and misnamed the sex, which in the Wilson original represents a male. Without a doubt also the odium in

[197] Compare *Ornithological Biography*, vol. iii, p. 203, where in Audubon's article on the Whooping Crane, there is this note: "Louisville, State of Kentucky, March, 1810. I had the gratification of taking Alexander Wilson to some ponds within a few miles of town, and of showing him many birds of this species, of which he had not previously seen any other than stuffed specimens. I told him that the white birds were the adults, and that the grey ones were the young. Wilson, in his article on the Whooping Crane, has alluded to this, but, as on other occasions, he has not informed his readers whence his information came."

[198] What appear to be the original legends, written on this drawing in ink, are as follows: "Chute de l'Ohio. July 1, 1808. No. 31. J. A. Que j'avais figuré [?] 12 pennes à la queue." Above were later added, also in ink, the names, "sylvia Trochilus delicata; Sylvia delicata, Aud."

this case must fall upon Audubon, but we are not at all certain that he was directly responsible for the theft. Audubon's plate of this species, which is finished in elaborate detail, was probably published towards the close of 1831, when he was in America. He furnished his engraver, we believe, with the drawing of the upper bird only, which he designated as a male, and the original still exists, with clearly written notes showing that it was executed in Louisiana in 1821.[199]

Audubon usually made up his drawings for the engraver with great care, but when pressed for time, Havell's skill was such that he often depended upon him to complete or change his figures, to fill in backgrounds, or even to combine several distinct figures into one plate, specific directions for all such changes being usually written on the drawing itself.[200] Inasmuch as no penciled directions whatever occur on this particular drawing, is it possible that Havell, in piecing it out to improve the composition, followed his own initiative, not fully appreciating the stigma that is rightly attached to such methods? The bird in the lower half of the plate, which was appropriated from Wilson, is misrepresented as a female, so that the composite, as it stands, is a remarkable product, supposedly depicting a pair but in reality showing two males. Although the apparent difference in sex in this bird was admittedly slight, it is improbable that so gross an error could have escaped the naturalist's eye had he been directly concerned with the result.

When Audubon was descending the Mississippi in December, 1820, he saw the kites busily engaged "in catching small lizards off the bark of dead cypress trees," but "having at that time no crayons or paper," he "did not draw one, and determined," as he then wrote in his journal, "never to draw from a stuffed bird." "I first saw the Mississippi Kite," he added, when "ascending in the steamboat Paragon, in June, 1819." Wilson, on the other hand, in his knowledge of this interesting bird was far in advance of his later rival, for his first observations were made in 1810, "in the Mississippi territory, a few miles below Natchez, on the plantation of William Dunbar, esquire, when the bird represented in the plate was obtained, after being slightly wounded; and the drawing made with great care from the living bird." "For several miles, as I passed near Bayo Manchak," Wilson continues, "the trees were swarming with a kind of cicada, or locust, that made a deafening noise; and here I observed numbers of the Hawk now before us sweeping about among the trees like Swallows, evidently in pursuit of these locusts; so that insects, it would appear, are the principal food of this species."[201] Wilson never succeeded in procuring the female of this graceful hawk, and his editor, George Ord, evidently continued the quest, for we find his correspondent, John Abbot, writing him from "Scriven County Georgia Mar. 1814": "Are you acquainted with the female yet of the Louisiana Kite?"[202]

We have entered into the detailed history of this plate because of the unfavorable comment which it has provoked, but it is easier to be critical than to be either just or correct, and without more definite knowledge than we possess, it would be unfair to censure Audubon too much or to shift the blame too completely upon the shoulders of another.

---

[199] On this drawing, which with Audubon's other originals is in the collections of the Historical Society of New York, the legends are as follows: "Mississippi Kite, Male, Falco mississippiensis; Drawn from nature by John J. Audubon, Louisiana, parish of Feliciana, James Perrie's Esq., Plantation. June 28th, 1821. Length 14 inches; Breadth 3 feet, ½ inches; Weight 10¾ ounces; Tail feathers, 12." It is drawn in his usual style of that period, in pastel, water color and pencil, and has been dismounted.

[200] See Vol. I.

[201] *American Ornithology*, vol. iii, p. 80.

[202] See Witmer Stone, "Some Letters of Alexander Wilson and John Abbot," *The Auk*, vol. xxiii, 1906.

To return again to the story of Wilson's diary, it is evident that Wilson would never have published his sentiments in the form in which they later appeared. They were perfectly characterized by a just critic of an early day,[203] who said that Wilson's words were without doubt written in a moment of keen depression and disappointment and were an exact description of his feelings, though, as we should also add, not of the facts. "A man who has given his heart to the accomplishment of an object, believing that he has no rival, must be somewhat more than human, if he be delighted to find that another is engaged in the same purpose, with equal energy and advantages far greater than his own." Barring his usual inaccuracies, it must be admitted that Audubon's account bears the thumbmarks of truth. He could not have known the bitter struggles of the proud spirit whose history we have briefly told; he saw only a stranger, an ardent devotee of nature, it is true, but a man of unbending disposition, who with a little more suavity of address could probably have won his friendship, if not his subscription. Of the literary quality of Wilson's work, now so well appreciated, he could have known nothing at all; after turning its pages in his Louisville store for the first time in 1810, he probably did not see it again for over ten years.

That Wilson was jealous of Audubon as a future rival is probable, but the real "rivalry" between these two pioneers was of later growth. It was fostered in this country chiefly by George Ord and some of his friends, together with others who were interested in the sale of Wilson's work. Ord, who seems to have felt that the mantle of this naturalist had fallen on his own shoulders, strove continually, and after 1826 with the aid of Charles Waterton in England, to hamper Audubon's progress, to discredit him as a man of integrity, and to break down his growing reputation as a naturalist. Though Ord was justified to some extent in his attacks upon Audubon which were made over Wilson's shoulders long after that estimable man was laid in the grave, the matter was carried too far. Neither of the rivals was wholly without fault, and a century is far too long to continue any quarrel, especially when one of those whose reputation was concerned was never a party to it.

Audubon, as we have seen, frankly attributed to personal vanity his failure to patronize Wilson's work, and added that "even at that time my collections were greater than his." But it should be noticed that money was far from plentiful with him at that moment. He was, in short, at the point of failure in the Louisville enterprise, and with Rozier was obliged to move down the river not long after the date of Wilson's visit. Audubon has been represented as at this time a well-to-do man of leisure, of fastidious tastes. Nothing could have been wider of the mark. He was still more of a sportsman than a naturalist, and when not occupied with drawing, he spent most of his time in the forest, to the neglect of his trade. We may be sure that he was quite as used to roughing it as any man on the frontier.

[203] In 1840, by W. B. O. Peabody, naturalist; author of a *Life* of Wilson.

Chapter 15

Experiments in Trade on the Frontier

At the beginning of the nineteenth century the banks of the Ohio River were but thinly settled, and over vast areas the virgin forest still reigned in undisturbed vigor and beauty. Yet traders were eagerly pushing westward in ever growing numbers, and by 1810 Audubon and Rozier found that competition at Louisville was already keen. This city, wrote Alexander Wilson in describing his experiences in the spring of that year, was as large as Frankfort, and possessed a number of good brick buildings and valuable shops; it would have been salubrious, he thought, "but for the numerous swamps and ponds that intersect the woods in its neighborhood," and the indifference of the people, whom he found too intent upon making money to give any heed to the drainage and sanitation of their town.

The prosperity of the partners, as already intimated, was shortlived. Audubon was doubtless right in admitting that his business abandoned him because he could not bear to give it the necessary attention. The conditions of life for the merchant-trader at that early day were at best far from easy, and an honest success, as then understood, required not only plenty of rough work but careful planning as well. His goods, purchased in the East, were laboriously transported across the State of Pennsylvania, and if they came from Philadelphia they must needs traverse the rough wagon roads that led through Bedford to Pittsburgh. There was an overland trail from Pittsburgh to Kentucky, but merchants with heavy loads would naturally take the easier river route. In going east to renew his stock in trade, it was a common practice to travel on horseback from as far west as St. Louis, but on returning the merchant would often sell his mount at Baltimore, Philadelphia or Pittsburgh, where a boat could be taken for the remainder of the journey.

The "ark" or flatboat was considered most convenient for the transportation of either passengers or merchandise down the Ohio, for any well-to-do traveler, while floating leisurely with the current, could make himself comfortable by fitting up snug sleeping quarters and a kitchen on deck, and could go ashore at will, with the certainty of satisfying his appetite for wild turkey, venison and other game in the season. Wilson, who descended the river in April, 1810, boarded and passed many of these "arks," which he described as built in the form of a parallelogram, from twelve to fourteen feet wide and from forty to seventy feet long, with a canopy to protect them from the weather; they were casually helped along by means of two oars in the bow, and steered by another and more powerful one in the stern. "Several of these floating caravans," said Wilson, "were loaded with store goods for the supply of the settlements through which they passed, having a counter erected, shawls, muslins," and the like, "displayed, and everything ready for transacting business. On approaching a settlement they blew a tin trumpet, which announced to the inhabitants their arrival." These "arks," he added, descended from all parts of the Ohio and its tributary streams, but in greatest numbers in the spring months. Although they cost originally about $1.50 per foot of length, when arrived at their destination they would seldom bring more than one-sixth of that amount. From forty to fifty days were commonly required to cover the entire distance of two thousand miles from Pittsburgh to New Orleans.

Another means of conveyance on the river, frequently used by Audubon, was the keel boat or barge, which, in some cases, was also roofed and would hold about two hundred barrels of flour.[204] When assisted by oars in the bow, it could reduce the time of a journey to New Orleans by ten or fifteen days. These barges were pushed up stream with the aid of setting poles at an average rate of about twenty miles a day, or, if loaded, they were laboriously "cordelled," or drawn by the hands of men who trudged along the banks pulling at the cordelle.

The chief pleasures which Audubon's business ventures in the West seem to have afforded him were his leisurely journey by river and long horseback rides to Philadelphia to buy goods, when he could roam through his "beautiful and darling forests of Kentucky, Ohio, and Pennsylvania," which gave him grand opportunities to make observations upon birds and animal life of every sort. He would seldom hesitate to swerve from his course to study his favorites, and has related how on one occasion, when driving before him several horses laden with merchandise and dollars, he quite lost sight of the pack saddles and the cash they bore, in watching the motions of a warbler. But few coaches, said Audubon, were available in those days, and the post roads were often unfit for lighter carriages. To cover the distance from Louisville to Philadelphia on horseback required about twenty days, and only a capable animal and rider could make forty miles a day; when steamer traffic on the Ohio[205] was well in hand this time was reduced to six or seven days, in performing a journey which the modern railroad has shortened to not far from as many hours.

Discouraged by the gloomy prospects which their business at Louisville presented, Audubon and Rozier determined in the spring of 1810 to move 125 miles down the river to Henderson.[206] Loading the residue of their stock on a flatboat, they resolutely set out for the new field, but great was their surprise to find, in place of the thriving settlement which their imaginations had pictured, only a cluster of log houses on the river bank, with a population of less than 200 people and a demand for little else than whisky, gunpowder and coarse woolen goods. When the partners arrived, the little town was eighteen years old, as the first log cabins were built there in 1792, but the whole country above and below them was, and for a considerable time remained, one vast canebrake. All the commodities known to the pioneer store were scarce, but the people of Henderson were friendly, and the new settlers had been

[204] Vincent Nolte, *Fifty Years in Both Hemispheres*.

[205] The first steamboat on the Ohio was the *Orleans*, a vessel of 200-400 tons, built at Pittsburgh in the summer and fall of 1811, by Robert Fulton and Robert M. Livingston, her first voyage, when she touched at Henderson, was signalized, as it seemed to many, by the great earthquakes of that year. The first Kentucky steamer was built at Henderson in 1817, the same year that a small vessel was constructed by Samuel Bowen and J. J. Audubon at the same place. Compare Edmund L. Starling, *History of Henderson County, Kentucky*.

[206] Known first as Redbank or Redbanks, to distinguish it from Yellowbank, or Owensboro, on a similar bend farther upstream; called also Hendersonville, but this term had no official standing. The population of Henderson in 1810 is given as 159, and that of the entire county, then larger than at present, as 5,000. See Starling, *op. cit.*

provident in bringing with them a goodly supply of flour and "bacon hams." Moreover, the Ohio, which was half a mile wide at that point, was well stocked with fish, and the woods and canebrakes were alive with birds, not to speak of larger and more important game. Not many years before, wild turkeys had been so plentiful that they were not sold but were given away, while a large buck deer could be bought in the season for fifty cents.

During their stay at Henderson, Rozier was in his habitual place behind the counter and attended to what little business was done, while Audubon with a Kentucky lad named John Pope, who was nominally a clerk, roamed the country in eager pursuit of rare birds, and with rod and gun bountifully supplied the table. Audubon's first abode in the town was, as he said, "a log-*cabin*, not a *log-house*," in which the richest piece of furniture was their child's cradle. He soon began to cultivate a garden, but his experience in horticulture must have been limited, for he naïvely remarks that the rankness of the soil kept the seeds they planted "far beneath the tall weeds which sprang up the first year."

Financial distress and hard times were already being felt in the Blue Grass State, and these conditions were not destined soon to improve. After experimenting for six months, or more, at Henderson, our two "rolling stones" determined to push still farther west and try their luck at a more promising point. They had hoped to reach St. Louis but finally went instead to Ste. Geneviève, then a small French settlement in Upper Louisiana, on the right bank of the Mississippi, a hundred miles north of the mouth of the Ohio.

This new venture promised to be both hazardous and uncertain, and as Mrs. Audubon and Rozier were not on the friendliest terms, Audubon decided to leave his family at Henderson, where a home for his wife and infant son could always be had under the hospitable roof of Dr. Adam Rankin, who became one of the naturalist's staunchest friends. If their stock in trade at this time actually consisted of "three hundred barrels of whisky, sundry dry-goods and powder," as Audubon affirmed, the keel boat which they then engaged was certainly calculated to bear a goodly load.[207] At all events the partners, with young Pope, their clerk, set out bravely, in a snow storm, in December, 1810. They floated with the current at a rate of about five miles an hour, while they helped their craft along by means of four oars in her bow and steered it with the aid of a slender tree trunk, "shaped at its outer extremity like the fin of a dolphin."

This journey of upwards of 165 miles lasted altogether more than nine weeks. It proved adventurous enough, but it was of no use to Audubon except in furnishing him with drawings of new birds and the raw materials for many "Episodes." The journal of his experiences on the great rivers during that eventful winter of 1810 and 1811 is interesting for the sidelights which it throws both upon his character and upon the state of the country at an elder day. Held up by the ice for several weeks at Cash Creek, near the mouth of the Ohio, to his own delight but to Rozier's sorrow, Audubon tramped the country and hunted wild swans and larger game with the friendly Shawnee Indians. "When one day's sport was over," he said, "we counted more than fifty of these beautiful birds whose skins were intended for the ladies of Europe. There were plenty of geese and ducks, but no one condescended to give them a shot." This was Audubon in 1810, when such "sport" was regarded as legitimate enough, and the feather-hunting of such Indians was not considered the nefarious trade that it proved to be. If we shift the scene to twenty years later, when William MacGillivray needed thousands of specimens of American birds for his studies upon their anatomy and variability, we find

[207] See Vol. I.

Audubon supplying him liberally, but he could not then bear to see them killed wantonly or for mere sport; more than once, out of compassion for individual birds that he chanced to be studying, whether in Florida or in Labrador, he would not permit them to be shot even when needed for his collections.

At the Shawnee Indian camp, to relate a characteristic anecdote, Audubon noticed that a squaw who "had been delivered of beautiful twins during the night" was busied on the next day at her usual task of tanning deer skins. "She cut two vines," his record reads, "at the roots of opposite trees and made a cradle of the bark, in which the new born ones were wafted to and fro with a push of her hand, while from time to time she gave them the breast, and was apparently as unconcerned as if the event had not taken place."

When at last our adventurers gained the Mississippi, the mighty volume of which was running three miles an hour, the patron ordered all hands ashore to pull at the bow rope. This characteristic remark of the naturalist is delightful, as showing the "single eye" which it has been declared of old shall be "full of light": "we made," said Audubon, "seven miles a day up the famous river; but while I was tugging with my back at the cordella, I kept my eyes fixed on the forests or the ground, looking for birds or curious shells."

Warping against the current was both difficult and dangerous, and though they rose two hours before the sun, they could make but one mile an hour or ten miles in the day. At night they would go ashore, light a good fire and cook their supper; then, after posting a sentinel to guard against unfriendly surprises, they would roll in their buffalo skins and sleep without further concern. Notwithstanding all their efforts, when they reached the Great Bend at Tawapatee Bottom, they were obliged to unship their cargo, protect their boat as best they could from being crushed in the growing pack, and await the final breaking up of the ice. "The sorrows of Rozier," at this dismal announcement, said Audubon, "were too great to be described; wrapped in a blanket, like a squirrel in winter quarters with his tail about his nose, he slept and dreamed his time away, being seldom seen except at meals." There was not a white man's cabin within twenty miles, but a new field opened to the naturalist, who tramped through the deep forests, and soon became acquainted with all the Indian trails and lakes in the neighborhood.

The six weeks spent at this camp passed pleasantly for Audubon, who devoted much of the time in studying the Osage Indians, whom he thought superior to the Shawnees, as well as in watching for wolves, bears, deer, cougars, racoons and wild turkeys, some of which were attracted by the bones and scraps of food thrown out for them: "I drew," said he, "more or less, by the side of our great camp-fire, every day." While detained at this point, they used for bread the breasts of turkeys, buttered with bear's grease, and opossum and bear's meat, until their stomachs revolted and they longed for a little Indian meal, which was procured only with the greatest difficulty.

When at last the ice broke up, splitting with reports like the thunder of heavy artillery, their prospects were dismal indeed, for their boat was immediately jammed by the rushing ice, and they were powerless to move her. "While we were gazing on the scene," to continue Audubon's record, "a tremendous crash was heard, which seemed to have taken place about a mile below, when suddenly the great dam gave way. The current of the Mississippi had forced its way against that of the Ohio, and in less than four hours we witnessed the complete breaking up of the ice." Having reloaded their goods, they were ready to start at a favorable moment, and taking leave of the friendly Indians, "as when brothers part," they pushed on through the floating ice, past Cape Girardeau, to Sainte Geneviève, a town which Audubon characterized as "not so large as dirty," declaring that the time spent there did not yield him half the pleasure he had felt at Tawapatee Bottom. It was near a granite tower which rose from a dangerous rock in the river below Ste. Geneviève that Audubon caught sight of what he afterwards described as "Washington's Eagle," a bird now believed to have been the true "bird of freedom," the "Bald-" or White-headed Eagle, but in an immature state.

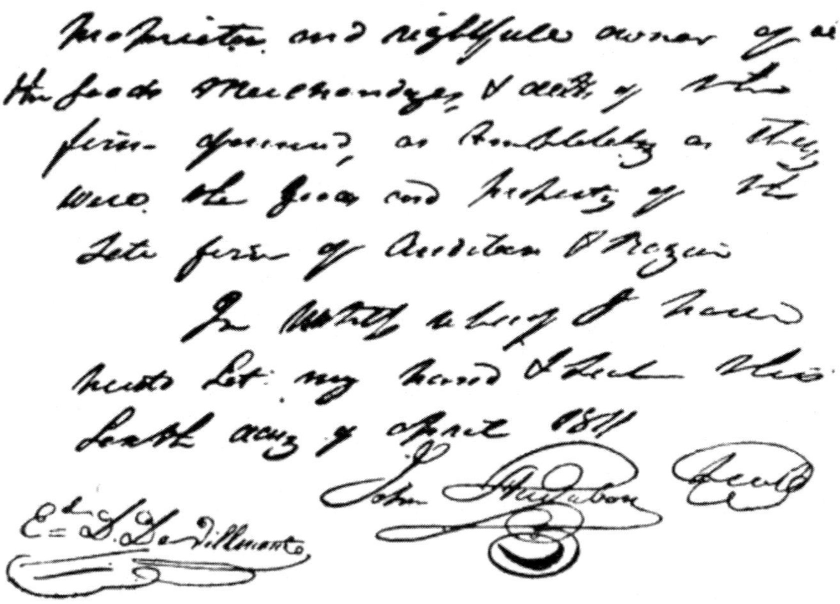

Figure 31. Signature to the Release Given by Audubon to Ferdinand Rozier on the Dissolution of Their Partnership at Ste. Geneviève, April 6, 1811.
From the Tom J. Rozier MSS.

Though their whisky was welcomed at Ste. Geneviève and what had cost the traders twenty-five cents, brought them two dollars, a gallon, Audubon heartily disliked the place and its people. Rozier, on the contrary, who had found plenty of Frenchmen with whom he could freely converse, was resolved to stay. Audubon accordingly proposed to sell out his share in the business, and the partnership was dissolved on April 6, 1811, Rozier paying part of the price in cash and the remainder in notes. In referring to the incident in his journal of 1820, Audubon wrote: "I parted with Mr. Rozier, and walked to Henderson in four days—165 miles"; but this does not agree with a later account, in which he spoke of having "purchased a beauty of a horse," and, happy in the prospect of again seeing his family, set out for Dr. Rankin's house in Kentucky. In the earlier record he also wrote that he once had a friend in trade, referring to Ferdinand Rozier, "with whom he did not agree, and so they parted forever"; but Audubon visited Ste. Geneviève in the autumn of 1811 and in the winter of 1812, probably for the purpose of collecting his money and settling his affairs, while the following letters of this period show that friendly relations with his old partner were not seriously impaired:[208]

208 See translations from copies of the originals, in French, in possession of the Louisiana Historical Society, New Orleans.

John James Audubon to Ferdinand Rozier

Louisville, 2d November 1811.

Mr. F. Rozier
St. Geneviève.
My Dear Rozier;

I reached here on the 31st of last month a little fatigued, as you can well imagine. Yesterday I wrote to T. W. Bakewell at New Orleans, and doubt not he is sending you regularly the prices current of the market there. I have found here a letter addressed to my brother-in-law from Benj. Bakewell, who complains of us, and says that we ought to settle with him in one way or another; write to him at Pittsburgh; I will be with him, possibly at the same time, and will speak with him; by the bill which he inclosed you will see that we are his debtor for 55$. I am leaving here in 2 or 3 days. I wish you health and prosperity, and with the respects of my wife, I am always your friend & Servant.

J. Audubon.

[Addressed] Mr. Fd. Rozier
Merchant
St *Genevieve*
u.L.

John James Audubon to Ferdinand Rozier

Shippingport. 10th August. 1812

My Dear Rozier;—

As it is quite likely that the present opportunity is safe, I take pleasure in writing you a few words.

Your letter sent to Philadelphia was duly received, and answered promptly; since I have heard news of you only by the most indirect means, I would be happy if you can give a few moments to your friends, if you would count me in their number, and would write me from time to time; I left Philadelphia last month with my wife and son; most of this time was spent in descending the Ohio, which is at present very low; we had the barge and crew of G[en]l. Clark, with the company of Mr. R. A. Maupin, and of Mrs. Galt, who had spent several months at New York & at Phila. I shall probably descend [the river] to New Orleans this autumn with N. Berthoud; [all kinds of] merchandise are extremely scarce and very dear, everywhere, but even more is this true of coarse woolens, which one does not find at all.

I have no doubt your lead is selling very well, this article having increased considerably [in value] since the war. In the latter part of my stay in the East I received a letter from my father, and one from your brother; all your family were then well, that is, four months ago; your brother is very anxious to hear from you; if peace should come at a day not far remote (and may it please God that this be so), I hope to get into communication with him.

I have written to him and I urge you to do the same; your letters can be delivered, if sent to New York, and from thence on the Cartel.[209] My wife is well and [so is] my son; may you be the same, and count among the number of your friends him who would esteem you always.

Adieu, J. Audubon.

[Addressed] Mrs F. Roziers
Mercht
St *Genevieve*
u.L.

Friendly relations with his former partner in trade were occasionally renewed by the naturalist in after life. At one of their last meetings, in 1842, Rozier, who had then returned from France, visited Audubon at his home on the Hudson, and both were entertained in New York by their mutual friend, Nicholas Berthoud.

Ferdinand Rozier, with whom we now part company, lived to enjoy abundant prosperity as a trader and merchant at Ste. Geneviève. Born in Nantes on November 9, 1777,[210] at the age of twenty-five he entered the French navy, at a time when Napoleon was contesting with England the supremacy of the sea. He made numerous voyages, and we hear of him at the Cape of Good Hope, the Island of France or Mauritius, at Cadiz, Teneriffe, and at the Island of Bartholomew. Eventually, on April 8, 1804, he embarked on the cutter *Experiment*, with Captain Upton in charge, bound for the United States, where he visited a number of American ports, including Philadelphia and Norfolk. In the following year he returned to France in the frigate *President*, Captain Gallic Lebrosse, and entered the harbor of Nantes on March 1, 1805.[211] In the spring of that year John James Audubon, as we have seen, had also returned to that city, and plans were eventually laid for their commercial aggrandizement in the New World which both had so lately visited. To what extent Audubon's dreams failed of realization may be gathered from the following chapters.

Having settled finally at Ste. Geneviève, Rozier, at thirty-six, married Constance Roy, a girl of eighteen, who bore him ten children, four of whom, all octogenarians, were living in 1905. Ferdinand Rozier's thrift and industry soon brought him substantial rewards. In his earlier days he is said to have made six journeys to Philadelphia on horseback to purchase merchandise, and these trading expeditions were uniformly successful. His trade extended over the whole of Upper Louisiana, and he lived to see the great growth of Missouri as a sovereign state, along with the development of the fabulous mineral wealth of the district.[212]

Rozier's old store at Ste. Geneviève, for long a landmark in that community and considered a pretentious building in its day, was undoubtedly built after the date of Audubon's visit.

---

209 Boat for the exchange of prisoners of war.

210 Compare Note, Vol. I above.

211 See Vol. I below.

212 For this characterization of Ferdinand Rozier I am indebted mainly to an account by his son, Firman A. Rozier, at one time mayor of Ste. Geneviève and member of the State Legislature; see his *History of the Early Settlement of the Mississippi Valley* (St. Louis, 1890).

Figure 32. Ferdinand Rozier; In His Eighty-Fifth Year (1862).

Figure 33. Ferdinand Rozier's Old Store at Ste. Geneviève, Missouri.
This and the above Published by Courtesy of Mr. Ruthven Deane.

The front was devoted to the service of customers and a large shed or stock room was placed at the rear, while the family lived in the main section, which was entered by a door not shown in our illustration.[213] When this building was demolished to make way for modern changes, the wooden pins used in joining the frame were treasured by many as souvenirs of pioneer times. Ferdinand Rozier, who outlived Audubon by thirteen years, died at Ste. Geneviève on January 1, 1864, at the age of eighty-seven years. If he were one of those who thought that Audubon was wasting his time in his ardent zeal for natural history, it should not surprise us, for their ideals were in conflict, and the naturalist's way of working was certainly not conducive to success in trade.

---

[213] For a photograph of the old Rozier store at Ste. Geneviève, as well as for the likeness of Rozier, made in 1862, when he was in his eighty-fifth year, I am indebted to the kindness of Mr. Ruthven Deane, who received them from a son of Ferdinand, Felix Rozier, in November, 1905, when the latter had attained his eighty-third year.

Chapter 16

AUDUBON'S MILL, AND FINAL REVERSES IN BUSINESS

The seven years which followed the outbreak of war with England in 1812 were the most disastrous in the naturalist's career. In many respects they were critical for the entire country, since hundreds who were not affected directly by the war were ruined by the financial troubles which followed in its wake. To Audubon reverses came at this time in rapid succession. Bereft of one and then another of his children,[214] with his family in straitened circumstances in France, and reduced to bankruptcy himself, he finally resolved to throw up trade, for which he was never fitted, and to make his avocation the real business of life. We shall see how, by the unstinted use of such talents as he possessed, through unremitting effort, and with the aid of his energetic and capable wife, he was able, at the age of forty-five, to turn failure into success.

After his return to Henderson in the spring of 1811, Audubon began to look for another opening in trade, living meanwhile with his family at the home of Dr. Adam Rankin, called "Meadow Brook Farm." Dr. Rankin was the first educated physician in his district, and was for many years an officer of the court. A doctor of the older school and a genuine lover of his kind, with a large heart and an open hand, he made his home a hostelry where anyone in need could find refuge without money and without price. No doubt he was attracted to the naturalist by kindred tastes, and it is known that they became life-long friends. The old house, to which Audubon refers in one of his "Episodes,"[215] was built of logs, and stood at some distance from the pike, about two miles from the village in a southeasterly direction. There were experienced in greatest frequency, in the winter of 1811 and 1812, the terrific earthquakes that repeatedly shocked the country at that time; there also Audubon's younger son, John Woodhouse, was born on November 30, 1812. The Rankin farm became at a much later day the site of the village of Audubon, which still later was to be incorporated in the growing city of Henderson, when most of the old landmarks had been obliterated. Dr. Rankin built a more commodious and pretentious brick house in the village itself, and was neighbor to the naturalist for many years, their houses being on the same or adjoining lots. He was thrice married and had many children, the eldest of whom, William Rankin, became

---

[214] While living at Henderson the Audubons lost their two daughters, Rosa and Lucy, both of whom died when very young.

[215] "The Earthquake," *Ornithological Biography* vol. i, p. 280.

Audubon's favorite companion in the field; together they ransacked the country for birds and animals of every sort.

Audubon's unfortunate business relations with his brother-in-law, Thomas W. Bakewell, began in the autumn or winter of 1811, when the naturalist was in the East and Bakewell was about to return to New Orleans in the employ of a firm of Liverpool merchants who dealt in cotton. Bakewell, who had seen much of the South since the failure of his uncle in New York, induced Audubon to join him in an independent commission business, with the assurance that his French nationality would help their undertakings. According to Vincent Nolte, when they were descending the Ohio in December, 1811, Audubon displayed a business card, showing the firm name of "Audubon and Bakewell," and indicating that they were to deal in such homely products as pork, lard and flour. Thomas Bakewell, we are told, taking with him all the disposable funds of Audubon, who continued to send him "almost all the money" that he could raise, opened their business at New Orleans in the winter or spring of 1812, just in time for the war, which broke out in June, to destroy it. When he returned north, in August of that year, Thomas Bakewell, said the naturalist, suddenly appeared one day at "Meadow Brook Farm," while he was making a drawing of an otter, and after bewailing their misfortune in trade, departed.

At the approach of spring in 1812 Audubon was hard pressed for funds, and Rozier's notes to him being then overdue he set out on foot for Ste. Geneviève to collect his money in person. He went out with a party of friendly Osage Indians, but returned, still afoot and unpaid, with his faithful dog as his only companion.[216] The prairies were then flooded and converted into vast lakes, but Audubon, anxious to reach his home, pressed on, walking, as he said, "one hundred and sixty-five miles in a little over three days, much of the time nearly ankle-deep in mud and water." It was probably on this journey, though it may have been in the previous year, that an incident occurred which he has related in "The Prairie,"[217] when, as he declared, for the first time in the course of his wanderings for upwards of a quarter of a century, his life was in actual danger from his fellow man.

When at last he had obtained some ready money, Audubon rode to Louisville, where he purchased on the half-cash, half-credit basis a small stock of goods, and again set up a retail shop at Henderson. This modest venture promised so well that he bought land with the intention of making that town his permanent home. "I purchased," said he, "a ground-lot of four acres, and a meadow of four more at the back of the first." On the latter, to follow this account, were several buildings and an excellent orchard, "lately the property of an English doctor, who had died on the premises and left the whole to a servant woman as a gift, from whom it came to me as a freehold": other land, he added, adjacent to the first, was later secured.

---

[216] This journey was probably made in February, though the date is given as April (see Maria R. Audubon, *Audubon and his Journals*, vol. i, p. 44), if the legends of four drawings of this time are to be trusted; all are labeled Pennsylvania, and bear the following dates: Swamp Sparrow, March, 1812; Spotted Sandpiper, April 22, 1812; White-throated Sparrow, April 24, 1812; and Whippoorwill, May 7, 1812.

[217] *Ornithological Biography*, vol. i, p. 81. In his biographical sketch of 1835 Audubon said that this occurred on his first return from Ste. Geneviève to Henderson (in 1811), a contradiction characteristic of his manner of dealing with biographical and historical details. For an account of this "Episode," see Chapter 18.

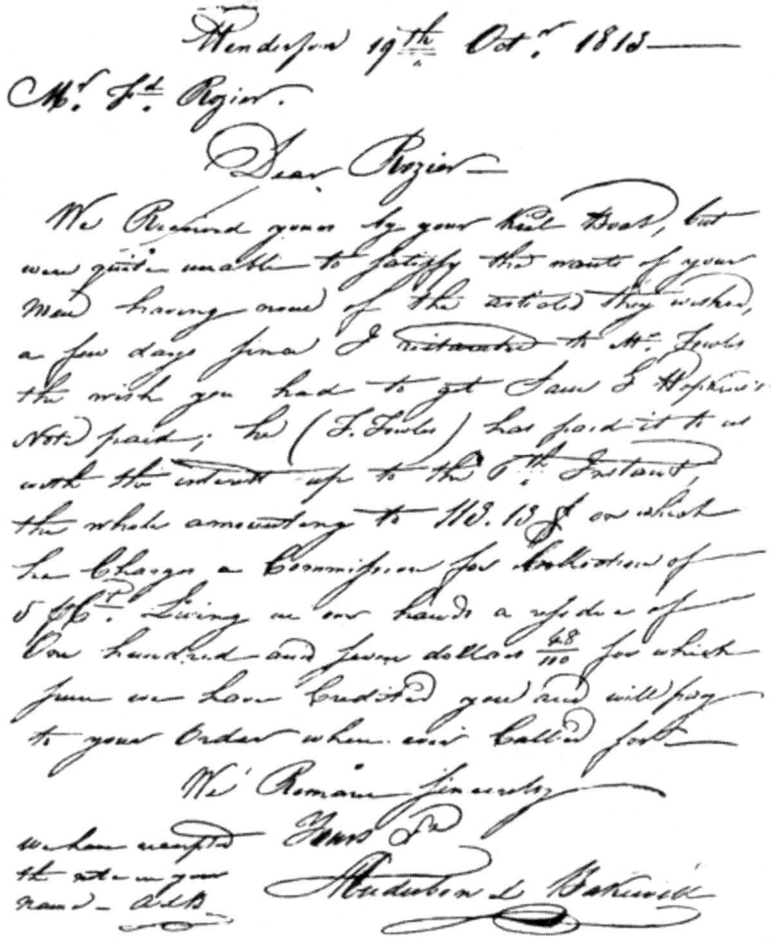

Henderson 19th Octr 1813—

Mr Fd Rozier.

Dear Rozier—

We Received yours by your Keel Boat, but were quite unable to satisfy the wants of your Man having none of the articles they wished, a few days since I mentioned to Mr Fowler the wish you had to get Saml G Hopkins's Note paid; he (F. Fowler) has paid it to us with the interest up to the 8th Instant, the whole amounting to 113.13 $ on which he Charges a Commission for Collection of 5 PCt Living in our hands a residue of One hundred and seven dollars 48/100 for which sum we have Credited you and will pay to your Order when ever Called for—

We Remain Sincerely

Yours &c

Audubon & Bakewell

Figure 34. Letter of Audubon to Ferdinand Rozier, Signed "Audubon & Bakewell," and Dated October 19, 1813, during the First Partnership under This Style.
From the Tom J. Rozier MSS.

These curiously embroidered statements regarding land transactions at Henderson in 1813 are not in harmony with the existing records of that frontier town. Henderson, as its historian[218] tells us, was laid out originally in 1797 into 264 one-acre lots, of which comparatively few had been sold at the time of which we speak, though nominal prices were asked and a few had been given away to encourage settlement.[219] Audubon is recorded as having purchased four one-acre lots from the town, two in 1813 and two in the following year, while a long lease was taken upon land adjacent to the river where later rose his famous mill.[220]

---

[218] For early references to Henderson I am indebted mainly to Edmund L. Starling, *History of Henderson County, Kentucky*, who had access to all the town and county records.

[219] In 1819, the year of Audubon's departure, 129 town lots had been sold, while 29 had been given to privileged persons or to prospective settlers.

[220] According to the town records, as quoted by Starling, on December 22, 1813, Audubon purchased lots numbers 95 and 96, which were one-half of the square lying on the west side of Third Street and between Green and Elm Streets, from General Samuel Hopkins, agent of the Messrs. Richard Henderson & Company; on September 3, 1814, he bought lots numbers 91 and 92, or one-half of the square on the west side of Second

The old Audubon store for general merchandise, built of hewn logs, in a single story, stood at the corner of Main and Mill Streets (now Second Street), fronting the latter, at a point where a modern departmental establishment has since risen. Adjoining this primitive store, on the main street, was his log dwelling,[221] of one and a half stories, with a square porch at the entrance. Immediately opposite, on the two-acre strip of land purchased in 1814, lay a small pond which Audubon is said to have stocked with turtles in order to gratify his special fondness for this delicacy.

Audubon's winning manners made him a popular figure among the early settlers of this region, and for the space of three years he enjoyed life as never before; "the pleasures," he said, "which I have felt at Henderson, and under the roof of that log-cabin, can never be effaced from my heart until after death." But in a community of exacting business men he could never have made a permanent success; he was too good a target not to be riddled by many who were ready to take advantage of his liberality and easygoing ways. Traveling from Frankfort to Lexington in 1810, Wilson complained that the people were all traders but no readers, even of the newspaper; every man, he said, had "either some land to buy or sell, some law-suit, some coarse hemp or corn to dispose of; and if the conversation does not to lead to any of these, he will force it."

Many stories, and no doubt much idle gossip, concerning Audubon's life and habits, were current at Henderson long after he left the village. It was said that he would often go into the woods in his pursuit of birds and remain from home for weeks at a time; that he was once known to have followed a hawk for three days in succession and in practically a straight course, swimming creeks when necessary, until it finally fell to his gun. When steamboats made their first appearance on the Ohio, they naturally excited the greatest interest, and a favorite pastime of many of the men and boys was diving from the side of a boat into the river. On one of these occasions Audubon is said to have made his appearance in the crowd of sightseers and to have astonished everyone by plunging from the bow and emerging from beneath the stern of the vessel after swimming under her entire length. According to traditional accounts, Mrs. Audubon, who was also an expert swimmer, would enter the river clad in a regular bathing costume and cross with ease to the Indiana shore.

In spite of the hard times Audubon managed to keep out of serious business troubles until he entered into another partnership with Thomas Bakewell, his brother-in-law. Their project in this second association was to erect a steam lumber and grist mill at Henderson, which of all mortal follies the naturalist considered in the retrospect to have been one of the worst. It is recorded that on the sixteenth day of March, 1817, John James Audubon and Thomas W. Bakewell, under the designation of "Audubon and Bakewell," applied to the trustees of the village for a ninety-nine year lease of a section of land on the river front. Their petition was granted, upon a consideration of $20 per annum, and the partners began to build their mill on the property and completed it within that year. Thomas W. Pears,[222] a former fellow-clerk of both Audubon and Bakewell in New York, early joined the enterprise, which was regarded at

---

Street, between Green and Elm. The mill site on the Ohio River was a part of the land given to Henderson by the Transylvania Company, the original owners of a large part of Kentucky; this site was leased for 99 years to J. J. Audubon, was sold and resold, but reverted to the city of Henderson in 1915. In the latter year the project was broached of obtaining the original mill site, together with adjoining property along the river, and converting the whole into a public park dedicated to Audubon.

[221] At a somewhat later time the naturalist occupied a one-story frame house, built in 1814, which stood at the corner of Fourth and Main Streets; see Starling, *op. cit.*

[222] See Vol. I.

the time as one of considerable magnitude. Their mill, which stood for ninety-five years, became famous in the annals of the Ohio Valley.[223] Said the historian of Henderson County, writing in 1879:

Figure 35. Audubon's Mill at Henderson, Kentucky, Since Destroyed, as Seen from the Bank of the Ohio River. After a Photograph of 1894, Published by Courtesy of Dr. R. W. Shufeldt.

> The weather boarding, whip-sawed out of yellow poplar, is still intact on three sides. The joists are of unhewn logs, many of them over a foot in diameter, and raggedly rough. The foundation walls are built of flat, broken rock and are four and a half feet thick. Mr.

---

[223] A Henderson correspondent of Joseph M. Wade, under the signature of "W. S. J.," August 8, 1883, gave the following account of the structure. The original mill covered forty-five by sixty-five feet, and consisted of four stories and basement; the basement walls of stone stood four feet thick, while at the third story the thickness was three feet; the three upper stories were in frame. The studding measured three by six, and the rafters four by eight, inches. Many of the large timbers that could then be seen were sound and apparently good for a century or more. Parts of the old machinery that had been used in the grist mill were lying about under the eaves; the building was then used as a tobacco stemmery. See Joseph M. Wade, *Ornithologist and Oölogist*, vol. viii, p. 79 (1883).
The old Audubon mill in more recent times was incorporated into a warehouse for the storage of leaf tobacco; it was burned to the ground on March 18, 1913.

Audubon operated the mill on a large scale for those times. His grist-mill was a great convenience, and furnished a ready market for all of the surplus wheat raised in the surrounding country. His saw-mill also was a wonderful convenience, doing the sawing for the entire county.

Mr. and Mrs. Pears, who had no liking for Henderson, early withdrew and sold their interest in the mill[224] to Audubon and Bakewell, thus adding to their financial embarrassment. The engines, which seem to have given no end of trouble, were constructed by David Prentice, an intelligent Scotch mechanic; since his first work after coming to this country was to erect a steam threshing mill at "Fatland Ford," his services were probably secured by William Bakewell, who afterwards helped to establish him at Philadelphia. While at Henderson he is said to have fitted a small engine and paddlewheels to a keel boat, which was christened the *Pike*, and to have taken it up the river to Pittsburgh. Prentice seems to have entered the partnership and to have retired with Bakewell.

In order to extend the sphere of their operations, Audubon is said to have purchased at this time a tract of 1,200 acres of government land,[225] and to have engaged a band of stalwart Yankees to fell and deliver the timber. According to one account, they were a party of emigrants who had come to Henderson with their families and encamped on the river bank. For a time all went well, but one day when they failed to deliver their usual supply of logs, it was found that they had decamped and fled down the river towards the Mississippi, taking on their flatboat Audubon's draft oxen and in fact all the plunder that they could lift. Nothing was ever recovered and but one of the fugitives was ever seen again; this man boarded a river boat on which the naturalist happened to be traveling, and it is said that upon being recognized he jumped into the river and swam to the shore like a frightened deer.

When Bakewell finally withdrew, Audubon appears to have been left stranded, and the business was taken over by a new set of men, including another brother-in-law, Nicholas Berthoud, and Benjamin Page of Pittsburgh, who continued it under the name of J. J. Audubon & Company.[226] Agents were also secured at various points on the Ohio and Mississippi rivers. Excepting, as we must assuredly do, his ever staunch friend, Nicholas Berthoud, Audubon believed that he was "gulled by all of these men."

In 1818 a new era of building and general prosperity seemed to dawn in the valley of the Ohio. A new bank was chartered at Henderson, and the woodwork of its brick structure was furnished by Audubon's mill.[227] This bank, however, failed in the course of two years, and

[224] The mill is supposed to have cost about $15,000; of this sum Thomas Pears is said to have contributed from $3,000 to $4,000, and William Bakewell a similar amount in the interest of his son, while Audubon presumably furnished the balance.

[225] Maria R. Audubon, *op. cit.*, vol. i, p. 47.

[226] In his journal of 1820 Audubon said that after the withdrawal of Bakewell, "men with whom I had long been associated offered me a partnership. I accepted, and a small ray of light appeared in my business, but a revolution occasioned by a numberless quantity of failures put all to an end."

[227] One of J. J. Audubon & Company's bills is here reproduced from Starling, *op. cit.*

"To the President and Directors of the Bank of Henderson to Henderson steam mill:

"To three pieces of scantling, 56 feet, 4½ c	$2.52
"To ten pieces of scantling, 34 feet	——
"To sixty rafters, 714 feet, at 4 c	28.56
"To five pieces scantling, 40 feet, at 3 c	1.20
"To fifteen joists [?], 278½ feet, at 6 c	16.71
"J. J. Audubon & Co."	$48.99

forty others scattered throughout that section broke in rapid succession, after having done little more than add to the flood of worthless paper notes that was demoralizing business and sending hundreds into bankruptcy.

The mill was in operation barely two years. The machinery, of which a wooden bolting shaft and wooden cog wheels remained as a curiosity to recent times, seems to have worked badly from the start. But aside from the inexperience of the builders and the financial troubles of the day, the enterprise was foredoomed to failure in a district which raised but little wheat, and in which the demand for lumber was then comparatively slight. "How I labored," said Audubon, "at that infernal mill! But it is over now; I am old, and try to forget as fast as possible all the different trials of those sad days."

In the course of the Audubon and Bakewell partnership[228] the naturalist became involved in a personal quarrel with a man whose initials are given as "S—— B——." It seems that in 1817 Audubon's mechanic, David Prentice, had built for him a small steamboat, though for what purpose is not known. When their interests were severed, we are told, Mr. B—— purchased this steamer, but paid for it in worthless paper. The captain of the craft ran her down to the Mississippi and thence to New Orleans, and Audubon, who was determined to arrest this man if necessary, started in pursuit in a skiff. He failed, however, to overhaul the fugitive, and reached New Orleans only to find that his vessel had been surrendered to another claimant. This was probably in May, 1819, for in his journal of the following year, under date of November 23, when he was again moving down the rivers but in more leisurely fashion, he speaks of two large eagle's nests, one of which he remembered having seen as he "went to New Orleans eighteen months" before.

Through the researches of a later historian I am now able to give a more exact account of this affair. The purchasers of the steamboat were William R. Bowen, Samuel Adams Bowen, Robert Speed, Edmund Townes, Obediah Smith, George Brent and Bennett Marshall, who immediately sued Audubon in the sum of $10,000, on the plea that he had maliciously taken out an attachment upon the vessel in New Orleans, where it had been detained. They represented to the judge of the circuit court, Henry P. Broadnax, that Audubon was about to leave Kentucky, and a warrant was issued to arrest him; he was taken into custody, said the narrator whom I am following, "but executed a bail bond in the sum of $10,000 with Fayette Posey as surety, and was released." Convinced that a trial at Henderson would lead only to a defeat of justice, Audubon now served notice that he would apply for a change of venue to another county. "That notice together with the other papers in the action, is among the records of the Daviess circuit court, at Owensboro, Kentucky. It was written and signed by Audubon. Application for a change of venue was made at Hardinsburg and the case was transferred to the Daviess circuit court." When the case was called, the plaintiffs asked for a continuance, and it was granted them, but when the case was called again at the next term of court, the plaintiffs failed to appear, and the action was finally dismissed.

Returning home, Audubon was obliged to walk from the mouth of the Ohio River to Shawnee Town. Upon reaching Henderson he found that Mr. Bowen had anticipated him. Acting upon advice, he was prepared for an encounter with this man, who as his neighbors declared, had sworn to kill him, and "whose violent and ungovernable temper was only too

---

[228] According to W. G. Bakewell, *Bakewell-Page-Campbell*, Thomas Bakewell sold his interest in the store and mill to Audubon in 1817, but this is contradicted by other accounts. For the incident which follows, see Maria R. Audubon, *op. cit.*, vol. i, p. 34.

well known." The anticipated encounter ensued. Audubon, who was then carrying his right hand in a sling from a recent injury received in his mill, waited, as he said, until he had received twelve severe blows from his assailant's bludgeon; then with his left hand he drew a dagger and struck in his own defense. His assailant was felled to the ground, but happily the wound inflicted was not mortal. Mr. Bowen was carried away on a plank, and when the affair was settled in the judiciary court, according to a Henderson tradition, Judge Broadnax gravely left the bench, approached the man who had been under charge of assault, and said: "Mr. Audubon, you committed a serious offense—an exceedingly serious offense Sir—in failing to kill the d—— rascal."[229] "Thomas Bakewell," added the naturalist, "who possessed more brains than I, sold his town lots and removed to Cincinnati, where he has made a large fortune, and I am glad of it.[230]

When the mill was finally closed and the company dissolved in 1819, Audubon as usual was the heaviest loser. Arrested and sent to the Louisville jail for debt, he was able to obtain release only by declaring himself a bankrupt in court. "I paid all I could,"[231] he said in his journal of the following year, "and left Henderson poor and miserable in thought. My intention to go to France and see my mother and sister was frustrated, and at last I resorted to my poor talents to maintain you and your dear mother, who fortunately became easy at her change of condition, and gave me a spirit such as I really needed, to meet the surly looks and cold reception of those who so shortly before were pleased to call me their friend." "I parted," to revert to his later account, "with every particle of property I held, to my creditors, keeping only the clothes I wore on that day, my original drawings, and my gun." Without a dollar in his pocket he left Henderson and walked to Louisville alone; "this," he said on reflection, "was the saddest of all my journies, the only time in my life when the Wild Turkeys that so often crossed my path, and the thousands of lesser birds that enlivened the woods and the prairies, all looked like enemies, and I turned my eyes from them, as if I could have wished that they never existed."

Passing down the Ohio in the following year Audubon made these entries in his diary:

> *November 2nd, 1820.* Floated down slowly within two miles of Henderson. I can scarcely conceive that I stayed there eight years, and passed therein comfortably, for it is undoubtedly on the poorest spot in the country, according to my present opinion.

---

[229] See Dixon L. Merritt, "Audubon in Kentucky," *The Taylor-Trotwood Magazine*, vol. 10 (1909), p. 293.

[230] Thomas Bakewell later became a successful builder of steamboats, first at Pittsburgh, and after 1824 at Cincinnati, where he was an important factor in the rising commerce of the Ohio Valley, and where he left his mark on the history of that city. As a theoretical mechanic in iron and wood he is said to have had no superior; his business was nearly destroyed in the panic of 1837, and he never regained his financial position. To his credit also it must be added that in 1860, at the age of seventy-two, he began at the bottom of the ladder again by engaging as a clerk with a paper company at Cincinnati, and, refusing the proffered aid of his children, he did not give up work until his eightieth year, seven years before his death in 1874. See W. G. Bakewell, *Bakewell-Page-Campbell.*

[231] Audubon was not so accurate when in his biographical sketch of 1835 he said: "Finally I paid every bill, and at last left Henderson probably forever...," for when at Charleston with Bachman in 1834, one of his former creditors attempted to sue him for debt and apparently carried his case to court. When Bachman asked for an explanation, Audubon wrote from New York, April 5, 1834, as follows: "Respecting the suit let me tell you ... that I went to Gaol at Louisville after having given up all to my creditors, and that I took the benefit of the act of insolvency at the Louisville Court House, Kentucky, before Judge Fortunatus Crosby & many witnesses, and that a copy of the record of that step can easily be had from that court.... I wish friend Donkin to do all he can to put a *Conclusion*—stop to this matter, for it makes me sick at heart." The lawyer here referred to was probably Judge Dunkin, friend of Bachman and distinguished in his profession, who had a plantation at Waccamaw, near Charleston, South Carolina

Nov. 3rd, 1820. We left our harbor at daybreak, and passed Henderson about sunrise. I looked on the mill perhaps for the last time, and with thoughts that made my blood almost run cold bid it an eternal farewell.

Chapter 17

THE ENIGMA OF AUDUBON'S LIFE AND THE HISTORY OF HIS FAMILY IN FRANCE

Lieutenant Jean Audubon, as already recorded, died at Nantes in 1818, at a time when his son's financial troubles in America were culminating, and left an estate, then none too large, for the sole enjoyment of his widow during her lifetime. The naturalist, so far as is known, never received a penny in payment of bequests made by either his father or stepmother, but the reasons for this fact were far different from those which his biographers have assigned.

We have referred to the curious wording which appears in the six different wills that were executed by Lieutenant Jean Audubon and Anne Moynet, his wife, between the years 1812 and 1821.[232] The first four of these documents[233] were of a mutual nature, and were so drawn that the survivor should enjoy the entire property of the other during his or her lifetime, but this eventually was to be divided between their two children, or heirs of the latter should any exist. In Jean Audubon's last will, made at Couëron on the 15th of March, 1816, he added the provision that in case his "dispositions in favor of Jean Rabain and Rose Bouffard, wife of Loyen du Puigaudeau, should be attacked and annulled," he bequeathed his entire estate, without exception, to his wife, Anne Moynet, for her sole use. His fears, as already intimated, were well grounded, and his will was immediately contested by four nieces, Mme. Lejeune de Vaugeon of Nantes, Mme. Jean Louis Lissabé, whose husband was a pilot, and Anne and Domenica Audubon, seamstresses at Bayonne.[234] This trial dragged on in the courts for a long time, and served further to impoverish Madame Audubon, who was obliged to dispose of most of her valuable effects, but it was finally settled by a compromise in 1820. In that year, at the age of eighty-five, she left "La Gerbetière" to live with her daughter and son-in-law at "Les Tourterelles" close by, where she remained until her death on October 18, 1821.

It seems incredible that Audubon should not have heard of the death of his foster mother, since he had been devotedly attached to her in his youth and was moreover a beneficiary under her will. Yet on August 6, 1826, he wrote in his journal: "My plans now are to go to Manchester, to Derbyshire to visit Lord Stanley, Birmingham, London for three weeks, Edinburgh, back to London, and then to France, Paris, Nantes, to see my venerable stepmother, Brussels, and return to England." On September 30 of the same year he wrote

---

232 See Chapter 9.

233 For complete text of these wills, in the original, See Appendix I, Documents 13-18.

234 See Note 4, Vol. I. The suit brought by these plaintiffs was based upon a French law, which at that time debarred a natural child from inheriting property.

from Liverpool: "I long to enter my old garden on the Loire and with rapid steps reach my mother,—yes, my mother! the only one I truly remember; and no son ever had a better, nor more loving one."[235] Again in 1828 he spoke of this estimable woman as if she were then alive, although she had been dead seven years.

In Madame Audubon's last will, which was made in the July preceding her death, she left her property to be equally divided between her two adopted children, "Mr. Jean Audubon, called Jean Rabin, husband of Lucy Bakewell, and who I believe is at present in the United States of America, and to Rose Bouffard, wife of M. Gabriel Loyen du Puigaudeau, my son-in-law, who is living at Couëron"; she also took care to guard against the pretensions of any spurious heirs, and to make provision for her grandchildren in case of the death of either or both of her heirs direct.

Figure 36. An Old Street in the Couëron of To-Day.

Having given the precise, if somewhat prosaic, recorded facts of the case, we will quote the story narrated by the naturalist's biographers, who never could have seen the legal documents and who thus had only hearsay and conjecture on which to build:

> At this juncture [of critical business affairs at Henderson], the father of Audubon died; but for some unfortunate cause he did not receive legal notice for more than a year. On becoming acquainted with the fact he traveled to Philadelphia to obtain funds, but was unsuccessful. His father had left him his property in France of La Gibitère [Gerbetière], and seventeen thousand dollars which had been deposited with a merchant in Richmond, Virginia. Audubon, however, took no steps to obtain possession of his estate in France, and in after years, when his sons had grown up, sent one of them to France, for the purpose of legally transferring the property to his own sister Rosa. The merchant who held possession of the seventeen thousand dollars would not deliver them

[235] Maria R. Audubon, *Audubon and his Journals*, vol i, pp. iii and 130.

up until Audubon proved himself to be the son of Commodore Audubon. Before this could be done the merchant died insolvent, and the legatee never recovered a dollar of his money.[236]

Figure 37. "Les Tourterelles," Couëron, Final Home of Anne Moynet Audubon, and the Resting Place of Exact Records of the Naturalist's Birth and Early Life.

A key to the origin of the fictitious seventeen thousand dollars is probably to be found in the letters of Jean Audubon to Francis Dacosta, written in 1805,[237] where he refers to certain unsettled business claims against his former partners, Messrs. Formon and Ross, who had been respectively interested with him in two vessels, *Le Comte d'Artois* and the *Annette*, the history of which has already been noticed.[238] They were also engaged at a later time in certain iron-works above Richmond, Virginia, but with these Lieutenant Audubon was not directly concerned. Formon, his partner in Santo Domingo trade, who was charged with having drawn $1,650 in excess of his share, had died without making any final settlement of their accounts; another associate, Edward, had died in London leaving an unsettled claim of $300; while David Ross, who was owing a certain sum, had also died without liquidating his debt. The amount of the latter claim probably was not large, since Dacosta was instructed to use this sum for his needs in developing the mine at "Mill Grove" should he be so fortunate as to collect it; "when you receive my papers from Miers Fisher," said Lieutenant Audubon in his letter of the 22d of June, 1805, "you will find a promissory note of Mr. Samuel Plaisance of Richmond, for the business of the widow Ross. If there were justice there this sum should be paid to me with the costs."

---

[236] Lucy B. Audubon, ed., *The Life of John James Audubon* p. 55.
[237] See Chapter 8.
[238] See Chapter 2.

Lieutenant Audubon was never able to collect these different amounts, which probably did not much exceed $2,000, but an echo of one of these transactions appeared as late as 1819, when Audubon's brother-in-law sent him a document referring to the claim on the Ross estate, in the hope that some money might still be forthcoming, writing as follows:[239]

> In turning over some letters I have found a letter of Mr. David Rost [Ross], and a memorandum that I thought pointed to what was referred to in it. As I have sometimes heard it said that this Mr. David Rost owed a considerable sum, it should be possible that this letter, which is in English, might be of use to you. I cannot say anything about it, not knowing your language, and not having ventured to get it translated, from fear of compromising us, I am sending it to you, [and] you will judge of its importance. Should chance will that it bring you money, send me some of it, I beg you, for I am in great need of it.

The same biographer whom we have just quoted said in reference to "La Gerbetière": "This estate was left by Commodore Audubon to his son John James, who conveyed it to his sister without even visiting the domain he so generously willed away." We have now seen what provisions were actually made for the disposition of this property under the terms of the various wills of Lieutenant Audubon and his wife. We need only add that not long after his father's death, the naturalist lost touch with his family in France; his one-half interest in his stepmother's estate, which was heavily encumbered, was never claimed, and at a much later day was informally relinquished in favor of his sister and her family.

During his Henderson period Audubon was in communication with his brother-in-law, Gabriel Loyen du Puigaudeau, who kept him informed in regard to all that transpired in their French home; on July 26, 1817, the naturalist had given him a power of attorney, the curious wording of which has already been noticed.[240] Whether deterred by the legal complications which soon followed, displeased by the mode of settlement, or for what other cause now unknown to us, Audubon seems to have severed all relations with his family at Couëron, or to have written to them only after long lapses of silence. On New Year's Day, 1820, Gabriel du Puigaudeau dispatched to him a friendly letter[241] of greeting:

> I take the opportunity at the renewal of the year, to offer you the good wishes of the entire family. Our every desire is that you, your beloved wife, and dear children may be happy, that you may prosper, that you may enjoy good health, and this is the wish of your nieces also. But, awaiting the pleasure of seeing you all, by what fatality during the past eighteen months have I not had any news of you, why no reply to at least twenty letters that I have written to you? Can I have been so unfortunate that someone has given you any report that would prejudice you against me? I do not believe that there could exist anyone who would be able to do this, at least with truth; if someone has really sought to estrange your friendship for me, act with frankness, and tell me your suspicions. I do not believe it would be difficult to destroy them, and I even promise that I would offer you no reproach for having momentarily believed it, should this after all have occurred. For what concerns our business affairs, I refer you to my letters which have preceded this.

---

[239] From G. L. du Puigaudeau's copy of his letter to John James Audubon (at Henderson), dated "Couëron, August 15, 1819," translated from the French. (Lavigne MSS.)

[240] See Vol. I.

[241] This, and the letter to follow, translated from Gabriel du Puigaudeau's copies. (Lavigne MSS.)

This letter was sent to Henderson, Kentucky, more than a year after the naturalist had finally left that state; at the moment it was written he was making his way down the Ohio River to New Orleans in a flatboat, "the poorest man aboard," as he thought at the time. Writing in his journal on December 26, 1820, when they had touched at Natchez, Audubon said that on that day he had received letters from his wife, who was then at Cincinnati, written on November 7 and 14, and that the last "contained one from my brother, G. Loyen Dupuigaudeau, dated July 24, 1820." If the month in this instance was misnamed, this might have been the following letter, which was written at Couëron on the twenty-fourth of June, 1820, and sent to Henderson like the last.

> Two years have passed without our having any news of you. What a long lapse of time, and in what anxiety are we plunged! In God's name give us some news about yourself, if it be but a word to set us at rest in regard to your condition. I should not know how to persuade myself that you were not on friendly terms with me, since I have given you no cause [for grievance]; if it is so, be generous enough to relieve me from this anxiety. The business matters of Mr. Audubon are at last concluded, and I await only the return of the papers from Cayes to set them in order with justice [to all].[242]
>
> Profiting by an opportunity for New York, I have only time to refer to my letters of 15 September, 30 October, 19 December, 1818, 1st February, 15 April, 15 May, 3d August, 1819, in all their contents.
>
> Madam Audubon is coming to live with us; she found herself isolated at "La Gerbetière," and was very dull there; I wish that she may be contented here. She does not cease to speak of you, and is as much astonished as I am that we receive no news of you.

The naturalist's elder son, Victor, visited Couëron about the year 1835, when his cousin, Gabriel Loyen du Puigaudeau the second, who was nearly of the same age, returned from military service to meet him. He was disappointed at the appearance of his father's old home, "La Gerbetière," which had not been occupied by the family for fifteen years.[243]

Rosa Audubon du Puigaudeau, the naturalist's sister, died at "Les Tourterelles" after August 3, 1842, leaving a daughter, Rose du Puigaudeau, who died without issue, October 20, 1881, and, if we are correctly informed, one son, Gabriel Loyen du Puigaudeau the second, who died at "Les Tourterelles," Couëron, June 23, 1892, when past his eightieth year; a daughter of this only son was married to Monsieur L. Lavigne, notary at Couëron. At the time of her uncle's death, his property, including the personal records of Lieutenant Jean Audubon, passed into the hands of Madame Lavigne, who is a grand step-niece of the naturalist, and who aside from her children, so far as known, is the only surviving member of his family in France.

---

[242] This reference is evidently to the litigation over Lieutenant Audubon's will and the final disposition of his estate.

[243] It was thought that Victor had come to settle the family's financial affairs, and his uncle and aunt asked if this were the case; he replied that it was not, that the children of Jean Audubon who were in America had taken their [share of the] property in that country, while those in France had theirs in France; he considered that all was settled, but if Rosa's children wished for any money, they had but to ask for it, and the heirs in America would send them what they desired; the subject was then dropped. A considerable correspondence followed this visit, but the letters were all destroyed about twenty-five years ago by Monsieur du Puigaudeau, when putting his effects in order. This account is given on the authority of Monsieur Lavigne.

At this point we must examine a little more carefully the peculiar status of what Audubon referred to as the "enigma" of his life. In some of his private journals and letters[244] he dramatically declared that a mystery had surrounded his early existence, which he was bound by a solemn oath exacted by his father never to reveal, and that this secret must be carried by him to the grave. If it be the duty of a biographer to make the true character of his subject known, the passage of time would now seem to sanction reference to many personal matters which a century ago should have been more rigidly guarded. I enter upon this task solely with the view of placing Audubon's character in a truer and fairer light.

The essential facts regarding Audubon's birth and early years have now been given, and this is the true, though possibly not the complete, story. Anything which we now add, however, can be regarded as little better than speculation. Audubon is said to have received through his father a large sum of money from an unknown or unnamed source,[245] but as such stories are apt to be exaggerated, especially when an ocean intervenes between a testator and his heir, the statement may be erroneous; we have seen that Lieutenant Audubon was not in a position to make such gifts himself had he been so disposed. If the report were true, the money may have come from the estate of his mother, and through the agency of the mysterious "Audubon of La Rochelle," who is said to have been a politician.[246] In some of the passages which we do not quote, the naturalist would have his family believe that he was of noble birth, that his adoptive father was not his true father, and that both he and Lieutenant Audubon had received irremediable injury through the treachery of the mysterious uncle, "Audubon of La Rochelle." Now these strange statements of the naturalist, though not in accord with the facts as they are known to us, should be interpreted, I believe, in the light of possible stories that may have come to him in the glamour of his youth; his mind may have been diverted by them, he may have believed them, but of this nothing now can positively be known. To continue our conjectures, it is possible that the plain conflict between these supposititious tales and the facts that were revealed at his adoption, his baptism, and in the wills of his father and stepmother, as well as by the lawsuit which followed the former's death, all led him to resort to "enigma." We should also remember that the naturalist, who was careless of dates and historical facts, had finally left his home at the age of twenty, when young men as a rule are not curious about their family history, and that he reached the reminiscent stage late in life. It seems probable that the wording of his father's will and the later attempt to annul it finally induced him to wash his hands of the whole matter, even to breaking off relations with his family in France. Feeling, as undoubtedly he did, that public knowledge of those conditions, for which he was in no way responsible, might be a bar to all future aspirations, he was not loath to let the matter rest, so far as he and his immediate family were concerned, under a cloak of mystery. If such were in truth the case, I think few would find cause to blame him.

---

[244] These passages, which were shown to me by his granddaughter, Miss Maria R. Audubon, in 1914, but not for publication, occur in his journals under the following dates; June 4, 1826, at sea; March 15, 1827, at Edinburgh, after describing a visit of Lady Selkirk and her daughter; again on the 18th of March of the same or the following year; and on October 8, 1828, when writing to his wife from Paris and reflecting on the advisability of visiting his old home at Nantes. While these extraordinary passages are not quoted, out of deference to the wishes of his granddaughters, it seems only just to Audubon, in view of the revelations that have already been made, to add this brief reference to the incidents in question.

[245] This statement was made to me in 1914 by Miss Maria R. Audubon.

[246] See Vol. I.

When we view the whole subject in this double light, of a duty owed to his family and of the possibility that conflicting stories had come to him at an earlier day, any embroidery or confusion which appears in many of his statements of a personal nature can be better understood. Such an explanation would be quite convincing if payments had actually come to him from his own mother's estate.

We will only add that Mrs. Audubon, who seemed to have shared her husband's intimate thoughts, apparently believed to the last in his high birth. When her younger son, John Woodhouse Audubon, lay at the point of death, in February, 1862, she was summoned to his bedside, but reached it too late to see him alive; upon entering the room Mrs. Audubon is said to have exclaimed: "Oh, my son, my son! to think that you should have died without having known the secret of your father's early life!" When asked by members of her family to what she then referred, she turned their questions aside, saying only that such remarks were common in moments of intense grief and excitement.

Chapter 18

EARLY "EPISODES" OF WESTERN LIFE

Audubon's sketches of life and scenery in America, which he designated as "Episodes," were interspersed in his *Biography* of birds[247] to brighten the narrative and beguile the reader. Extending to the number of sixty, and dealing mainly with events between the years 1808 and 1834, they abound in tales of adventure and graphic pictures of pioneer life which for their personal charm, local coloring, and human interest are worthy of high praise. Some of these sketches have been copied widely and some have been translated into Audubon's native tongue; some have even found their way into schoolbooks. While they have deservedly won the naturalist many readers, not a few have subjected him to harsh criticism on the score of too vivid coloring or historical inaccuracy, a fault to which he was particularly prone. Whenever Audubon went directly to nature to exercise his pencil or brush or wrote with his subject before him, he was truth itself, but in writing offhand and from memory of past events he was wont to humor his fancy, disregarding dates as readily as he did the accents on French words. This tendency is particularly apparent in the accounts of some of his early adventures in the western country, such as "Louisville in Kentucky" (1808-10), "The Prairie" (1812), "A Wild Horse" (1811-13), and "The Eccentric Naturalist" (1818), the history of which is detailed in the following chapter. We shall examine some of these stories at this point, though their composition belongs to a later period, in order to reach a just conclusion in regard to the author's method, as well as for the intrinsic interest of the narratives themselves.

During Audubon's early life in Kentucky, as we have seen, he frequently visited the East, whether in the interest of birds or business, traveling by way of the river and the forest roads. Incidents of these journeys frequently occur in the "Episodes," but since dates commonly are omitted and the order of events is liable to be blended or confused, they cannot be trusted always for historical accuracy. Thus, "The Wild Horse" episode[248] professes to be an account of a single journey from Henderson, in Kentucky, to Philadelphia and back again, whereas some of the events recorded occurred in reality at least two years apart, such as the meeting with Nolte at the Falls of the Juniata River in December, 1811, and the naturalist's return from

[247] In the first three volumes only of the *Ornithological Biography*, being omitted from the last two on account of the exigencies of space.

[248] *Ornithological Biography*, vol. iii, p. 270.

Pennsylvania with the proceeds of "Mill Grove," which could not have been earlier than 1813, the date of its sale to Mr. Samuel Wetherill, Junior.[249]

Audubon visited Philadelphia in November, 1811, and returned to Kentucky in December of that year, but whether it was upon this or some other journey that he rode a wild horse through seven states in going from his home at Henderson to the Quaker city, or whether such a journey ever occurred, is immaterial to the interest of the narrative. In this instance, however, we have the advantage of comparing the notes of a fellow traveler, Vincent Nolte, then a merchant at New Orleans.[250] First to follow Audubon's account, as given in his "Episode," we are told that he rode a wild mustang, named "Barro," that had never known a shoe, having been recently captured near the headwaters of the Arkansas. In going east he diverged from the beaten track to extend his knowledge of the country and of its bird life. From Henderson he passed through the heart of Tennessee to Knoxville, thence to Abington, the Natural Bridge, and Winchester in Virginia, crossed the corner of West Virginia to Harper's Ferry, then to Frederick, Maryland, and on through Lancaster to Philadelphia; there, he said, he remained four days, and returned by way of Pittsburgh, Wheeling, Zanesville, Chillicothe, Lexington and Louisville, to Henderson. He estimated the whole distance traversed at "nearly two thousand miles," and at a rate of "not less than forty miles a day." Much is said in praise of his favorite bay horse, and its food and daily treatment are duly recorded. This horse was very docile, and would wade swamps, swim rivers, and clear a rail fence like an elk; corn blades as well as corn and oats entered into his daily ration, to which a pumpkin and fresh eggs, when procurable, were occasionally added.

It was upon his return journey that the naturalist met with Vincent Nolte, who twelve years later did his chance acquaintance a good turn, when the latter was about to sail for England in 1826.[251] Nolte, said Audubon,

> was mounted on a superb horse, for which he had paid three hundred dollars, and a servant on horseback led another as a change. I was then an utter stranger to him, and when I approached and praised his horse, he not very courteously observed that he wished I had as good a one. Finding that he was going to Bedford to spend the night, I asked him what hour he would get there: "Just soon enough to have some trouts ready for our supper, provided you will join when you get there." I almost imagined that Barro understood our conversation; he pricked up his ears, and lengthened his pace, on which Mr. Nolte caracolled his horse, and then put him to quick trot, but all in vain; for I reached the hotel nearly a quarter of an hour before him, ordered the trouts, saw to the putting away of my good horse, and stood ready at the door to welcome my companion. From that day to this Vincent Nolte has been a friend to me.
>
> Audubon added that they rode together as far as Shippingport, now a part of Louisville, where his brother-in-law, Nicholas Berthoud, was then living.

We shall now follow the equally circumstantial but widely divergent account of this meeting and the subsequent journey as given by the other traveler. Nolte had sailed from Liverpool in September, 1811, and landed in New York after a perilous voyage of forty-eight

[249] While the object of this visit is not mentioned in the "Episode," it is stated in the second biographical sketch; the ambiguities connected with the sale of this farm, in which others besides Audubon were then interested, are discussed in Chapter XI.

[250] Vincent Nolte, *Fifty Years in Both Hemispheres*.

[251] See Chapter 21.

days. He had no servant, but was accompanied by a young Englishman, named Edward Hollander, whom he had engaged in a business capacity while in London and with whom he was making his way to New Orleans. Hollander had been sent in advance to Pittsburgh to purchase two flatboats, for in addition to their horses they had planned to carry 400 barrels of flour, from the sale of which in the South they expected to defray the expenses of their journey. Having purchased a fine horse in Philadelphia, Nolte left that city in December, and with saddle-bags strapped to his horse's back, rode on "entirely alone." He crossed the highest point of the Alleghany ridge at ten o'clock of a winter's morning and later in the same day reached a small inn "close by the Falls of the Juniata River." "The landlady," to quote his narrative, "showed me into a room, and said, I perhaps would not mind taking my meal with a strange gentleman, who was already there." This stranger, who immediately struck him as "an odd fish," "was sitting at a table, before the fire, with a Madras handkerchief wound around his head, exactly in the style of the French mariners, or laborers, in a seaport town." In the course of the conversation which then ensued he declared that he was an Englishman, but Nolte was the last person to be deceived on a question of nationality and remarked at once that his speech betrayed him. "He showed himself," to quote our senior traveler again, "to be an original throughout, but at last admitted that he was a Frenchman by birth, and a native of La Rochelle. However, he had come in his early youth to Louisiana, had grown up in the sea-service, and had gradually become a thorough American." When asked how this account squared with his earlier statement, said Nolte, "he found it convenient to reply in the French language: 'when all is said and done, I am somewhat cosmopolitan; I belong to every country.' This man," to conclude, "who afterwards won for himself so great a name in natural history, particularly in ornithology, was Audubon, who, however, was by no means thinking, at that time, of occupying himself with natural history."

In the interview as thus far recorded, Audubon was clearly chaffing his new acquaintance, for not one of the statements attributed to him was true, if we accept the fact of his French extraction. Nolte, to be sure, writes as a somewhat vain and garrulous man, and after a lapse of forty-three years, but he professes to speak the truth and there is no reason to suppose that his narrative is pure invention. Nolte further informs us that Audubon's father-in-law, Mr. Bakewell, "formerly of Philadelphia," was "then residing and owning mills at Shippingport," which was not the case. To continue, finding that Audubon, who was bound for Kentucky, was a companionable man and devoted to art, a field which he had cultivated himself, Nolte proposed that they should travel together, and offered the naturalist a berth on one of his flatboats.

> He thankfully accepted the invitation, and we left Pittsburgh in very cold weather, with the Monongahela and Ohio rivers full of drifting ice, in the beginning of January, 1812. I learned nothing further of his traveling plans until we reached Limestone, a little place in the southwestern corner of the State of Ohio.[252] There we had both our horses taken ashore, and I resolved to go with him overland, at first to visit the capital, Lexington, and from there to Louisville, where he expected to find his wife and parents-in-law.... We had hardly finished our breakfast at Limestone, when Audubon, all at once, sprang to his feet, and exclaimed in French; "Now I am going to lay the foundation of my establishment." So saying, he took a small packet of address cards from his pocket, and

[252] Limestone or, as it was later called, Maysville, was on the left bank of the river, in Kentucky, and about a hundred miles east of Cincinnati.

some nails from his vest, and began to nail up one of the cards to the door of the tavern, where we were taking our meal.

Later they rode on together as far as Lexington, where they appear to have parted company.

The discrepancies between these accounts could hardly be greater, and they serve to illustrate the liberties which Audubon sometimes took with facts in composing his "Episodes." The travelers met, not on horseback, but at the supper table of a country inn; Nolte was then alone and had but one horse, while the greater part of the return journey was made by flatboat with Audubon as his guest; corn blades, pumpkins and trout suggest any other season than midwinter, with heavy snows on the mountains and rivers choked with ice. Audubon in this instance, as already explained, combined the incidents of two different journeys and colored the narrative to suit his fancy. There was no apparent motive to mislead the reader, and one of his readers he must have known would probably be Vincent Nolte, though he was not a subscriber to *The Birds of America*; Nolte did read the story, and was pleased with the "flattering acknowledgment of the little service" that he was able to render Audubon at that time as well as later in his career.

Both travelers felt the great earthquakes while making this journey, but probably not until they had parted company at Lexington. Audubon has given a vivid account of this experience in a characteristic sketch, but as usual there are no dates.[253] He was overtaken, as he said, while "traveling through the Barrens of Kentucky ... in the month of November," when he thought his terrified "horse was about to die, and would have sprung from his back had a minute more elapsed, but at that instant all the shrubs and trees began to move from their very roots; the ground rose and fell in successive furrows, like the ruffled waters of a lake." For "November" he should have written "January" of the year 1812.[254]

This series of memorable earthquakes was followed in 1813 by a hurricane, more terrific than destructive, which swept the lower part of Henderson County, Kentucky, and cut a wide swath through the virgin forests, without causing any loss of life. Audubon's account of this event[255] is that of a close observer who escaped destruction by a hair's breadth and who related only what he himself had experienced. Critics inclined to be supercilious have complained that he exaggerated the importance of a merely local event and stretched the course of the storm some 800 miles until it had covered several states. "Sir," said Waterton, in pointing a dart through Audubon to another target, "this is really too much even for us Englishmen to swallow, whose gullets are known to be the largest, the widest, and the most elastic, of any in the world." What Audubon said was: "I have crossed the path of this storm, at a distance of a hundred miles from the spot where I witnessed its fury, and, again four hundred miles farther off, in the State of Ohio. Lastly, I observed traces of its ravages on the

---

253 "The Earthquake," *Ornithological Biography*, vol. i, p. 239.

254 These historic earthquakes, which were most destructive of life and property in the lower Mississippi Valley, began on December 16, 1811, and therefore before Audubon and Nolte had reached the western country. They were noted for their remarkable frequency and persistence, 221 shocks having been recorded in a single week at Henderson, Audubon's home at that time; though their force was mostly spent after the first three months, they did not wholly die away in the Ohio Valley until December 12, 1813, when the last feeble vibration was recorded by Dr. Daniel Drake at Cincinnati; the worst shocks at this point were experienced on December 16, 1811, on January 23 and February 7, 1812. See Daniel Drake, *Natural and Statistical View of Cincinnati, and the Miami Valley; with an appendix, containing observations on the late Earthquakes*, (Cincinnati, 1815); and Edmund L. Starling, *History of Henderson County, Kentucky*.

255 "The Hurricane," *Ornithological Biography*, vol. i, p. 262.

summits of the mountains connected with the Great Pine Forest of Pennsylvania, three hundred miles beyond the place last mentioned. In all these different parts, it appeared to me not to have exceeded a quarter of a mile in breadth." Audubon was doubtless mistaken in his hasty inference that marks of forest devastation observed at such widely separated points were due to the same storm, but this would only illustrate a lack of caution which he sometimes displayed.

A contemporary writer[256] declared that Audubon's account of "Mason," the outlaw, whose name we are told should be spelled "Meason," was altogether fabulous; that he was not killed by a regulator party, nor was his head stuck upon a tree in the way described.[257] The same critic further discredited the naturalist's account of Daniel Boone, whom he had characterized as follows:[258] "The stature and general appearance of this wanderer of the western forests approached the gigantic. His chest was broad and prominent; his muscular powers displayed themselves in every limb; his countenance gave indication of his great courage, enterprise, and perseverance." "Boone," said this writer, "was under six feet high, probably not more than five feet, ten inches, and of that round, compact build, which makes little show. Though very active, he had the appearance of being rather slender and did not seem as large as he really was." In the case of the outlaw, Audubon no doubt retold a story that had passed from mouth to mouth, but he later learned to be wary of second-hand information, which in matters of natural history sometimes led him into more serious difficulties. In his description of Boone there was no more apparent motive to deceive than in the case of his own father, to whom his imagination had added nearly half a foot in stature.[259]

When Audubon was returning from Ste. Geneviève in the spring of 1812, an incident occurred in which, for the first time in the course of his wanderings for upwards of twenty-five years, he felt his life to be in danger from his fellow man.[260] Overtaken by night on the prairie, he approached the hearth fire of a small log cabin, which at first was mistaken for the campfire of some wandering Indians. On craving shelter, he was admitted by a tall, surly woman in coarse attire, who displayed both an evil eye and a repellent countenance; but she offered him a supper of venison and jerked buffalo meat and bade him to make his bed upon the floor. When she espied his gold watch and chain, her demeanor suddenly changed and she asked to take them in her hand; she put the chain around her brawny neck and by her manner betrayed every token of covetous desire. Meanwhile, a young Indian stoic, who was nursing a recent arrow wound, had been sitting in silence by the fire; though he spoke not a word, he cast an expressive glance in Audubon's direction whenever the woman's back was turned, and having drawn his knife from its scabbard, expressed in pantomime what the confiding stranger might eventually expect.

Audubon's suspicions were at last thoroughly aroused. He asked for his watch, and under pretense of forecasting the weather, took up his gun and sauntered out of the cabin; in the darkness outside he slipped a ball in each of the barrels of his gun, scraped the edges of his flints, renewed the primings, and returned with a favorable report of his observations. Then laying some deer skins on the floor in a corner and calling his faithful dog to his side, he lay down and to all appearances was soon asleep. Presently sounds of approaching voices were

---

[256] James Hall, *Western Monthly Magazine*, vol. ii (1834).
[257] "The Regulators," *Ornithological Biography*, vol. i, p. 105.
[258] "Colonel Boone," *ibid.*, vol. i, p. 503.
[259] See Chapter 5.
[260] "The Prairie," *Ornithological Biography*, vol. i, p. 81.

heard, and at length two sturdy youths, who were evidently the woman's sons, appeared bearing a dead stag, which they had slung to a pole; they asked at once about the stranger, and called loudly for whisky. Audubon tapped his dog, who showed by eye and tail that he was already alert. Observing that the whisky bottle was paying frequent visits to the mouths of the trio, he hoped that they would soon be reduced to a state of helplessness, but the woman was seen to take in her hands a large carving knife and go deliberately outside to whet its edge on a grindstone; then, calling to her drunken sons, she asked them to settle the stranger and bade them do their bloody work without delay. Audubon cocked both barrels of his gun, touched his dog again, and was resolved to shoot at the first suspicious move. At this dramatic moment the door suddenly opened and two burly travelers with rifles on their shoulders entered the cabin. Audubon sprang to his feet, and welcoming the strangers with open arms, lost no time in making known to them his desperate position. No parley was necessary, for, said he, they were regulators, who then and there took the law into their own hands. The woman and her sons were promptly secured, bound, and left until morning to sober off; they were then led into the woods and shot. "We marched them into the woods off the road," said Audubon, "and having used them as Regulators were wont to use such delinquents, we set fire to the cabin, gave all the skins and implements to the young Indian, and proceeded, well pleased, towards the settlements." Would you believe, he added, that not many miles from where this happened, "and where fifteen years ago, no habitation belonging to civilized man was expected, and very few ever seen, large roads are now laid out, cultivation has converted the woods into fertile fields; taverns have been erected, and much of what we Americans call comfort is to be met with? So fast does improvement proceed in our abundant and free country."

I have given a paraphrase of this "Episode" as a further illustration of Audubon's tales of adventure. There is doubtless a certain amount of invention, and it reads like the setting of a dime novel incident, but we see no reason to doubt the substantial truth of either the local coloring or the fact. In answer to the question of a recent commentator,[261] "Did remote prairie cabins have grindstones and carving knives?" we would reply that the knife and the ax have followed man to the frontier posts of civilization everywhere, and without the grindstone the ax is useless. As a concrete instance in point, compare this minute entered in the Proprietors' Book of Records of Perrytown, afterwards Sutton, New Hampshire,[262] for the third day of September, 1770: "Voted a grindstone of about 8 shillings to be sent up to Perrystown, for the use of the settlers there"; the first settler had entered that wilderness but three years before, and at the time this vote was taken the number was five.

---

[261] John Burroughs, *John James Audubon*, p. 37.

[262] "See *History of Sutton, New Hampshire*, compiled by Augustus Harvey Worthen, pt. I (Concord, 1890).

Chapter 19

AUDUBON AND RAFINESQUE

Audubon's humorous sketch of "The Eccentric Naturalist" has often been quoted, and it presents a picture which is amusing, however short of the truth it may fall or however it may fail in doing justice to its subject. Though his real hero is not named, no doubt as to his identity has ever been entertained. This episode occurred at Henderson in the late summer of 1818, and was published thirteen years after in the *Biography* of birds.[263] Since the story was not fully told then and the after-effects were productive of much harsh criticism, it cannot be overlooked if we would do justice to both the writer and his subject.

When walking one day by the river, to follow Audubon's story, he saw a man landing from a boat with what appeared like a bundle of dried clover on his back; he concluded from his appearance that the stranger must be "an original," a term which had been applied also to himself. A meeting followed, and the stranger, who had inquired for Mr. Audubon's house, explained that he was a naturalist, and had come to see Audubon's drawings of birds and plants; he bore also a letter from a friend, introducing "an odd fish" which might "prove to be undescribed." The visitor was made welcome in Audubon's Henderson home, where, to quote the naturalist,

> at table his agreeable conversation made us all forget his singular appearance.... A long loose coat of yellow nankeen, much the worse of the many rubs it had got in its time, and stained all over with the juice of plants, hung loosely about him like a sac. A waistcoat of the same, with enormous pockets, and buttoned up to the chin, reached below over a pair of tight pantaloons, the lower parts of which were buttoned down to the ankles. His beard was as long as I have known mine to be during some of my peregrinations, and his lank black hair hung loosely over his shoulders. His forehead was so broad and prominent that any tyro in phrenology would instantly have pronounced it to be the residence of a mind of strong powers. His words impressed an assurance of rigid truth, and as he directed the conversation to the study of the natural sciences, I listened to him with as much delight as Telemachus could have listened to Mentor.

All had retired for the night when of a sudden a great uproar was heard in the visitor's room. To his great astonishment, Audubon found his guest running about the apartment naked, holding the "handle" of his host's favorite violin, the body of which had been battered

---

[263] "The Eccentric Naturalist," *Ornithological Biography*, vol. i, p. 455.

to pieces against the walls in the attempt to secure a number of fluttering bats which had entered by an open window. "I stood amazed," said Audubon, "but he continued jumping and running round and round, until he was fairly exhausted, when he begged me to procure one of the animals for him, as he felt convinced they belonged to 'a new species.' Although I was convinced to the contrary, I took up the bow of my demolished Cremona, and administering a sharp tap to each of the bats as it came up, soon had specimens enough." Other incidents of this visit, which Audubon said lasted three weeks, are fully recorded. The eccentric naturalist collected an abundance of plants, shells, bats and fishes. One evening he failed to appear, and after a prolonged search was nowhere to be found; nor were the Audubons wholly assured of his safety until some weeks later they received a letter with due acknowledgments of their hospitality.

The "M. de T." of this episode was Constantine Samuel Rafinesque, in many respects the most singular figure that has ever appeared in the annals of American science. Although young in years, for Rafinesque was then but thirty-five, he was already old in experience and that of the bitterest sort; and although already known to many in both hemispheres, he had few friends. It is certain that neither Audubon nor anyone else in that part of Kentucky had ever heard of him before.

Born in Constantinople, of a father who was a French merchant from Marseilles and of a mother with a German name who by nativity was Greek, Rafinesque had known life in many lands, and was destined, as he said, to be a traveler from the cradle to the tomb.[264] His first voyage, made with his parents on their return to France, by way of Scutari in Asia, Smyrna, and Malta, led to his first discovery, when he was a year old, for he was able to announce that "infants are not subject to sea-sickness." At eleven he read Latin and collected plants; at thirteen he wrote his first scientific paper, "Notes on the Apennines," which he had seen when traveling from Leghorn to Genoa. His father, who set out for China in 1791, fell in with pirates, but managed to reach America; he died of the yellow fever in Philadelphia in 1793. To escape the Reign of Terror in France, Rafinesque's mother fled with her children to Italy, where four years were passed at Leghorn. There Constantine studied with private tutors, but his education was never formal and he was allowed to follow his omnivorous tastes, reading, as he said, ten times more than was taught in the schools. His writings are mainly in French, Italian, and English, and his facility with languages was no doubt remarkable, even if we discount his egotized estimate of his own attainments: "I have undertaken to read the Latin and Greek, as well as the Hebrew, Sanskrit, Chinese, and fifty other languages, as I felt the need or inclination to study them."

In 1802 Rafinesque was sent with his brother to America and became a shipper's, clerk at Philadelphia, where he spent all of his spare time in the study of nature, plants being his first and greatest love. Here he was befriended by Dr. Benjamin Rush, and during this period he made the acquaintance of many pioneer naturalists in the United States. In 1805 the offer of a lucrative situation in Sicily lured him back to the Old World and to a country already known to him. There he soon discovered the medicinal squill, of ancient repute and thought to be an antidote, which in the form of syrup was long the bane of childhood; this and other medicinal

[264] For the characterization of Rafinesque given in the present chapter I am chiefly indebted, aside from his own writings, to his two most sympathetic biographers, Richard Ellsworth Call and T. J. Fitzpatrick, as well as to David Starr Jordan. Fitzpatrick gives photographic reproductions from Rafinesque's exceedingly diversiform and scattered works; his bibliographic titles extend to 939, and "Rafinesquiana" to 134.

drugs he exported to the European and American markets in such quantities that before the secret of his trade became known to the jealous Sicilians, he had reaped from it, in conjunction with his other enterprises, a small fortune. During the ten years that were spent in Sicily we find him the manager of a successful whisky distillery, the chancellor or secretary of the American Consulate at Palermo, editor, writer, and correspondent of learned men in Europe, as well as traveler and explorer in every part of the island, which he proposed to monograph with all of its contents. At Palermo Rafinesque met the English naturalist, William Swainson, his lifelong correspondent; together they tramped over the island and together they worked for a number of years on the fishes of the western coast.[265] Swainson, who became the friend of Audubon, was one of the few who later defended Rafinesque.

Rafinesque espoused a Sicilian woman of the Catholic faith, and had by her two children, of whom a daughter lived to maturity; this experience seems to have embittered him against the sex, for no other woman excepting his mother, to whom his *Life of Travels* was dedicated, was ever mentioned in his writings, and this one was disinherited in his extraordinary will. Through fear of being drafted into the French wars, he assumed for a time his mother's family name of Schmaltz, and finally left Sicily in disgust; taking with him his fortune and "fifty boxes of personal goods," he set out again for America in 1815. Sicily, he declared in epigram, offered "a fruitful soil, a delightful climate, excellent productions, perfidious men, deceitful women."

This second voyage to the New World began late in July but did not end until 100 days later, when, on the night of November 2, his ship ran on the Race Rocks near New London, at the western end of Long Island Sound, and eventually went down within sight of land with all his possessions. "I had lost everything," he said, "my fortune, my share of the cargo, my collections and labors for twenty years past, my books, my manuscripts, my drawings, even my clothes ... all that I possessed, except some scattered funds, and the insurance ordered in England for one third of the value of my goods." "I have found men," he continued, "vile enough to laugh without shame at my misfortune, instead of condoling with me! But I have met also with friends who deplored my loss, and helped me in need." One of these friends was Dr. Samuel L. Mitchell of New York, who had given a helping hand to Audubon,[266] and it was probably through him that Rafinesque obtained a position as private tutor in a family living on the Hudson. Traveling up and down the country, collecting objects in natural history, writing, with frustrated attempts at business, occupied a number of the following years; meanwhile he had aided in founding the Lyceum of New York and had become a member of the Literary and Philosophical Society. At Philadelphia he found another friend in Mr. John O. Clifford, of Lexington, Kentucky, who encouraged him to visit the West, and in the spring of 1818 he descended the Ohio in an "ark" in company with several others who had joined him in the enterprise. At Shippingport he was welcomed by the Tarascon brothers, flour merchants, formerly of Marseilles and Philadelphia, and it was through them, possibly, that he first heard of Audubon's drawings of birds.

---

[265] "At Palermo," said Swainson, "I had the pleasure of meeting ... Rafinesque Schmaltz, whose first name is familiar to most zoölogists. In the society of such congenial minds, I passed many happy hours, and made many delightful excursions ... by the inducement of the latter, I was led to investigate the ichthyology of the western coast."

[266] See Vol. I.

Such was the "odd fish" who a little later greeted Audubon on the river bank at Henderson. Had Audubon known the true history of his visitor either then or at a later time, he would not, we believe, have held him up to ridicule in the "Episode" quoted above, and could he have foreseen the unpleasant consequences that ensued, his conduct would assuredly have been different. A part of the episode, which Audubon does not relate, was supplied by another naturalist at a much later day.[267] Audubon, it seems, was at that time a good deal of a wag, and whether to vent his dislike of species-mongers, to avenge the loss of his violin, or to gratify some spirit of mischief, he played upon the credulity of his guest, in a way that could be deemed hardly creditable, in giving him detailed descriptions and even supplying him with drawings of sundry impossible fishes and mollusks. Rafinesque took the bait eagerly, duly noted down everything on the spot, and, what was more unfortunate for American zoölogy, a year later began to publish the results. The fictitious species of fish, to the number of ten, "communicated by Mr. Audubon," first appeared as a series of articles in a short-lived and long forgotten western magazine,[268] but in 1820 they were gathered into a little volume [269] now considered so quaint and rare that it has been reproduced in its entirety. In this pioneer work on the ichthyology of the Ohio River and the great Middle West, 111 kinds of American fresh-water fishes are briefly described. Those ten "new species," representing apparently a number of new genera, "so like and yet so unlike to anything yet known," long remained a stumbling block to American zoölogists; naturally they tended to discredit the work of Rafinesque.

---

[267] See David Starr Jordon, *Popular Science Monthly*, vol. xxix (1886). "The true story of this practical joke was told me by the venerable Dr. Kirtland, who in turn received it from Dr. Bachman;" the latter, I might add, was the friend and correspondent of the "Sage of Rockport" after a visit at his home near Cleveland in the summer of 1852. In the private notebooks of Rafinesque copies of Audubon's drawings are still to be seen, and "a glance at these," said Dr. Jordon, "is sufficient to show the extent to which science through him has been victimized."

Audubon was also responsible for a number of extraordinary "new species" of birds, the most notorious of which was the Scarlet-headed Swallow, of which Rafinesque published the following account in 1820: "*Hirundo phenicephala*. Head scarlet, back gray, belly white, bill and feet black. A fine and rare swallow seen only once by Mr. Audubon near Henderson, Kentucky...." See Samuel N. Rhoads, "Constantine S. Rafinesque as an Ornithologist," *Cassinia*, No. XV (Philadelphia, 1911).

[268] *The Western Review and Miscellaneous Magazine*, Lexington, 1819-20.

[269] *Ichthyologia Ohiensis, or Natural History of the Fishes inhabiting the River Ohio and its tributary Streams, preceded by a physical description of the Ohio and its branches.* By C. S. Rafinesque, Professor of Botany and Natural History in Transylvania University, Author of the Analysis of Nature, &c. &c. Member of the Literary and Philosophical Society of New-York, the Historical Society of New-York, the Lyceum of Natural History of New-York, the Academy of Natural Sciences of Philadelphia, the American Antiquarian Society, the Royal Institute of Natural Sciences of Naples, the Italian Society of Arts and Sciences, the Medical Societies of Lexington and Cincinnati, &c. &c.

"The art of seeing well, or of noticing and distinguishing with accuracy the objects which we perceive, is a high faculty of the mind, unfolded in few individuals, and despised by those who can neither acquire it, nor appreciate its results."

Lexington, Kentucky: printed for the author by W. G. Hunt. (Price one dollar.) (Pp. 1-90. Lexington, 1820.)

Fitzpatrick gives a list of 14 copies of this work, the whereabouts of which are known; we can add another from the library of Dr. Jared P. Kirtland, now in the collections of Western Reserve University; it is bound up with Dr. Kirtland's notebook on birds and fishes, and labeled "Scraps of Natural History. My Note Book;" a written notice on the inside of the cover, imploring the finder to return the volume to its owner if lost, is signed by Dr. Kirtland and dated "Cleveland, O., Oct. 16th, 1839." Probably fewer than 20 original copies of the work now exist. It was reproduced in a limited edition, with a sketch of Rafinesque's life and works by Richard Ellsworth Call, published by the Burrows Brothers' Company of Cleveland in 1899.

Figure 38. Early Unpublished Drawings of American Birds: Above, "L'alcion D'amerique Septentrionale Buffon. King Fisher. Chute De L'ohio July 15, 1808. Belted King Fisher A. W.-J. A. Alcedo Alcion. Drawn by J. J. Audubon. No. 110;" Below, "Passenger Pigeon—A. W. Columba Migratoria. Chute De L'ohio. December. 11, 1809. 12 Pennes À La Queue Tres Étagee. Appelé Ici Wild Pigeon. J. Audubon. No. 109." Published by Courtesy of Mr. Joseph Y. Jeanes.

As a specimen of these spurious fish stories, which were previously published in both America and Europe, we reproduce a part of Rafinesque's description of the "91st. Species. Devil-Jack Diamond-fish. Litholepis adamantinus":

> This may be reckoned the wonder of the Ohio. It is only found as far up as the falls, and probably lives also in the Mississippi. I have seen it, but only at a distance, and have been shown some of its singular scales. Wonderful stories are related concerning this fish, but I have principally relied upon the description and figure given me by Mr. Audubon. Its length is from 4 to 10 feet. One was caught which weighed four hundred pounds. It lies sometimes asleep or motionless on the surface of the water, and may be mistaken for a log or a snag. It is impossible to take it in any other way than with the seine or a very strong hook, the prongs of the gig cannot pierce the scales which are as hard as flint, and even proof against lead balls! Its flesh is not good to eat. It is a voracious fish: Its vulgar names are Diamond fish, (owing to its scales being cut like diamonds) Devil fish, Jack fish, Garjack, &c.... The whole body covered with large stone

> scales laying in oblique rows, they are conical, pentagonal, and pentædral with equal sides, from half an inch to one inch in diameter, brown at first, but becoming of the colour of turtle shell when dry: they strike fire with steel! and are ball proof!

While we cannot defend Audubon in his treatment of Rafinesque, it would be hardly fair to judge such incidents wholly in the light of after events, for, as our narrative will show, it is unlikely that he ever saw Rafinesque or heard of him again until long years after this incident, certainly not until after his "Episode" was published in 1831.[270] Rafinesque evidently enjoyed this sketch of himself, for he gave unstinted praise to the work in which it was published. As late as 1832, when the appearance of *The Birds of America* seems to have stimulated him to even more grandiose conceptions of his own merits than was usual, he declared that his discoveries were counted by the thousand, and that he had traveled twenty thousand miles, always collecting and drawing. In view of the fact that drawing was a talent which nature had unequivocally denied him, it is interesting to read this boast that an unfriendly critic drew forth: "My illustrations of 30 years' travels, with 2,000 figures will soon begin to be published, and be superior to those of my friend Audubon, in extent and variety, if not equal in beauty. I shall study and write as long as I live, in spite of all such mean attempts against my reputation and exertions, trusting in the justice of liberal men."[271]

After leaving Audubon at Henderson in the summer of 1818, Rafinesque passed down the Ohio into the Mississippi, pausing only to pay his respects at the famous communistic settlement of New Harmony, by the mouth of the Wabash in Indiana, then the abode of Thomas Say, David Dale Owen, and Charles Le Sueur, all of whom have left bright and honored names in the annals of American science. He eventually returned to Philadelphia by way of Lexington, Kentucky, where he was induced to settle and teach natural history and the modern languages in the Transylvania University, at that time the most important seat of learning in the West. After closing up his business affairs in Philadelphia, Rafinesque entered upon his new labors at Lexington in the autumn of 1819. He was probably the first teacher of these subjects west of the Alleghanies, and certainly the first in that section of the country to use the present object method in the elucidation of natural history. The lot of a pioneer in education has never been a sinecure, and the post which Rafinesque then filled was not a "chair" but a hard "settee." In those days the classics were in the saddle and "rode mankind," while the natural sciences, when tolerated at all, were given short shrift; yet this eccentric foreigner held his position for seven years and accomplished an extraordinary amount of work. As usual he spread his energies over the whole field of knowledge, lecturing, writing and publishing on almost every subject, but concentrating upon none. Meanwhile, he roamed far and wide and made extensive collections.

While at the Transylvania University Rafinesque seems to have applied for the master of arts' degree, but was at first refused, as he said, "because I had not studied Greek in a college, although I knew more languages than all of the American colleges united, but it was granted at last; but the Doctor of medicine was not granted, because I would not superintend anatomical dissections."

---

[270] Probably not before October of that year, when Audubon first met John Bachman, at Charleston, South Carolina.

[271] Reply to a criticism of G. W. Featherstonhaugh (*The Monthly American Journal of Geological Science*), in Rafinesque's *Atlantic Journal and Friend of Knowledge*, No. 3, p. 113 (Philadelphia, 1832). Rafinesque occasionally spoke of meeting "my friend Audubon," who, he declared, had invited him to join his expedition to Florida in 1831-32.

One of his many projects, as meritorious as it was impractical, at that time, was a Botanic Garden with a Library and Museum for Lexington, which was then but a small village; though land was actually secured and a start in tree planting begun, the project of course came to nothing and had to be abandoned. Rafinesque also invented, as he believed, the present coupon system of issuing bonds, the "Divitial Invention," as he called it; in 1825 he set out for Washington in order to secure his patent rights, but his journey and idea never brought him any returns. On the contrary, the incident marked the culmination of his troubles with the president of the University and its governing board, whom he seems to have constantly nettled by his independent ways and roaming habits. Upon returning from Washington he found that Dr. Holley, who, he said, "hated and despised the natural sciences" and wished to drive him out altogether, had broken into his rooms during his absence, and had "given one to the students, and thrown all my effects, books and collections in a heap in the other," besides depriving him of certain other privileges. "I took lodgings," he continued, "in town and carried there all my effects; thus leaving the college with curses on it and Holley; who were both reached by them soon after, since he died next year at sea of the yellow fever, caught at New Orleans; having been driven from Lexington by public opinion; and the College has been burnt in 1828 with all its contents."

After this unpleasant experience Rafinesque returned to Philadelphia, where he spent the last and saddest part of his checkered career. His insistent ideas, which were undoubtedly the index of an unbalanced mind, increased, especially his mania for describing "new species" of animals and plants; this mania perverted everything that he wrote, especially toward the end of his life, and made him a thorn in the side of every naturalist who tried to verify his work. A non-conformist and a respecter of no authority but his own is never popular, though a part of the antagonism which Rafinesque aroused was due to the conservatism of his age. He boldly advocated organic evolution when almost the whole world believed that species were fixed and unchangeable things, and in many other respects was fifty years ahead of his time; but nothing was ever carefully worked out in his fertile mind, with the consequence that the world paid no heed to his crude and undigested ideas.

The great mass of Rafinesque's books and monographs, his "tracts," broadsides, and ephemeral papers of all sorts, extending to nearly a thousand titles, must have gone into paper rags, when not used to kindle fires, for he was generous in their distribution, and they are now exceedingly rare. He touched nearly everything, it is true, but little that he touched, especially in this later period of his life, did he ever truly ornament. His best pioneer work, in the opinion of competent students, was that done upon the fishes of Sicily and the natural history of the Ohio Valley; his *Medical Flora*, in two volumes (1828 and 1830), is also admitted to have possessed real value; but his writings are now sought after as literary or scientific curiosities, and as such they are unique.

No doubt Rafinesque was often treated unjustly, either through ignorance or intent, while many naturalists were exasperated by the barbed arrows which he shot into the air or direct at the mark. Others through sheer inability to follow him gave up the attempt, one writer[272] saying that such an attitude was justified when it appeared that he had made six species out of one, not to speak of several different genera and two sub-families. If anyone still believes that Rafinesque has been misjudged, says Günther,[273] let him read his letters to Swainson, from

[272] Isaac Lea, in *A Synopsis of the Family of Naiades*, pp. 8-9 (Philadelphia, 1836).

[273] See Bibliography, No. 204.

1809 to 1840, fifty-three in number, covering 178 closely written quarto or folio pages, now in possession of the Linnæan Society of London. "Rafinesque," continues this critic, "was a man deeply to be commiserated, not merely on account of the unfortunate circumstances which left him in his youth to himself, without teacher or guide, but still more on the ground of that natural disposition by which his universal failure in life was brought about. He was possessed of a feverish restlessness which entirely disqualified him from serious study of any of the multitudinous subjects which attracted his mind in rapid succession."

Rafinesque, bereft of friends and fortune, unknown even to his neighbors, by whom he seems to have been regarded as a harmless herb doctor, was left to struggle on alone, without recognition and without sympathy or support. Reduced finally to abject poverty, he concocted and sold medicines which were advertised much like quack remedies at the present day, especially his "Pulmel," which without a doubt he thought had cured him of the pulmonary consumption. To advertise this he wrote a little treatise, hoping to realize something from its sale and at the same time to avoid any undue appearance of empyricism.

Toward the very end of his life, Rafinesque projected a savings bank, and, strangely enough, this seems to have been a success, though just how is not clear, since it both borrowed and loaned money at six per cent. He had already attempted to secure rights on a "steam-plough," a "submarine boat," "incombustible houses," and similar novelties which abler inventors have later perfected. For a long time he led the life of a perfect recluse in a garret in a poor quarter of Philadelphia, in the midst of his collections, his books and his manuscripts, never the world forgetting but ever by the world forgot. There, in the direst misery, he died in 1840, at the age of fifty-six, without a word of cheer or a tear of regret. His body was barely saved from the dissecting table and given decent burial through the loyalty and promptitude of one of his few remaining friends, Dr. William Mease, who with undertaker Bringhurst, broke into the room where his body lay and let it down through a window by ropes.[274] Even his will was ruthlessly violated, and all of his effects, in eight dray-loads, were hurried off to the public auction rooms and sold in bargain lots, his books and all else bringing but a mere pittance, not even enough to pay his landlord and the administrator of his estate.

Thus died the "eccentric naturalist" whom Audubon had portrayed, and for whom the world in general had shown scant sympathy. Rafinesque, nevertheless, possessed a mind of extraordinary acumen and an energy and versatility little short of marvelous. He dipped into every field of knowledge, looking for precious metal, but much that he brought to the surface was dross. His restless versatility alone would probably have ruined him, for nothing short of an analysis of the globe with all of its contents would have satisfied his ambitious spirit. His was the ardor of the traveler and the explorer, with a passionate love for nature seldom equaled, but without the incentive and the patience of the investigator or a balance-wheel in the judgment. His ambition in early life was to become the greatest naturalist of his age; had his early training and environment been suited to his needs, and had fortune favored him more consistently with her smiles, this ambition possibly might have been realized, but we suspect that in this case nature would have proved stronger than nurture, and that he would have been Rafinesque to the end.

---

[274] The landlord, to whom Rafinesque had been in arrears for rent, had locked his body in the room and refused permission for its burial, thinking to find a market for it in one of the medical schools of the city. Rafinesque was buried in a little churchyard, then outside of the limits of the city, known as Ronaldson's cemetery, now at Ninth and Catharine Streets. See Call and Fitzpatrick, Bibliography, Nos. 198 and 228.

Chapter 20

AUDUBON'S ÆNEID, 1819-1824: WANDERINGS THROUGH THE WEST AND SOUTH

Audubon's failure at Henderson was the crucial turning point in his career. For the five years that immediately followed he led a peripatetic existence in the southern and western states, seldom tarrying long at one point, often leaving his family for months at a time, living from hand to mouth, but ever bent on perfecting those products of his hand and brain, his life studies of American birds and plants.

At this crisis Audubon could have accomplished nothing but for the intelligent devotion of his capable wife. Generous, emotional, inclined to be self-indulgent, Audubon needed both the example and the spur of a strong character such as his wife possessed, and at this time Lucy Audubon furnished both the motive power and the balance-wheel that were requisite for the development of her husband's genius. Without her zeal and self-sacrificing devotion the world would never have heard of Audubon. His budding talents eventually would have been smothered in some backwoods town of the Middle West or South. For the space of nearly twelve years, Mrs. Audubon, now as the head of a small private school, now as a governess in some friendly family who appreciated her worth, practically assumed the responsibility for the support and education of their children in order that her husband's hands might be free, and with her hard-earned savings was able to aid him materially in the prosecution of his labors. When relatives or friends upbraided him for not entering upon some form of lucrative trade, she recognized his genius and always came to his support, being fully persuaded that he was destined to become one of the great workers of the world. Whatever others may have said or done at that time, both Audubon and his wife were confident of the ultimate success of his mission. In short, the work in which the naturalist was engaged became a family interest, in which every member was destined sooner or later to bear a part.

Audubon recalled a somber incident of this time which he thought might furnish a lesson to mankind, and he shall relate it in his own words:

> After our dismal removal from Henderson to Louisville, one morning when all of us were sadly desponding, I took you both, Victor and John, from Shippingport to Louisville. I had purchased a loaf of bread and some apples; before you reached Louisville you were hungry, and by the river side we sat down and ate our scanty meal. On that day the world was with me as a blank, and my heart was sorely heavy, for scarcely had I enough to keep my dear ones alive; and yet through those dark days I was

being led to the development of the talents I loved, and which have brought so much enjoyment to us *all*....

At Shippingport Audubon was welcomed by his brother-in-law, Nicholas A. Berthoud. Wasting no time in vain regrets, he began doing portraits in crayon, and with such success that he was able to rent a modest apartment and have his family about him again. From no charges for his tentative efforts the price was gradually raised until he received five dollars or more a head; with the spread of his fame orders filled his hands, and he was called long distances to take likenesses of the dying or even of the dead. Audubon's facility in portraiture was a valuable resource, and it kept him from the starving line at many a pinch in later years.

Through the influence of friends the naturalist was offered a position as taxidermist at a museum which had just been started at Cincinnati; here his family joined him in the winter of 1819-20, and here he remained for nearly a year. The published accounts of this Cincinnati experience are strangely confused and have led to aspersions of bad faith which were, we believe, quite undeserved. "I was presented," said Audubon, "to the president of the Cincinnati College, Dr. Drake, and immediately formed an engagement to stuff birds for the museum there, in concert with Mr. Robert Best, an Englishman of great talent," adding that his salary was large; so industrious were they, to continue his account, "that in about six months we had augmented, arranged, and finished all that we could do," but they found to their sorrow "that the members of the College museum were splendid promisers and very bad paymasters."[275] It has been stated that Audubon got nothing from Dr. Drake, but that "Mrs. Audubon afterwards received four hundred dollars, of the twelve hundred due," and that the remainder was never paid.[276] This matter can now be fully cleared up, and it will appear that the Cincinnati College was in no way involved; Dr. Drake was not its president, although he drew its charter and was one of its trustees; the Museum in which the naturalist worked was an independent foundation; and Mrs. Audubon was probably paid in full for the service which her husband had rendered.

Audubon wrote in his journal in 1820, when this experience was fresh in his mind, that owing to his talent for stuffing fishes he entered the service of the Western Museum at a salary of $125 a month; he made no complaint at that time of any lack of pay. Moreover, on the day before he started on his cruise down the Ohio River on the 11th of October of that year, the Rev. Elijah Slack gave him a letter of introduction in which he said that Audubon had "been engaged in our museum for 3 to 4 months, and that his performances do honor to his pencil." Since Mr. Slack, like Dr. Drake, was one of the managers of the Western Museum, he must have known of Audubon's term of service. We are convinced that Dr. Daniel Drake,[277] whose character was above reproach and who was a keen naturalist himself,

---

275 Maria R. Audubon, *Audubon and his Journals*, vol. i, p. 36.

276 *Ibid.*, vol. i, p. 49.

277 Dr. Daniel Drake (1785-1852) was one of the most versatile and prolific writers on medicine which the West has ever produced, and Cincinnati owed to him much, for he was instrumental in organizing in that city a church, a literary society, a museum, a hospital, a college, and a school of medicine, while he enjoyed a large medical practice, lectured on botany, and was a partner in two mercantile establishments. We might also add that his "Notice concerning Cincinnati" (pp. 1-28, i-iv. Printed for the author at Cincinnati, 1810), of which only three copies are known to exist, is the earliest and rarest published record of that city. This little pamphlet included a "Flora" of the city for 1809, and from it we transcribe this interesting extract (p. 27):
"May 10. Black locust in full flower.
"It is highly probable that the flowering of this beautiful tree, the Robinia pseudococacia of Linnæus, indicates the proper time for planting the important vegetable the Indian corn. For several successive years I have

was Audubon's good friend, and that no misunderstanding ever rose between them. In writing offhand from memory, years after the events, Audubon misstated the facts but evidently without design.

In 1818 Dr. Drake organized the Western Museum Society, of which he said: "I have drawn up the constitution in such a manner as to make the institution a complete school for natural history, and hope to see concentrated in this place, the choicest natural and artificial curiosities in the Western Country." The first meeting of the Society was held in the summer of 1819, not long before Audubon was engaged to work for it. The membership fee was $50, a considerable sum for that period, but the enterprise was well patronized. It was in charge of a board of whom Dr. Drake was the moving spirit; another member, as we have seen, was Rev. Mr. Slack, who became the first president of the Cincinnati College, which was organized in 1818-19. The collections of the Museum were placed in one of the buildings of the College in order better to serve the students and public, which would account for some of the confusion noted above.

Dr. Drake's hands at this time were more than full; in October, 1819, he wrote to a friend: "The ties which bind me to the world at large seem every day to increase in strength and numbers. The crowd of mankind with whom I have some direct or indirect concern, thickens around me, and I see little prospect of more leisure, nor any of retirement and seclusion." At this juncture also, when Audubon and Best were working for his Museum, Dr. Drake was experiencing the first disastrous check in his energetic career. In January, 1820, in spite of the opposition and intrigue of professional rivals, he succeeded in organizing the Medical College of Ohio, and Robert Best became the assistant in chemistry and the curator of the Western Museum. Opposition did not abate, but instead of strangling the College which he had founded, the marplots succeeded in expelling the Doctor from its staff. At last, feeling obliged to leave the city, Dr. Drake accepted in 1823 a position in the rival medical school of Transylvania University, and thus became a colleague of Constantine Rafinesque. It will be seen that Audubon's engagement at Cincinnati fell in a troubled era, and the annoyance which he may have felt at lack of pay was probably no fault of the harassed doctor.

While at Cincinnati Audubon was obliged to resort to his crayon portraits; and he also started a drawing school, but it required all of Mrs. Audubon's skill in management to keep the family out of debt. In 1820 he began for the first time seriously to consider the possibility of publishing his drawings, and under the spur of this incentive began to exert himself as never before. He planned a long journey through the Middle West and South, his intention being to descend the Ohio and Mississippi Rivers, explore the country about New Orleans, and then proceed as far east as the Florida Keys; he wished also to ascend the Red River, cross Arkansas, and visit the Hot Springs, before returning again by river to Cincinnati. Lack of ready money was no drawback, for he was now confident of being able to live by his talents alone.

Accordingly, he left his wife to care for their two boys, and on October 12, 1820, started down river in a flatboat, bound for New Orleans. His companions on this journey were

---

observed our farmers generally to plant corn during some stage of its flowering. This from the 10th to the 20th of May."

For the privilege of examining one of the original copies of this paper, I am indebted to Mr. Wallace H. Cathcart of the Western Reserve Historical Society of Cleveland.

Captain Cummings,[278] an engineer who had been in the government service, to whom Audubon became much attached; Joseph R. Mason, a promising artist of eighteen, in the *rôle* of pupil-assistant, and his dog "Dash." Although Audubon had no funds, he was careful to provide himself with letters to or from men of mark who could be of assistance to him and this custom was followed to good effect at a much later day. On this occasion he bore recommendations from William H. Harrison, who afterwards became President, to Governor Miller of Arkansas, and from Henry Clay, as well as his letter from Rev. Elijah Slack, in which it was stated that the naturalist was traveling to complete his collection of the birds of the United States which he intended to publish at some future time. Audubon also wrote a personal letter to Governor Miller, fully outlining his plans, and asking for information; he told the Governor that he had been working fifteen years, and that his drawings of birds and plants were all from nature and life-size, showing that the idea of publication which was afterwards realized was then fixed in his mind. Audubon kept a careful journal on this journey, which extended over a year, the last entry being for the close of 1821.[279]

As their flatboat stopped at many towns and plantations on the rivers, Audubon could hunt game and birds to his heart's content. Having resolved, as he said, never to draw from a stuffed specimen, he worked at every new bird with the greatest diligence. It seems almost incredible that he should never have met with the Hermit Thrush before this journey, yet under date of "Oct. 14, 1820," there is this entry: "We returned to our boat with a Wild Turkey, a Telltale Godwit and a Hermit Thrush, which was too much torn to make a drawing of it; this was the first time I had met with this bird, and I felt particularly mortified at its condition."[280]

Their visit to Natchez furnished Audubon with materials for at least two of his "Episodes."[281] This incident of his generosity may be taken as characteristic; finding that one of his companions was down at the heel and as short of ready money as himself, he sought out a shoemaker and offered to do a portrait of the man and his wife for two pairs of boots; the proposal was accepted forthwith, and he set to work; the sketches were finished in the course of two hours, and Audubon and his companion, having selected their boots, went on their way rejoicing.

Audubon left Natchez on December 31, 1820, on a keel boat belonging to his brother-in-law, Nicholas A. Berthoud, who accompanied him, and at one o'clock the steamer *Columbus* hauled off from the landing and took them in tow. Towards evening, when they were looking up their personal belongings, the naturalist found to his dismay that a portfolio containing all

---

[278] See Audubon's letter to Thomas Sully. In his Ohio and Mississippi Rivers Journal Audubon wrote on April 5, 1821: "Cap. Cumming left us on the 10 for Phila; the poor man had not *one* cent with him."

[279] This early journal fills a large unruled book, measuring about 13 by 8 inches, of 201 pages, beginning with Oct. 12, 1820, and closing with December 31, 1821; it forms a part of the John E. Thayer collection of Audubon and Wilson manuscripts and drawings in possession of Harvard University, having been once included in the estate of Joseph M. Wade. The collection embraces four early drawings by Audubon, presumably at one time in the hands of Edward Harris; 73 of Audubon's original letters, comprising largely his correspondence with Dr. John Bachman; 60 letters by Victor G. Audubon; and a few by other members of the naturalist's family. See the *Annual Report of the Curator of the Museum of Comparative Zoology for 1910-1911*.
Through the courtesy of Professor E. L. Mark, and the Director of the Museum, Dr. Samuel Henshaw, I have been permitted to examine these numerous documents. In any direct or casual reference to this valuable material, I have endeavored not to overstep the bounds of propriety, in view of the fact that the University contemplates publishing copious extracts from it at an early day. It should be noticed that excerpts from this journal have already appeared in print. See following Note.

[280] See Ruthven Deane, *The Auk*, vol. xxi, pp. 334-338.

[281] "Natchez in 1820" and "The Lost Portfolio," *Ornithological Biography*, vol. iii, pp. 529 and 564.

of the drawings that he had made on the voyage down the river was missing. Letters were despatched to Natchez friends, but it was not until the 16th of March that his anxiety was relieved; the missing portfolio had been found and left at the office of *The Mississippi Republican*, whence it was forwarded on his order, and reached his hand on the 5th of April. "So very generous had been the finder of it," he said, "that when I carefully examined the drawings in succession, I found them all present and uninjured, save one, which had probably been kept by way of commission."

On New Year's Day, 1821, they came to at Bayou Sara, at the mouth of the inlet of that name, which later saw much of Audubon and his family. On the following day he made a likeness of the master of their craft, Mr. Dickenson, for which he was paid in gold; he also outlined two warblers by candle-light in order to have time to finish them on the morrow. The captain of their steamer in his anxiety to make haste had set them adrift at this point, and they were obliged to make their way as best they could, by aid of the current and oars, to the port of New Orleans, which was finally entered on Sunday, January 7, 1821.

Audubon landed at New Orleans without enough money to pay for a night's lodging, for someone had relieved him of the little he possessed, and he was obliged to pass several nights on the boat while looking for work. Undismayed by his financial straits, his first visit at daybreak on Monday was to the famous markets of the southern city, where he found dead birds exposed for sale in great numbers—mallard, teal, American widgeon, Canada and snow geese, mergansers, tell-tale god-wits, and even robins, bluebirds and red-wing blackbirds; he added that the prices were very dear.

Upon leaving Cincinnati Audubon had resolved upon making one hundred drawings of birds; this was actually accomplished, but only after repeatedly modifying his plans and working in more humble capacities than he was at first inclined to consider. On the 12th of January he wrote in his diary of meeting an Italian painter at the theater, and of showing him his drawing of the White-headed Eagle[282] at the rooms of Mr. Berthoud; "he was much pleased," and took him "to his painting apartment at the theater, then to the directors, who very roughly offered me one hundred dollars per month to paint with Monsieur l'Italien. I believe really now that my talents must be poor," said Audubon. His refusal of this offer in view of his straitened circumstances, and the entry which followed, were characteristic: "Jan. 13th, 1821. I rose up early, tormented by many disagreeable thoughts, again nearly without a cent, in a bustling city where no one cares a fig for a man in my situation." The following day Audubon applied to a self-taught portrait painter, John W. Jarvis, and after showing his drawings, was engaged to assist him in finishing the "clothing and ground"; but this artist's manners were declared to be so uncouth and the pay so poor that he left him in disgust.

When he had made a hit, as he said, with the likeness of a well known citizen, orders came to him, and he was able to resume his drawing of birds. On February 22 he recorded that he had spent his time in "running after orders for portraits, and also in vain endeavors to obtain a sight of Alexander Wilson's 'Ornithology,' but was unsuccessful in seeing the book, which is very high priced." Later, however, he appears to have succeeded in this quest, for on the 17th of that month he was able to send his wife twenty drawings of birds, eight of which were marked as "not described by Willson." Among them were the originals of some of the

[282] The original of this admirable drawing had been shot at New Madrid, on the Ohio, on November 23, and Audubon, who immediately began to work on it, recorded his conviction that the White-headed or Bald Eagle and the "Brown Eagle," which he later called "The Bird of Washington," were two different species; he thought that the young of the former, which was also brown, was much smaller in size.

most famous of his plates, such as the Great-footed Hawk, the White-headed Eagle, and the Hen Turkey.[283] Having seen in a newspaper a notice of an expedition which the Government was about to send to the Pacific Coast, to survey the boundary of the territory that had been recently ceded by Spain, Audubon became much excited over a possible appointment as draughtsman and naturalist. He sat down at once and wrote a personal letter to President Monroe, while hundreds of imaginary birds of new and interesting kinds seemed to come within the range of his gun; on the 31st of March he was still pondering on the project, and although it is not likely that his letter ever reached the eye of the President, he did receive a recommendation from Governor Robertson of Louisiana. It was with this expedition in view that he sought an interview with John Vanderlyn,[284] an eminent painter of historical subjects, then working in New Orleans; according to one version Vanderlyn treated him as a mendicant, and ordered him to lay down his portfolio in the lobby, but ended by giving him a very complimentary note, in which he praised his drawings without stint, particularly his studies of birds.

During the five months spent at New Orleans in 1821, Audubon attempted to support himself and his companion by means of their artistic talents, while he was pushing forward his ambitious design of figuring all of America's birds and most characteristic plants. That he received scant encouragement but many rebuffs is not surprising. They did succeed in obtaining a few pupils in drawing, and Audubon made a number of rapid portraits, but after living for a time on Ursuline Street, near the old Convent, and later shifting from one quarter to another, their finances had reached so low an ebb by the beginning of June that a move was imperative. Audubon then decided to go to Shippingport, Kentucky, and on the 16th of June, with young Mason, he again boarded the steamer *Columbus*, John D'Hart, captain, and started

---

[283] These drawings were as follows:

- "Common gallinule; Not described by Willson;
- Common gull; Not described by Willson;
- Marsh hawk;
- Boat tailed grackle; Not described by Willson;
- Common Crow;
- Fish Crow;
- Rail or Sora;
- Marsh Tern;
- Snipe; Not described by Willson;
- Hermit Thrush;
- Yellow Red poll Warbler;
- Savannah Finch;
- Bath Ground Warbler; Not described by Willson;
- Brown Pelican; Not described by Willson;
- Great Footed Hawk;
- Turkey Hen; Not described by Willson;
- Cormorant;
- Carrion Crow or Black Vulture;
- Imber Diver;

White Headed or Bald Eagle."

[284] Vanderlyn, like Audubon, had been a pupil of David at Paris; he produced historical paintings of merit, as well as panoramas, then coming into vogue; some of the latter were exhibited in the "Rotunda" which he erected for that purpose in City Hall Park, New York, but this enterprise failed, and his building was seized by the city for debt. Vanderlyn died in absolute want in 1852. See Samuel Isham, *The History of American Painting* (New York, 1915).

up river. An incident now occurred which affected the naturalist's whole after life by introducing him to one of the most favored spots in Louisiana, if not in the whole country, for the study of bird life, not to speak of the impressions which the charm of new scenery, a rich flora, and natural products of the most varied description must have then made on his mind. Mrs. James Pirrie, wife of a prosperous cotton planter of West Feliciana Parish, happened to be their fellow-passenger. Doubtless her curiosity was piqued by the winning manners and flowing locks of the artistic traveler, whose Gallic accent at once betrayed his nationality. Whether Audubon had made her acquaintance previous to this journey or not is not known, but before it was ended his fine enthusiasm and ambitious plans had found a sympathizer, and he was engaged as tutor to Mrs. Pirrie's daughter at $60 a month. To further his ornithological pursuits it was understood that he and his companion should live at "Oakley," her husband's plantation, five miles from St. Francisville, on Bayou Sara, and that one-half of his time should be absolutely free for hunting and drawing. Thus, on June 18, 1821, was forged the link that bound the heart of Audubon to the State which was first in his affections, and which he would fain believe might have been the scene of his nativity. Well may the Louisianians of today adopt him as their son, for from that early time he cherished their State as in a peculiar sense his own.

It was a hot and sultry day when our wanderers landed at Bayou Sara,[285] a small settlement at the junction of the sluggish stream which bears that name and the Mississippi, and proceeded to climb to St. Francisville, the village a mile away on the hill. Mrs. Pirrie, who seems to have preceded the travelers by carriage, sent some of her servants to relieve them of their luggage, which Audubon said they found light. They rested in the village at the house of Mr. Benjamin Swift, where they were invited to stay to dinner, then at the point of being served, but feeling somewhat ill at ease, they thanked their host and again took to the road. Following their leisurely guides, they now traversed a country so new, so strange, and so enchanting, that the five miles to the Pirrie house seemed short indeed. "The rich magnolias, covered with fragrant blossoms, the holly, the beech, the tall yellow poplar, the hilly ground, and even the red clay," to quote Audubon's record made at the time, "all excited my admiration. Such an entire change in the face of nature, in so short a time, seems almost supernatural, and surrounded once more by numerous warblers and thrushes, I enjoyed the scene."

In passing up the Mississippi from New Orleans, the topography of the country suddenly changes at about this point; in the parish of West Feliciana the alluvial lowlands of the river valley give place to beautiful highlands, which still harbor as rich and distinctive a flora and fauna as in Audubon's day. Following Audubon's course in June, 1916, or ninety-five years later, Mr. Arthur found the region about St. Francisville wonderfully rich in birds, and there noted seventy-eight resident kinds which were seen on the same day, from shortly before noon to seven o'clock in the evening.

---

[285] "Bayou," in Louisiana, is a term commonly applied to any slow-running stream. According to the tradition gathered on the spot by Mr. Stanley C. Arthur, both stream and settlement were formerly called "New Valentia," while the present name was derived from an old woman called "Sara," who many years ago lived at the mouth of the Bayou, where she practiced some sort of spurious physic. St. Francisville, on the hill, received its name from the circumstance that the brothers of St. Francis, who had a mission at Pointe Coupée, on the opposite bank, were in the habit of ferrying their dead over the river, in order to bury them on the high ground; "Bayou Sara" and "St. Francisville" are used interchangeably by the inhabitants. See S. C. Arthur, *Times-Picayune*, New Orleans, August 6, 1916.

Figure 39. Bayou Sara Landing, West Feliciana Parish, Louisiana, at the Junction of Bayou Sara and the Mississippi River.

Figure 40. Scene on Bayou Sara Creek, Audubon's Hunting Ground in 1821.
This and the above after Photographs by Mr. Stanley Clisby Arthur, 1916.

Upon reaching the plantation house, Audubon and his companion were kindly received by the Scotchman, James Pirrie, who introduced to them his daughter, Eliza, then a beautiful and talented girl of seventeen—"my lovely Miss Pirrie, of Oakley," as Audubon once characterized her in his journal—who was to become his pupil in drawing, and who, as after events proved, was destined to a romantic and checkered career.

The "Oakley" house, which by a strange turn of fortune's wheel thus became the naturalist's home in the summer of 1821, has changed but little since that time, but the century that has nearly sped its course has added strength and beauty to the moss-hung oaks which now encompass it and temper the heat of the southern sun in the double-decked galleries which adorn its whole front. Built of the enduring cypress, as my correspondent remarks, the house stands as firm and sound as the gaunt but living sentinels of that order which tower from the brake not far away.

Audubon spent nearly five months at the Pirrie estate. He worked with great ardor at his *Ornithology* and produced the originals of many of his plates that were afterwards published, while his assistant, Joseph Mason, who had followed him from Cincinnati, labored with equal diligence at the plants that were chosen as a setting for the birds.[286] An early drawing of the Chuck Will's Widow is dated "Red River, June, 1821," and it is probable that he followed this stream into Arkansas, for on leaving Cincinnati in the autumn of the previous year, he had planned to enter that State, and later references in his journals clearly imply that this object was attained. Another favorite hunting ground was Thompson's Creek, and he often recalled its heated banks, where, on a Fourth of July, he once satisfied his hunger by "swallowing the roasted eggs of a large soft shelled turtle."

On August 11, 1821, while Audubon was living at "Oakley," he made this entry in his journal:

> Watched all night by the dead body of a friend of Mrs. P——; he was not known to me, and he had literally drunk himself to an everlasting sleep. Peace to his soul! I made a good sketch of his head, as a present for his poor wife. On such occasions time flies very slow indeed, so much so that it looked as if it stood still, like the hawk that poises over its prey.

In the same journal also, for August 25, occurs a record which throws light on one of Audubon's most discussed and questionable pictures, that of the mocking-birds defending their jessamine-embowered nest from the sinister designs of a rattlesnake;[287] little did he think at the time how much discord this venomous reptile, when coiled in the branches of a tree, could later breed.[288] The entry was:

---

[286] On the original drawing of the Pine-creeping Warbler, *The Birds of America* (Plate cxl), the following legends appear in Audubon's autograph: "Drawn from Nature by John J. Audubon, James Pirrie's Plantation, Louisiana, July 10, 1821. Plant, J. R. Mason."
Sixteen of Audubon's originals, which still bear the designations of time and place, were produced during this interval, in the year 1821; they embrace the Mississippi Kite (Plate cxvii), June 28; Yellow-throated Vireo (Plate cxix), July 11; Red-cockaded Woodpecker (Plate ccclxxxix), July 29; American Redstart (Plate xl), August 13; Summer Red-bird (Plate xliv), August 27; Prairie Warbler (Plate xiv), Sept. 3; and the Tennessee Warbler (Plate cliv), Oct. 17.

[287] *The Birds of America*, Plate xxi.

[288] See Chapter 28.

> Finished drawing a very fine specimen of a rattlesnake, which measured five feet and seven inches, weighed six and a quarter pounds, and had ten rattles. Anxious to give it a position most interesting to a naturalist, I put it in that which the reptile commonly takes when on the point of striking madly with its fangs. I had examined many before, and especially the position of the fangs along the superior jaw-bones, but had never seen one showing the whole [of the fangs] exposed at the same time.

He then described the generous provision which nature has made to keep the rattlesnake in fighting trim, by giving it a dental arsenal on which it can draw in case of loss; he added that the heat of the day was such that he could devote only sixteen hours to the drawing.

At this time Audubon was a handsome and attractive man; his pupil, who did not enjoy the best of health, was attended by a young physician who was also her lover. It is not surprising therefore to learn that jealousy on the part of the doctor led to a misunderstanding, and that the naturalist suddenly made his departure and returned to New Orleans. In recording this incident Audubon could not repress his amusement at the prescription of the physician, who ordered the young lady to abstain from all writing and drawing for a period of four months, but meanwhile permitted her to eat anything which pleased her fancy, in spite of the relapses of fever that occasionally occurred. Audubon was allowed to see her only at appointed hours, as if, he said, he were an extraordinary ambassador to some distant court, and was obliged to preserve the utmost decorum of manner; he expressed the belief that he had not once laughed in the presence of the young lady during the entire term of his tutorial engagement, which lasted from the 18th of June to the 21st of October. Later, in December of the same year, when his former pupil passed him without recognition in the streets of New Orleans, he indulged in the reflection that she had apparently quite forgotten the great pains with which at her own request he had done her portrait in pastels, but, thanks to his talents, he thought that he could run the gauntlet of the world without her help.[289]

At New Orleans Audubon soon found new pupils, particularly through the aid of Mr. R. Pamar and Mr. William Braud,[290] who came to his assistance, Mrs. Braud and her son paying him at the rate of three dollars for a lesson of one hour. On November 10, 1821, he wrote:

> Continued my close application to my ornithology, writing every day, from morning until night, omitting no observations, correcting, re-arranging from my notes and measurements, and posting up; particularly all my land birds. The great many errors I found in the work of Wilson astonished me. I try to speak of them with care, and as seldom as possible, knowing the good will of that man, and the vast many hearsay accounts he depended on.

Again, on the 25th of that month is this entry:

---

[289] The vivacious Miss Pirrie did not marry the young doctor, but eloped to Natchez with the son of a neighboring planter, who died within a month in consequence of a cold, said to have been contracted when he waded a deep stream with his lady-love in his arms. Audubon's pupil was thrice married, and bore five children; she died April 20, 1851, and her ashes now rest by the side of her second husband, the Reverend William Robert Bowman, the parish minister at St. Francisville. See Arthur, *loc. cit.*

[290] Mistakenly written "Brand" by Audubon's biographers, according to Mr. Stanley C. Arthur, who writes that "Braud" is a very common name in New Orleans.

Figure 41. Road Leading from Bayou Sara Landing to the Village of St. Francisville, West Feliciana Parish, Louisiana.

Figure 42. "Oakley," James Pirrie's Plantation House Near St. Francisville. This and the above after Photographs by Mr. Stanley Clisby Arthur, 1916.

> Since I left Cincinnati I have finished 62 drawings of birds and plants, 3 quadrupeds, 2 snakes, fifty portraits of all sorts, and the large one of Father Antonio,[291] besides giving many lessons, and I have made out to send money to my wife sufficient for her and my Kentucky lads, and to live in humble comfort with only my talents and industry, without *one cent* to begin on. I sent a draft to my wife, and began to live in New Orleans with forty-two dollars, health, and much anxiety to pursue my plan of collecting all the birds of America.

The close of the year 1821 found Audubon teaching a few pupils at New Orleans, where, he said, his style of work and the large prices he received caused him the ill will of every artist in the city. The figure which he cut in the streets, with his loose dress of nankeen and long, flowing locks, made him wish to appear like other people, and he was soon able to rejoice in a new suit of clothes. Though still in need of work, when he was asked to aid in painting a panorama of New Orleans, he refused, begrudging the time, saying that he did not wish to see any other perspective than that of the last of his drawings.

Having been from home for over a year, Audubon now wished to have his family about him again.[292] His plan did not appeal to his practical wife, who had many friends at Cincinnati, where she was assured of a good income through her teaching; Mrs. Audubon also felt that to be constantly shifting about was anything but favorable to the education of their children. Her reluctance, however, gave way, and in December she joined her husband in New Orleans, but only to find that the city could afford them no settled means of support. The situation of the Audubon family during the winter of 1821-22 became precarious in the extreme, and for two months Audubon gave up his habit of journalizing, one reason being that he could not afford the paltry sum necessary to buy a blank book for this purpose.

Compelled at last to make a new move, Audubon started for Natchez, on the 16th of March, 1822, paying for his passage on the steamer *Eclat* by doing a crayon portrait of the captain and his wife. It was while going up the river at this time that he opened a chest containing two hundred of his drawings to find them sadly damaged by the breaking of a bottle of gunpowder, but the loss then sustained was apparently slight in comparison with that which he had experienced in an earlier disaster. To follow his account of this earlier and better known incident, when leaving Henderson for Philadelphia, he carefully placed all of his drawings in a wooden box and entrusted them to the care of a friend, with injunction that no harm should befall them; upon returning several months later, his treasure chest was opened, but only to reveal that "a pair of Norway rats had taken possession of the whole, and had reared a young family amongst the gnawed bits of paper, which but a few months before represented nearly a thousand inhabitants of the air." The heat that was immediately felt in his head, said the naturalist, was too great to be endured, and the days that followed were days of oblivion to him; but upon recuperation he took up his gun, his notebook and his pencils, "and

---

[291] Father Antonio de Sedella, popularly known as "Père Antoine," after 1791 pastor of St. Louis Cathedral; an idol of the people, but execrated by historians.
"This seditious priest is a Father Antoine; he is a great favorite of the Louisiana ladies; has married many of them, and christened all their children; he is by some citizens esteemed an accomplished hypocrite, has great influence with the people of color, and, report says, embraces every opportunity to render them discontented under the American Government." *Executive Journal of Governor Claiborne*. See Charles Gayarré, *History of Louisiana*, vol. iv, pp. 154-155 (New Orleans, 1903).

[292] This item occurs in Audubon's journal for October 25; "Rented a house in Dauphine street at seventeen dollars per month, and determined to bring my family to New Orleans."

went forth to the woods as gaily as if nothing had happened"; after a lapse of three years his portfolio was again filled, and the earlier work replaced by better. Audubon's drawings and plates were also repeatedly ravaged by fires, but this was at a much later day.

While Audubon was engaged in teaching French, music, or drawing, now to private pupils at Natchez, now in a school at Washington, Mississippi, nine miles away, the summer of 1822 passed with the outlook as ominous as ever. On August 23 he wrote: "My friend, Joseph Mason, left me today, and we experienced great pain at parting. I gave him paper and chalks to work his way with, and the double barrelled-gun ... which I had purchased in Philadelphia in 1805." Mason, who, for a year and nine months, was Audubon's aid and constant companion, seems to have settled eventually as an artist in Philadelphia, where we hear of him in 1824 and again in 1827.[293]

In the following December Audubon received a fresh impetus towards the goal of his ambition by the arrival at Natchez of a traveling portrait painter, named John Stein, who gave him his first lessons in the use of oils; his initial attempt was the copy of an otter from one of his own drawings. Audubon and Stein together later painted a full-length portrait of Father Antonio which was sent to Havana. Artists who have worked long in one medium are not always successful in another, but those who have seen some of Audubon's later and better works in oil, such as his large canvas of the Wild Turkeys,[294] must admit that he attained a high degree of skill. As will be seen, this acquisition was a strong string to his bow; when in England his brush helped largely to pay for the issue of his early plates.

Mrs. Audubon, who joined her husband in New Orleans on December 8, 1821, soon felt obliged to seek employment. She engaged as nurse or governess in the family of Mr. Braud, presumably the same whose wife and son had received instruction in drawing from the naturalist the previous autumn, and remained with that family until September, 1822, when the death of the child that was placed in her charge left her free to follow her husband to Natchez. After attempting a similar position in the home of a clergyman there and finding it impossible to obtain her salary, in January, 1823, she was invited by the Percys to West Feliciana,[295] then a prosperous cotton district, at the apex of the salient made by the neighboring state of Mississippi and bordered on two sides by the great river. Her worth was evidently appreciated, for she was encouraged to establish a private school on the Percys' plantation, which she conducted successfully for five years.

Captain Robert Percy, who before coming to America in 1796 had been an officer in the British Navy, was living at this time with his wife and five children at their plantation of "Weyanoke," on Big Sara Creek, fifteen miles from St. Francisville; this town, owing to its large shipments of cotton, was then at the height of prosperity, and its population no doubt exceeded that of the present day; it now stands at about one thousand souls. Letters and journals of the period constantly refer to "Beechwoods," which was not the mansion house, though it undoubtedly belonged to the Robert Percy estate. There it was that the wife of the naturalist lived, and there she started her school, for the benefit not only of the Percy boys and girls, but also of a limited number of children of their wealthy neighbors; her own son, John

[293] See Audubon's letter to Sully.

[294] Now in the collection of Mr. John E. Thayer, Lancaster, Mass.

[295] Mr. Stanley C. Arthur, whose recent visit to this region has already been noticed, gathered there from the lips of old residents, some of whom were descendants of those who had known the Audubons, a store of reliable data by which the history of the naturalist at this important phase of his life is revealed in its true light; to him I am indebted for a series of excellent photographs of the region, its historic houses and people, as well as for much needed information. See Arthur, *loc. cit.*

Woodhouse Audubon, then eleven years of age, at this time received instruction at her hands. The parish of West Feliciana, at this early period, was one of the richest cotton-producing sections of the entire State; its care-free planters led an easy life until the "king" was unceremoniously dethroned by a small, but not insignificant insect which has proved mightier than either fire or sword, namely, the boll-weevil; now many a fine old estate which has languished under the influence of the pest could probably be bought for a song. "Beechwoods," thus devoted to educational purposes, later came into the hands of Thomas Percy, but the house, like that of "Weyanoke," was long since burned to the ground.

While Mrs. Audubon was establishing her rules and authority at the Percy school, the naturalist was painting with Stein at Natchez, and he remained there with his elder son until the spring of 1823. At this period he wrote in his journal: "I had finally determined to break through all bonds, and follow my ornithological pursuits. My best friends solemnly regarded me as a madman, and my wife and family alone gave me encouragement. My wife determined that my genius should prevail, and that my final success as an ornithologist should be triumphant."

In March, 1823, Audubon and friend Stein bought a horse and wagon, and in the hope of raising money through their joint efforts as itinerant portrait painters, set out with Victor on a tour of the Southern States. This venture, however, did not succeed, and after visiting Jackson and a number of other towns, they disbanded at New Orleans. Audubon then started north with his son for Louisville, but upon paying a visit to his wife at the "Beechwoods" school, he was invited by the Percys to remain there for the summer and "teach the young ladies music and drawing." According to a tradition which has survived among the Percy descendants, Audubon spent most of his time in roaming through the woods, but he also taught his wife's pupils to swim in the large spring house at "Weyanoke," where the water could be deepened at pleasure. It was also said that he painted the Wild Turkeys in the woods of Sleepy Hollow nearby, but I have already given Audubon's own record in regard to one of these pictures, and, as Mr. Arthur remarks, the places in Louisiana where he drew these famous subjects are as numerous as the beds in which Lafayette slept when at New Orleans.

Audubon remained with the Percys during the greater part of the summer, or until some misunderstanding arose, when he was again adrift and upon a sea of difficulties. While visiting a plantation near Natchez, both he and Victor were stricken with fever; his faithful wife hastened to them, and after nursing both back to health, she returned with them to the Percy plantation, where they remained from the 8th to the 30th of September.

In the autumn of 1823 Audubon was determined to visit Philadelphia, in the hope of finding a sponsor for his "Ornithology." Although the work was then far from ready for publication, he felt that at least he might better his condition, and with this end in view he sent his drawings from Natchez to that city; a hasty visit was made also to New Orleans, for the purpose, no doubt, of obtaining credentials to possible patrons in the East. At last, on October 3, he started with Victor on the steamer *Magnet*[296] for Louisville. Low water quickly held them up after entering the mouth of the Ohio, and they were obliged to disembark at the little village of Trinity, at the mouth of Cash Creek, the scene of Audubon's misadventures with Rozier thirteen years before. The remoteness of the situation and the state of their funds, which corresponded with that of the river, left no alternative but to walk, and they undertook

[296] One of the early steamboats on the Ohio that had been built at Pittsburgh, in 1821, by Thomas W. Bakewell, his brother-in-law and former partner.

to reach Louisville, several hundred miles distant, afoot. Two other travelers joined them, and with Victor, then a lad of nearly fourteen, the party left the creek at noon on October 15 and struck across country through the forests and canebrakes. At Green River, which was reached on the 21st, Victor gave out from sheer exhaustion,[297] and the remainder of the journey was finished in a Jersey wagon. At length, said Audubon, "I entered Louisville with thirteen dollars in my pocket." At Shippingport, then an independent town at the Falls of the Ohio, he was obliged to settle down for the winter. A place for Victor was found in the counting-house of Nicholas A. Berthoud, while the father undertook anything that came to hand, painting portraits, landscapes, panels for river boats, and even street signs, so hard pressed was he at times to eke out a subsistence for them both. Yet Audubon was as sanguine as ever, and on November 9 he recorded the resolution "to paint one hundred views of American scenery," and added: "I shall not be surprised to find myself seated at the foot of Niagara," a prediction which was fulfilled in the following year.

During the winter spent at Shippingport, Audubon lost a gentle friend in Madame Berthoud,[298] the mother of Nicholas. In his journal for January 20, 1824, we read his emotional words:

> I arose this morning by the transparent light which is the effect of the moon before dawn, and saw Dr. Middleton passing at full gallop towards the white house; I followed—alas! my old friend was dead!... many tears fell from my eyes, accustomed to sorrow. It was impossible for me to work; my heart, restless, moved from point to point all round the compass of my life. Ah Lucy! what have I felt to-day! ... I have spent it thinking, thinking, learning, weighing my thoughts, and quite sick of life. I wished I had been as quiet as my venerable friend, as she lay for the last time in her room.

---

[297] See "A Tough Walk for a Youth," *Ornithological Biography*, vol. iii, p. 371; and "The Hospitality of the Woods," *ibid.*, vol. i, p. 383.

[298] This lady had a remarkable history. She was the widow of the Marquis de Saint Pie, and was at one time a *dame d'honneur* of Queen Marie Antoinette; like many others of noble birth, she had fled from Paris during the Revolution, and emigrated to America, where with her husband she assumed the name of Berthoud. Her son, Nicholas Augustus, had married Mrs. Audubon's sister, Eliza Bakewell, in 1816.

Chapter 21

DÉBUT AS A NATURALIST

In 1824 after five hard years of struggle and embarrassment, Audubon decided that the time had come to bring his labors to the light of day. At thirty-nine, he read and spoke two languages but was without adequate training in either; he had never written a line for publication, and to the scientific world he was a stranger. Though without a definite plan, he cherished the ardent hope of presenting the birds of his beloved America as he had depicted them, to the size of life, and with all the added interest and zest that a natural environment could give them.

To Philadelphia the naturalist now turned his steps, for that city was then a Mecca for scientific men. Leaving Shippingport in March, he reached the Quaker capital on the fifth day of April. There he purchased a new suit of clothes, and, dressed "with extreme neatness," paid his respects to Dr. William Mease, the one friend there whom he had known intimately in his younger and more prosperous days. It was primarily through this excellent man's interest that Audubon met the leading artists and scientific men of the city, including Thomas Sully, Robert and Rembrandt Peale, Richard Harlan, Charles Le Sueur, and Charles L. Bonaparte, the latter then a rising young ornithologist of one and twenty. It was Bonaparte who introduced Audubon to the Academy of Natural Sciences, where his drawings were exhibited and generally admired. Among his critics on that occasion was George Ord, who from their first interview seems to have looked upon the new luminary with jealous eyes. Whether this was true or not, there is no doubt that Ord became one of his few really bitter and implacable adversaries, and not many days elapsed before Audubon came to feel that many in Philadelphia would be glad to see him return to the backwoods of the Middle West, from which, like an apple of Sodom, he seemed suddenly to have dropped into their midst. Those who were most interested in the continued sale and success of Wilson's *Ornithology*, he declared, advised him not to publish anything, and threw not only cold water but ice upon all his plans. Thus began that unseemly rivalry, fostered for many years by George Ord in this country, between the friends of Alexander Wilson and those of John James Audubon, the dead embers of which are occasionally stirred even to this day.[299]

Ord, who was about Audubon's own age, was a quiet, persistent, and unassuming worker, held in high esteem by many of his associates. Audubon seems to have done his best to conciliate him then and at a later day, but all to no purpose; Dr. Harlan once advised him to

[299] See Chapter 14.

give up the attempt, since Ord, he declared, had no heart for friendship, having been denied that blessing by nature herself. Ord, as we have seen, had edited the eighth and written the ninth, or concluding, volume of Wilson's *American Ornithology*, as well as a life of its author; the appearance of a new star in the ornithological horizon may not have been a welcome sight. At all events, we soon find him engaged upon a new edition of Wilson's work.[300] Ord had objected to Audubon's method of combining plants and other accessories with his drawings of birds, a criticism that in the case of purely technical works could be easily sustained, and some of his later charges, though carried too far, were not wholly without foundation.[301]

Bonaparte,[302] on the other hand, was captivated by Audubon's drawings and anxious to secure his services for his own work, then well in hand. This was the *American Ornithology*, for which Titian R. Peale was then making the drawings, and Thomas Lawson, who had been Wilson's engraver, was engaged on the plates; though quite distinct in itself, this was much in the style of Wilson's earlier work, of which it was virtually a continuation. When Bonaparte introduced Audubon to these men, it is not surprising that the meeting was not productive of the best of feeling on either side. Peale's stiff and rather conventional portraits of birds naturally failed to awaken enthusiasm in "the trader naturalist," as some who looked upon him as a rival rather contemptuously called him. The interview with Lawson, if correctly reported by his friend,[303] shows that his interest could not have been of the most disinterested sort. "Lawson told me," said this reporter, "that he spoke freely of the pictures, and said that they were ill drawn, not true to nature, and anatomically incorrect." Thereupon Bonaparte defended them warmly, saying that he would buy them and that Lawson should engrave them. "You may buy them," said the Scotchman, "but I will not engrave them ... because

---

[300] This was the third edition of the *American Ornithology*, issued by Messrs. Collins & Company in New York and by Harrison Hall of Philadelphia, in three octavo volumes, with an atlas of 76 plates colored by hand, in 1828-9. Mr. Hall, who appears to have been the person most interested financially in this edition, was a brother of James Hall, author of a notorious review in which this work was praised at the expense of Audubon, who was viciously attacked. Friends of Audubon repeatedly asserted that as soon as his popularity and success began to check the sales of Wilson's work, Ord and a few others, aided by interested publishers, began a systematic series of attacks, some notice of which is taken in Chapter 28.

[301] See Chapter 14.

[302] Charles Lucien Jules Laurent Bonaparte, Prince of Canino and Musignano, the eldest son of Lucien, and nephew of Napoleon, Bonaparte, was born at Paris in 1803, and died there in 1857. At this time he was settled with his uncle and father-in-law, Joseph Bonaparte, former King of Spain, at Philadelphia, and there and at Bordentown, New Jersey, where Joseph had an estate, he undertook the study of American birds. His best known scientific works are: *American Ornithology, or the Natural History of the Birds of the United States, not Given by Wilson*, 4 volumes, quarto, with 27 colored plates, Philadelphia, 1825-1833; and *Iconographica della Fauna Italica*, Rome, 1833-1841. In 1828 he retired to Italy, where he was devoted to literary and scientific pursuits. He was an early subscriber to Audubon's *Birds of America*, but their relations were somewhat strained on the publication of the *Ornithological Biography* in 1831. Bonaparte later entered politics in Italy, and was leader of the republican party at Rome in 1848 and 1849; after having been expelled from France by the order of Louis Napoleon, he was permitted to return in 1850, and became director of the Jardin des Plantes in Paris.

He was a closet naturalist rather than a field student, but did much for the reform of nomenclature. In his *Ornithology* the number of American birds was raised to 366, nearly one hundred having been added since the work of Wilson was revised by Ord, but he added only two that were new, Cooper's Hawk, (*Accipiter cooperi*), named after William Cooper of New York, and Say's Phœbe (*Sayornis saya*), dedicated to Thomas Say, and first procured by Titian R. Peale in the Rocky Mountain districts of the Far West. Perhaps his most important technical work, the *Conspectus Generum Avium*, begun in 1850, was incomplete at the time of his death.

[303] William Dunlap, *History of the Rise and Progress of the Arts of Design in the United States*, vol. ii, p. 402 (New York, 1834).

ornithology requires truth in the forms, and correctness in the lines. Here are neither." Other meetings are said to have followed, but to have ended only in mutual dislike. Nevertheless, one of Audubon's drawings was engraved by Lawson and appeared in Bonaparte's work,[304] but most of the figures in Bonaparte's concluding volumes were by the hand of a German named Alexander Rider. It was doubtless a fortunate circumstance that the prejudice and obstinacy of this overbearing Scot was a bar to any further absorption of Audubon's talents.[305]

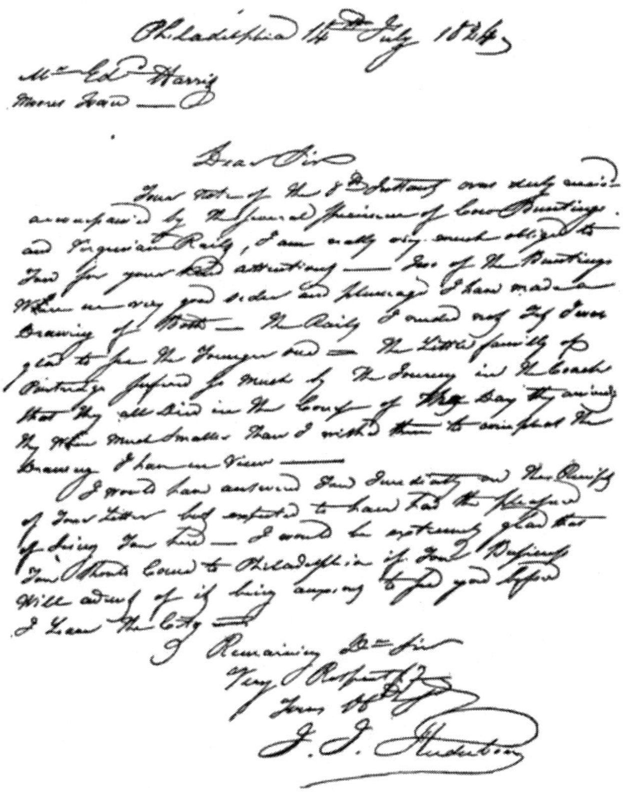
Philadelphia 14th July 1824

Mr Edd Harris
Morris Town —

Dear Sir
Your note of the 8th Instant was duly receiv-
accompanied by the several specimens of Cow Buntings
and Virginian Rail, I am really very much obliged to
You for your kind attention — Two of the Buntings
were in very good order and plumage I have made a
Drawing of Both — The Rail I would not try, I was
glad to see the Younger one — the little family of
Partridges suffered so much by the Journey in the Coach
that they all died in the Course of the day they arrived
they were much smaller than I wished them to compleat the
Drawing I have in View —
I would have answered You immediately on the Receipt
of Your Letter but expected to have had the pleasure
of seeing You here — I would be extremely glad that
You should Come to Philadelphia if Your Business
Will admit of it being anxious to see you before
I Leave the City —
I Remaining Dr Sir
Very Respectfully
Yours &c
J. J. Audubon

Figure 43. An Early Letter to Edward Harris, Dated July 14, 1824 at the Beginning of Their Life-Long Friendship. From the Jeanes MSS. Audubon's Last Letter to Edward Harris, from the Same Source, Is Reproduced in Volume II.

Audubon met at this time a more appreciative engraver in Mr. Fairman, who urged him to take his drawings to Europe and have them engraved in a superior style; on July 12 the naturalist wrote that he had drawn "for Mr. Fairman a small grouse to be put on a banknote belonging to the State of New Jersey." By some lucky chance this incident brought him the acquaintance of Edward Harris,[306] whom he met that summer in Philadelphia, and who

---

304 The Boat-tailed Grackle, vol. i, plate iv.

305 He seems, however, to have supplied Bonaparte liberally with notes, for after devoting fifteen pages to the biography of the Wild Turkey, Audubon said: "A long account of this remarkable bird has already been given in Bonaparte's American Ornithology, volume I. As that account was in a great measure derived from notes furnished by myself, you need not be surprised, good reader, to find it often in accordance with the above." *Ornithological Biography* vol. i, p. 16.

306 Edward Harris was born at Morristown, New Jersey, in 1799, where he died in 1863. Without the incentive to earn money or the ambition to acquire fame, he lived the life of a gentleman of leisure, devoted to natural

became one of his most constant and disinterested friends. It was Harris who a few days after their meeting took all of the drawings which Audubon had for sale and at the artist's own prices;[307] who for years was continually sending him rare or desirable specimens of birds; who accompanied him through the Southern States to Florida in 1837 and on the famous Missouri River Expedition in 1843. Edward Harris became a patron of science through his friendship with scientific men, and many besides Audubon were indebted to him for judicious advice as well as more substantial benefits.

The Academy of Natural Sciences, founded in 1812, was well established at this time, and its rapidly growing Museum was already the largest and most valuable in the New World; ornithology was a favored subject, and the Academy's roll embraced every American pioneer worker of note in the entire field of the natural sciences. The following account of a meeting of the Academy, held on October 11, 1825, when Ord presided, has been preserved in a letter of the period:[308]

> A few evenings since I was associated with a society of gentlemen, members of the Academy of Natural Sciences. There were present fifteen or twenty. Among the number were Le Sueur, Rafinesque, Say, Peale, Pattison, Harlan, and Charles Lucien Bonaparte.
>
> Among this collection life was most strikingly exemplified: Le Sueur, with a countenance weather-beaten and worn, looked on, for the muscles of his ironbound visage seemed as incapable of motion, as those on the medals struck in the age of Julius Cæsar. Rafinesque has a fine black eye, rather bald and black hair, and withal is rather corpulent. I was informed that he was a native of Constantinople; at present he lives in Kentucky. Dr. Harlan is a spruce young man.... Peale is the son of the original proprietor of the Philadelphia Museum, and one who visited the Rocky Mountains with Major Long; he is a young man, and has no remarkable indications of countenance to distinguish him. Say, who was his companion in the same expedition, is an extremely interesting man; to him I am particularly obligated for showing me their Museum and Library. I think he told me that their society had published nine volumes.... Bonaparte is the son of Lucien Bonaparte and nephew to the Emperor Napoleon; he is a little set, black-eyed fellow, quite talkative, and withal interesting and companionable.

Among the working naturalists at Philadelphia Dr. Richard Harlan was possibly one whose friendship was most valuable to Audubon; the artist from whom he received most encouragement was Thomas Sully, the portrait painter, who took him into his studio and gave him lessons in the use of oils. Sully was one of those who saw the good side of Audubon's

history, to sport and to the cultivation of his paternal acres. He had the gift of friendship, was widely traveled, wrote charming letters, and kept careful records of his observations, but rarely published anything. The breeding of fine stock was one of his hobbies, and as a result of a journey to Europe in 1839, when he visited a horse fair in Normandy, he is credited with having first introduced the Norman breed into America. "The beneficent results of his quiet, unobtrusive life," says an appreciative biographer, "reach down to our time, and, after half a century, we are glad that Edward Harris lived." See biographical sketch by George Spencer Morris, in *Cassinia*, vol. vi (Philadelphia, 1902).

[307] See Chapter 12.

[308] Written by Dr. Edmund Porter of Frenchtown, New Jersey, to Dr. Thomas Miner of Haddam, Connecticut, on October 25, 1825. See Witmer Stone, "Some Philadelphia Ornithological Collections and Collectors, 1784-1850," *The Auk*, vol. xvi (New York, 1899).

character, discerned his talent, and predicted for him a great future; at a later day Sully was able to rejoice in finding his prediction amply fulfilled.[309]

Convinced that the advice which Fairman and Bonaparte had given him was sound, Audubon decided to look to Europe for a publisher of his *Birds*, and with this end in view, set hard to work at his drawings. "I had some pupils offered," he said, "at a dollar per lesson; but I found the citizens unwilling to pay for art, although they affected to patronize it. I exhibited my drawings for a week, but found the show did not pay, and so determined to remove myself." Audubon remained in Philadelphia until August, and while in doubt as to what step he should take next, he was cheered by a visit to "Mill Grove," made in the carriage of his Quaker friend, Reuben Haines. To quote his journal:

> As we entered the avenue, which led to the farm, every step brought to my mind the memory of past years, and I was bewildered by the recollections until we reached the door of the house, which had once been the residence of my father as well as of myself. The cordial welcome of Mr. Wetherill, the owner, was extremely agreeable. After resting a few moments, I abruptly took my hat and ran wildly to the woods, to the grotto where I first heard from my wife that she was not indifferent to me. It had been torn down, and some stones carted away; but raising my eyes towards heaven, I repeated the promise we had mutually made. We dined at Mill Grove, and as I entered the parlor I stood motionless for a moment on the spot where my wife and myself were forever joined.

In this dramatic rehearsal the naturalist clearly implies that he was married in the parlor of his own home, but his excellent wife, who was surely in this instance the better authority, explicitly states that their marriage took place in her father's house at "Fatland Ford." Since Audubon was in the habit of sending extracts from his journal to his family, it is clear that errors of this sort were the simple result of an impulsive temperament; the moment his imagination pictured his wedding as having taken place in his old abode, down went the jotting in the journal, which was written at odd moments anywhere, often at late hours, and with no care in revision or thought of future publication.

On August 1, 1824, Audubon recorded in his diary that he had left Philadelphia for New York on the day before, "in good health, free from debt, and free from anxiety about the future." Sully had given him glowing letters of introduction to Gilbert Stuart, Washington Allston and Colonel Trumbull, but then as now midsummer was not a propitious time to find city people at home, and he began to consider the advisability of visiting both Albany and Boston. Alternately elated or depressed by the prospects of the day or the hour, Audubon wrote on August 4 that he had called with a letter of introduction on Dr. Mitchell, who had

---

[309] Thomas Sully (1783-1872), Englishman by birth, who had come to America at an early age, and like Audubon had waged a bitter struggle before success was achieved, became one of the first portrait painters of the early American School.

In 1831 Sully wrote to Audubon that his success in England and France had charmed all of his friends in America, that it was like a personal triumph to them, and that it would soon silence his few remaining enemies; "Be true to yourself, Audubon," he added, "and never doubt of success." It has been said that when Audubon first came to Philadelphia in 1824 he applied to Sully for instruction, saying that he wished to become a portrait painter (see Dunlap, *op. cit.*); again that he was ready to sell his drawings to the highest bidder; but the records of his journals from 1820 onward are sufficiently consistent to show what his purpose really was.

given him "a kind letter to his friend Dr. Barnes." This hurriedly penciled note from the Nestor of American science of that day has been carefully preserved, and reads as follows:[310]

Dr. Samuel L. Mitchell to Dr. Barnes

Mr. A. who brings strong testimonial of excellence from our friends in Pha is now sitting with me—I have been delighted and instructed by a Display of his Port Folio containing Drawings Done from Life of North American Birds and illustrating the Connect, of ornithology with Botany. He has Superior attainments & skill in the natural sciences which he has cultivated for more than 20 y. he wishes to show his Elegant performances to the Members of the Lyceum and to be made a Member of that Society—it is his intention to Leave this City for Boston on Sunday morning. Meanwhile I recommend him to your good offices.

Yours Truly as ever
Sam, L. Mitchell
Aug t 4 t 1824

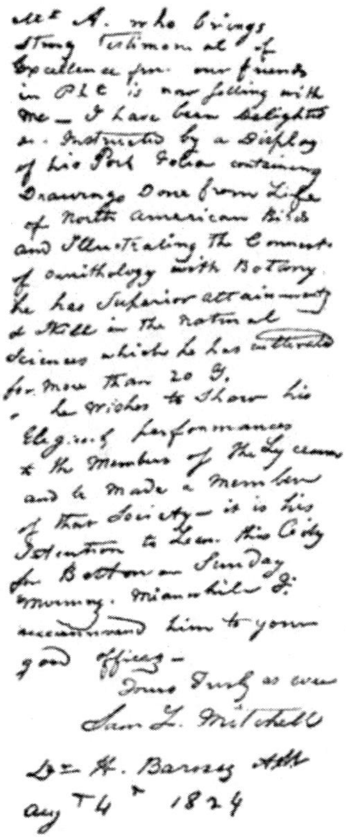
Mr A. who brings
strong testimonial of
excellence from our friends
in Pha is now sitting with
me — I have been delighted
and instructed by a Display
of his Port Folio containing
Drawings Done from Life
of North american birds
and Illustrating the Connect.
of ornithology with Botany.
he has Superior attainments
& skill in the natural
sciences which he has cultivated
for more than 20 y.
he wishes to show his
Elegant performances
to the Members of the Lyceum
and be made a member
of that society — it is his
Intention to Leave this City
for Boston on Sunday
morning. Meanwhile I
recommend him to your
good offices —
Yours Truly as ever
Sam L. Mitchell
Dr H. Barnes &c
Aug t 4 t 1824

Figure 11. Note of Dr. Samuel L. Mitchell, Written Hurriedly in Pencil, Recommending Audubon to His Friend, Dr. Barnes, August 4, 1824. From the Howland MSS.

[310] For the favor of examining a collection of interesting autograph letters written to Audubon in Europe and America, some of which are here reproduced, I am indebted to the kindness of Mr. Henry R. Howland, secretary of the Buffalo Society of Natural Sciences. This note was written on a narrow strip of manila-colored drawing paper.

Dr. Mitchell, who was the father and first president of the Lyceum of Natural History, had been a friend of young Audubon when he was clerking in New York in 1807.[311] His recommendation was accepted, and the naturalist was enrolled on the Lyceum's list of members; to justify his election, two papers, representing his first contribution to ornithology, were presented to the Society, and appeared in its *Annals* of that year.[312] Audubon visited the Lyceum with Dr. DeKay and exhibited his drawings, but said that he felt awkward and uncomfortable. On August 3 he called on John Vanderlyn, the artist, examined his pictures, and "saw the medal given him by Napoleon, but was not impressed with the idea that he was a great painter." Upon meeting Vanderlyn again a little later, he was asked to sit for a portrait of Andrew Jackson; his journal entry regarding the incident was as follows:[313]

August 10. My spirits low, and I long for the woods again; but the prospect of becoming better known prompts me to remain another day. Met the artist Vanderlyn, who asked me to give him a sitting for a portrait of General Jackson, since my figure considerably resembled that of the General, more than any he had ever seen. I likewise sketched my landlady and child, and filled my time.

The context shows that the sitting was given, and as Mr. Stanley C. Arthur remarks, Vanderlyn's portrait, which now hangs in the City Hall in New York, shows "Old Hickory" from the shoulders up, but from the shoulders down it is John James Audubon.

On the 14th Audubon wrote cheerfully to Sully:

Audubon to Thomas Sully

My reception in New York has surpassed my hopes. I have been most kindly [received], and had I seen Col. Trumball, I would have found him the gentleman you represented, but his absence at Saratoga Springs has deprived me of that pleasure.

New York is now an immense city. Strangers are received here with less reserve generally than at Philadelphia. I found the Academy well supplied with paintings, and sculptures of the Greek masters. The steam boats of the Sweet Ohio, with all their swiftness of motion and beautiful forms, do not interest the eye like those that are here tossing over the foaming billows with the grace of the wild swan. Were I a painter—ah could I, like ——, carry in my mind's eye all my mind feels when looking at the Battery at the moon's tender reflections on the farthest sails, forcing the vessel they move with the very wind's heart,—express as he does the quick moving tar hauling in a reef at the yard's end,—and make on the canvas a noble commander speak, as you have done; then, my dear friend, I could show you New York's harbor and all its beauties....

I cannot part with that Fair City [Philadelphia] this soon; I cannot help thanking Fairman, Peale, Neagle, Le Sueur, and many others besides Mc Murty for their attentions to me. Should you see honest Quaker Haines, beg him to believe me his friend; should you see Mr. Ord, tell him I never was his enemy. Think of me some time, and accept the truest best wishes of John J. Audubon.

I leave for Boston tomorrow. Should you please to write to me, direct to Care of Messrs. Anshutz & Co, Pittsburgh, where I shall be in about 40 days.

---

[311] See Chapter 11.
[312] See Bibliography, Nos. 15 and 16.
[313] See Lucy Audubon, ed., *Life of John James Audubon, the Naturalist*, p. 107.

The very next day Audubon changed his plans and sailed up the Hudson to Albany, where he hoped to meet De Witt Clinton, then at the height of his fame, who in the course of his great undertakings had found time to write letters on the natural history and antiquities of his State, and Dr. Beck, the botanist. Failing to find either at home, Audubon was compelled by the depleted state of his pocketbook to give up his plan of visiting Boston, and being determined to see Niagara Falls, he took passage on a canal boat to Buffalo instead. The Falls were reached on the 24th of August, and it was then, on recording his name at an hotel, that Audubon wrote underneath: "Who, like Wilson, will ramble, but never, like that great man, die under the lash of a bookseller."[314] Upon his first view of the Falls he was satisfied that Niagara never had been and never could be painted. He wanted to cross the bridge at Goat Island but was deterred by the necessity of economy. Visitors it seems, had already learned to venture under a small section of the American Falls, and Audubon said that while looking through the falling sheet of water, "at their feet thousands of eels were lying side by side, trying vainly to ascend the torrent." After strolling through the village to find some bread and milk, the naturalist recorded that he ate a good dinner for twelve cents, and that he went to bed "thinking of Franklin eating his roll in the streets of Philadelphia, of Goldsmith traveling by the aid of his musical powers, and of other great men who had worked their way through hardships and difficulties to fame, and fell asleep, hoping, by persevering industry, to make a name for himself among his countrymen."

The schooner from Buffalo to Erie, Pennsylvania, on which Audubon had taken deck passage, as he was unable to afford a berth in the cabin, was caught in a violent gale on the way and was obliged to anchor in the harbor of Presque Isle. "It was on the 29th of August, 1824," his diary reads, "and never shall I forget that morning." Captain Judd, of the United States Navy, had sent a gig with six men to its relief, and "my drawings," he continues, "were put into the boat with the greatest care. We shifted into it, and seated ourselves according to direction. Our brave fellows pulled hard, and every moment brought us nearer the American shore; I leaped upon it with elated heart. My drawings were safely landed, and for anything else I cared little at the moment."

At this point Audubon set out with a fellow traveler, who was also an artist, for Meadville, Pennsylvania. The earliest version of his journal[315] which gives an account of this experience reads as follows:

> On the shore of upper Canada, my money was stolen. The thief, perhaps, imagined it was of little importance to a naturalist. To repine at what could not be helped would have been unmanly. I felt satisfied Providence had relief in store. Seven dollars and a half were left to us, two persons, 1500 miles from home, at the entrance of Presque-Isle Harbor.

Five dollars was paid to their driver, and when they reached Meadville, and entered J. E. Smith's "Traveler's Rest," they had but one hundred and fifty cents between them. No time was to be lost, and Audubon at once started out with his portfolio and his artist friend to look for work:

---

[314] See Vol. I.

[315] Probably first published in a newspaper, and reprinted in pamphlet form, dated April 9, 1846.

> I walked up the Main Street, looking for *heads*, till I saw a Hollander gentleman in a store, who looked as if he might want a sketch. I begged him to allow me to sit down. This granted, I remained perfectly silent till he very soon asked: "What is in that portfolio"? This sounded well; I opened it. He complimented me on my drawings of birds and flowers. Showing him a portrait of my Best Friend, I asked him if he would like one of himself. He said "Yes, and I will exert myself to gain as many more customers as I can."

According to a story current at Meadville long after the event Audubon made the acquaintance of Mr. Benedict, a merchant, lately come from New Haven, whose attractive daughter, named Jennett,[316] was then one and twenty; his family lived at the village tavern, called the "Torbett House," in which Mr. Augustus Colson had a store. It was Mr. Colson, to whom Audubon probably refers, who responded generously to his appeal for work, and called in a number of his young friends as possible patrons. Among them was Miss Jennett Benedict, and the naturalist, attracted by her agreeable manners and pleasing appearance, asked permission to make a portrait-sketch, saying that he would pay for the privilege by presenting her with a copy. This was evidently good business enterprise, for, according to the story, a grain bin in the Colson store was soon converted into a studio, and Audubon was rewarded by a number of sitters. Here is his account from the record just quoted:

> Next day I entered the *artist's room*, by crazy steps of the store-garret; four windows faced each other at right angles; in a corner was a cat nursing, among rags for a paper-mill; hogsheads of oats, Dutch toys on the floor, a large drum, a bassoon, fur caps along the walls, a hammock and rolls of leather. Closing the extra windows with blankets, I procured a painter's light.
>
> A young man sat to try my skill; his phiz was approved; then the merchant; the room became crowded. In the evening I joined him in music on the flute and violin. My fellow traveller also had made two sketches. We wrote a page or two in our journals, and went to rest.
>
> The next day was spent as yesterday. Our pockets replenished, we walked to Pittsburgh in two days.

A month was spent at Pittsburgh, where Audubon searched the country for birds and continued his drawings. While there he made the acquaintance of the Reverend John Henry Hopkins, a man of superb appearance and rare conversational and oratorical powers, later known as the learned and versatile first Episcopal Bishop of Vermont. Audubon attended some of the ministrations of this remarkable man, through whose influence, he said, "I was brought to think, more than I usually did, of religious matters; but I never think of churches without feeling sick at heart at the sham and show of some of their professors. To repay evil with kindness is the religion that I was taught to practice, and this will forever be my rule."

---

[316] Miss Jennett Benedict in 1836 became Mrs. Butts; the crayon portrait which Audubon made at this time was carefully treasured by her daughter, the late Mrs. Frederick A. Sterling, of Cleveland, Ohio, to whose kindness I am indebted for the privilege of reproducing it. This original drawing, which is presumably a fair specimen of Audubon's itinerant portraiture, was made on a sheet of buff, water-marked paper, 14½ by 10½ inches in dimensions; it was outlined in pencil, and carefully finished in crayon-point; its legend "J. J. Audubon-1824," was inserted in pencil, in a very fine hand at the lower margin of the sketch. The Colson store was at the corner of Water Street and south of Cherry Alley. For an account of this incident I am indebted to Mrs. Sterling, and to an article in the *Tribune Republican*, of Meadville, for February 7, 1907.

Figure 45. Miss Jennett Benedict. An Example of Audubon's Itinerant Portraiture; Drawn at Meadville, Pennsylvania, in 1824. After the Original Crayon Drawing in Possession of Mr. Frederick A. Sterling.

Figure 46. Miss Eliza Pirrie. "My Lovely Miss Pirrie of 'Oakley,'" Audubon's Pupil in 1821. After a Portrait in Possession of Her Granddaughter, Miss Lucy M. Matthews. Published by Courtesy of Mr. Stanley Clisby Arthur.

In the autumn of 1824 Audubon planned another visit to the Great Lakes in search of new birds, and tried to induce his friend, Mr. Edward Harris, to accompany him. While wandering in the forests along those lakes he thought out the plan which was finally followed in the publication of his *Birds of America*:

> Chance, and chance alone, had divided my drawings into three different classes, depending upon the magnitude of the objects to be represented; and, although I did not at that time possess all the specimens necessary, I arranged them as well as I could into parcels of five plates—I improved the whole as much as was in my power; and as I daily retired farther from the haunts of man, determined to leave nothing undone, which my labor, my time, or my purse could accomplish.[317]
>
> Audubon's journal kept on the lakes has been lost, but that journey was fresh in mind when he wrote the following letter to Edward Harris.[318]

Audubon to Edward Harris

Beechwoods. Near Bayou Sara, La.
Jany. 31 1825.

> Surely I have not dismerited your esteem; when on the Lakes, both Ontario and Champlain, I wrote to you—again from Pittsburgh, all without any answer, and I am sorry to say that I have been either abandoned or forgotten by all those other persons who had promised to keep up a correspondence with me....
>
> The country I visited was new, in great measure, to me. I have been delighted with the tour, but will forever regret that your sister's indisposition could not allow you time to augment my pleasure by your company.
>
> [Audubon offers to send his friend shrubs and fruits from the South, and concludes;] In fact, my dear Mr. Harris, I am yet the same man you knew at the corner of 5th, and Minor Streets [in Philadelphia], and will continue forever the same.

After his tour of the Lakes Audubon returned to Pittsburgh, and on October 24, 1824, started down the Ohio in a skiff, intending to descend to the Mississippi and thence reach his family in Louisiana. Bad weather and lack of funds interfered with this plan, and ere long he was once more stranded in Cincinnati, where he was beset by claimants for payment upon articles ordered for the Western Museum five years before. Finding it difficult at this time to replenish an empty purse, Audubon felt that he must borrow fifteen dollars, but could not make up his mind how to ask the favor until he had several times walked past the house where he had once been known. Nevertheless, he succeeded in obtaining the necessary funds, took passage on a boat bound for Louisville, and slept cheerfully that night on a pile of shavings which he managed to scrape together on deck. "The spirit of contentment which I now feel," he wrote, "is strange; it borders on the sublime; and, enthusiastic or lunatic, as some of my relatives will have me, I am glad to possess such a spirit"; later he added: "I discover that my friends think only of my apparel, and those upon whom I have conferred acts of kindness prefer to remind me of my errors."

---

[317] *Ornithological Biography*, vol. i, p. xi.
[318] The Jeanes MSS.

Louisville was reached on November 20, and a number of days were spent in visiting his eldest son, Victor, who was then at Shippingport.[319] He finally arrived at Bayou Sara in late November, 1824. The captain of his vessel, which was bound for New Orleans, put him ashore at midnight, and he was left to grope his way to the village on the hill. St. Francisville, to his dismay, was nearly deserted, a scourge of yellow fever having driven most of its inhabitants to the pine woods. The postmaster, however, was able to assure him that his wife and son were well, and Mr. Nübling, a friendly German, whom he described as "a man of cultivation and taste, and a lover of Natural Science," gave him refreshment and a horse. In his eagerness to cover the fifteen miles to the Percy house as rapidly as possible, he tried to strike a straight course through the dark forest, but missed his way, and dawn found him on unfamiliar ground; he then learned from a negro that he was two miles beyond the place. When he arrived at last "with rent and wasted clothes, and uncut hair, and altogether looking like the wandering Jew," his wife was busily engaged in teaching her pupils. During his absence of nearly fourteen months she had prospered greatly, and she was not only ready but eager to place her earnings at her husband's disposal.

When he had finally decided to take his drawings to Europe for publication, Audubon set to work to increase his capital, and soon had pupils in French, music, and drawing, while a dancing class of sixty was organized in a neighboring town. His country lads and lassies proved rather awkward material, and he broke his bow and nearly ruined his violin in his impatience to evoke a single graceful step or motion; when, however, he consented to dance to his own music, he never failed to bring down thunders of applause. These efforts were continued for over a year, until he had realized a considerable sum. With this money in hand, supplemented by what his wife could spare, he determined to seek his fortunes in the Old World.

---

[319] "Shipping Port," as the village below the rapids or falls of the Ohio was then called, was joined to Louisville by the Louisville and Portland Canal, a channel two and one-half miles long, in 1830, two years after the city received its charter. The "Louisville" or "Portland" cement, a name now applied to the product of a considerable district, was first manufactured at Shipping Port, in 1829, for the construction of this canal.

Chapter 22

TO EUROPE AND SUCCESS

When Audubon had reached the age of forty-one, his fortunes were destined to undergo still further kaleidoscopic changes, but the patterns and hue were now of a more agreeable character. He had failed repeatedly in business ventures of various kinds; he had failed also to find either encouragement or support for his ambitious schemes of publishing his drawings in the United States. But there was still a chance for success in the Old World, and thither he was determined to go to try the hazard of fortune in either England or France. Accordingly, he left his family at St. Francisville and went to New Orleans, where he engaged passage on a cotton schooner bound for Liverpool, named the *Delos*, Captain Joseph E. Hatch. With his drawings, a few books, and a purse, if not ample, at least sufficient for his immediate needs, and fortified with numerous letters, he finally set sail on the 17th of May, 1826.

Figure 47. Early Unpublished Drawing of the "Frog-Eater," Cooper's Hawk; "Red Banks, Ky. November 29, 1810. Longeur Total 19 Pouces Poids 1lb. 6oz. Queue 12 Pennes." Published by Courtesy of Mr. Joseph Y. Jeanes.

Figure 48. Pencil Sketch from Audubon's Journal of His Voyage to England in 1826: "Shark 7 Feet Long, off Cuba, June 18th, 1826." Published by Courtesy of Miss Maria R. Audubon.

26 April 1826—

I Left My Beloved Wife Lucy Audubon and My Son John Woodhouse on Tuesday afternoon the 26th April, bound to England. remained at Doct. Pope's at St. Francisville untill Wednesday 5 o'clock P.M. in the Steam Boat Red River Capt. Kimble and having for Company Mr. A. Holl and John Holiday — reached New Orleans Thursday 27th at 12 — Visited many Vessels for my Passage and concluded to go in the Ship Delos of Kennebunk Capt. Joseph Hatch bound to Liverpool, loaded with Cotton entirely —

Figure 49. First Page of Audubon's Journal of His Voyage from New Orleans to Liverpool. Reproduced by Courtesy of the Misses Florence and Maria R. Audubon.

This voyage, like every other which the naturalist ever made, was turned to good account; the log book or journal kept on this occasion abounds in interesting observations upon the life of the sea, particularly on the fishes and birds which were encountered in the Gulf. The first page of this journal,[320] reproduced with orthographic exactness, reads as follows:

> 26 April 1826—
>
> I Left My Beloved Wife Lucy Audubon and My Son John Woodhouse on Tuesday afternoon the 26th April, bound to England, remained at Doctr Pope at St Francisville untill Wednesday 4 o'clock P. M.: in the Steam Boat Red River Cape Kimble—having for Compagnons Messrs D. Hall & John Haliday—reached New Orleans Thursday 27th at 12—Visited Many Vessels for My Passage and concluded to go in the Ship Delos of Kennebunk Cape Joseph Hatch bound to Liverpool, Loaded with Cotton entirely—
>
> The Red River Steam Boat left on her return on Sunday and I Wrote by her to Thee My Dearest Friend and forwardd Thee 2 Small Boxes of Flowering Plants—
>
> saw, spoke to & walked with Charles Briggs, much altered young man—
>
> Lived at New Orleans at G. L. Sapinot in Company with Costé—
>
> During My Stay at New Orleans, I saw my old and friendly acquaintances the familly Pamar; but the whole time spent in that City was heavy & dull—a few Gentlemen Calld to see My Drawings—I Generally Walked from Morning untill Dusk My hands behind me, paying but very partial attention, to all I saw—New Orleans to a Man who does not trade in Dollars or any other Such Stuffs is a miserable Spot ==
>
> fatigued and discovering that the Ship could not be ready for Sea for several days, I ascended the Mississipy again in the Red River and once more found Myself with my Wife and Child. I arrived at Mrs Percy at 3 o'clock in the morning, having had a Dark ride through the Magnolia Woods but the Moments spent afterwards full repaid me—I remained 2 days and 3 Nights, was a Wedding—of Miss Virginia Chisholm with Mr. D. Hall &c. I Left in Company With Lucy Mrs Percy house at Sun rise and went to Breakfast at My good [friend's, Augustin Bourgeat].

The captain and mates of the *Delos* were friendly, and whenever their vessel was becalmed, they would let down a boat so that Audubon could procure the stormy Petrel and numerous other birds which he was anxious to examine in the flesh or depict for his "Ornithology."

During his long voyage of sixty-five days our adventurous traveler was alternately elated or depressed by hopes or fears for the future, until land was at last reached on Friday, July 21, 1826. The appearance of Liverpool, said Audubon, "was agreeable, but no sooner had I entered it than the smoke became so oppressive to my lungs that I could hardly breathe." At the customs he was charged two pence on each of his drawings, "as they were water-colored,"

---

[320] Audubon's 1826 manuscript journal, which I examined through the courtesy of Miss Maria R. Audubon in 1914, was written, mostly in pencil, in a ruled blank book, of similar size and quality to that used on the Ohio River in 1820-21, and was illustrated with a number of pencil sketches, chiefly of fishes. On page 2 was a rough outline sketch of first mate Sam L. Bragdon, of Wells, Maine, reading in the booby hatch; to his kindness Audubon paid a written tribute; there was also a drawing of a "Balacuda [Barracouta] Fish, June 17, 1826;" of a "Shark, 7 ft. long; off Cuba, Jn. 18"; and of a "Dolphin; Gulph of Florida, May 28;" other sketches were of a line or "thread-winder," a Flying Fish, and outlines of the Cuban coast.

Audubon presented a sketch of the "Dolphin" to Captain Hatch, whose vessel, the *Delos*, went down on the Grand Banks of Newfoundland in the summer of 1831, but not until her crew and valuables had been transferred to another boat that stood by. (For this note I am indebted to Miss Maria R. Audubon.)

but on his American books he had to pay "four pence per pound," a circumstance in which he was possibly favored by the following letter which he had brought with him from a friend in New Orleans:

Edward Holden to George Ramsden

New Orleans, *May 26th., 1826.*

George Ramsden, Esq.
Dear Sir.

The present will be handed to you by Mr. J. J. Audubon of this city, whom most respectfully I beg to introduce to you.

The principal object of Mr. Audubon's visit to England is to make arrangements for the publication of an extensive and very valuable collection of his drawings in Natural History, chiefly if not wholly of American Birds, and he takes them with him for that purpose. Can you be of any assistance to him by letters to Manchester and London? If you can I have no doubt that my introduction of him will insure your best attention and services.—Mr. Audubon is afraid of having to pay heavy duties upon his drawings. He will describe them to you, and if in getting them entered Low at the Custom House, or if in any other respect you can further his views, I shall consider your aid as an obligation conferred upon myself. Pray introduce him particularly to Mr. Booth, who I am sure will feel great interest in being acquainted with him, were it only on account of the desire he has always expressed to be of service to the new Manchester Institution, to which Mr. Audubon's drawings would be an invaluable acquisition.

I am Dr. Sir
Yours truly,
Edward Holden.

Among the letters which Audubon carried on this occasion, but which apparently he did not deliver, was the following, addressed by a friend in New Orleans to General Lafayette:[321]

Louis P. Caire to General Lafayette

New Orleans, *15 May, 1826.*

My dear General,

Monsieur Audubon, after having spent twenty-two years in the United States, is returning to Europe in order to publish a work to which he has devoted his entire life. This distinguished ornithologist, who bears letters from the most eminent citizens of the Union, will find, I trust, the encouragement to which his talents and his perseverance so fully entitle him, and however flattering may be the recommendations which his friends are eager to give him, these are yet, my dear General, beneath his merits. I have presumed to assure him of your patronage, and in introducing him to you I am convinced that it will be agreeable to you both.

[321] Addressed "*General Lafayette*, Paris ou Lagrange."
Translated from the French original, kindly sent to me by Mr. Ruthven Deane.

Adieu my General: give my kind regards to all your family, and permit me to embrace you as I love you.

Louis P. Caire.

Before Audubon left New Orleans, an old acquaintance, Mr. Vincent Nolte[322] of that city, had also furnished him with credentials, in which it was stated that the naturalist was carrying with him four hundred original drawings, and that his object was "to find a purchaser or a publisher." "He has a crowd of letters," continued Nolte, "from Mr. Clay, De Witt Clinton, and others for England, which will do much for him; but your introduction to Mr. Roscoe and others will do more." This judgment was sound, but the most valuable letter which Audubon carried proved to be that of Nolte himself addressed to Richard Rathbone, Esq., of Liverpool, for it brought him into immediate friendly relations with an influential family of merchants which also included William Rathbone, a brother, as well as their father, William Rathbone, Senior, whose interest in birds had made him in his younger days an amateur collector and student. Seldom has the *rôle* of Mæcenas been played more effectively and with less ostentation than by those intelligent men of affairs, to whom Audubon, with his fine enthusiasm and bold literary plans, seemed to embody all the romance of the New World. They stood sponsor for his work and worth, and did all in their power to make their new discovery known. At the home of the senior Rathbone, called "Greenbank," three miles out of Liverpool, the naturalist was warmly welcomed, and his excellent hostess, Mrs. William Rathbone, the "Queen bee," as he called her, received from him lessons in drawing and became his first subscriber.

At this period Audubon often complained of shyness felt in meeting strangers, but his "observatory nerves," as he said, never gave way. He studied his English friends as closely as he had the birds of America, and the results of his shrewd observations were often turned to practical account. That he was as diffident as he declared himself to be may be doubted, for he seems to have met nearly everyone of prominence wherever he went, and a list of his acquaintance at the end of his sojourn abroad would read much like a "Blue Book" of the British Isles.

At Liverpool Audubon received much assistance also from Edward Roscoe, botanist and writer, Dr. Thomas S. Traill[323] and Adam Hodgson, who introduced him to Lord Stanley. When he came to write his *Ornithological Biography*, these early friends were all publicly called by name, and we thus had (though, as it afterwards appeared, in name only) the "Rathbone Warbler,"[324] "Stanley Hawk," "Children's Warbler," "Cuvier's Regulus," "Roscoe's

[322] For an account of Audubon's meeting with Nolte see Chapter 18.

[323] Dr. Thomas Stuart Traill, after whom one of our common flycatchers was named, was a founder of the Royal Institution at Liverpool, and later a professor of medical jurisprudence at Edinburgh. When the keepership of the Department of Natural History in the British Museum became vacant through the resignation of Dr. Leech in 1822, Dr. Traill supported William Swainson for the position; when George J. Children received the appointment, he was disinclined to accept defeat, and entered upon a crusade against the Museum's trustees in a series of anonymous articles contributed to the *Edinburgh* and *Westminster* Reviews. Traill's exposure of the neglect which the natural-history collections had suffered in the custody of the British Museum paved the way to a separate Department of Zoology, which in the able hands of John E. Gray, and later in those of Sir Richard Owen, led to the present great Museum of Natural History at South Kensington.

[324] In dedicating the *Sylvia rathbonia* Audubon said: "Were I at liberty here to express the gratitude which swells my heart, when the remembrance of all the unmerited kindness and unlooked-for friendship which I have received from the Rathbones of Liverpool comes to my mind, I might produce a volume of thanks. But I must content myself with informing you, that the small tribute of gratitude which it is alone in my power to pay, I now joyfully accord, by naming after them one of those birds, to the study of which all my efforts have been

Yellow-throat," "Selby's Flycatcher," and still possess "Bewick's Wren," "Traill's Flycatcher," "Henslow's Bunting,"[325] "MacGillivray's Finch," and "Harlan's Hawk," to cite a few instances of this form of acknowledgment.

Within barely a week after landing at Liverpool a total stranger, Audubon was invited to show his drawings at the Royal Institution. The exhibition, which lasted a month, was a surprising success; 413 persons, as he recorded, were admitted on the second day, and it netted him one hundred pounds although no charge for admission was made during the first week.

Everyone, said the naturalist, was surprised at his appearance, for he wore his hair long, dressed in unfashionable clothes, rose early, worked late, and was abstemious in food and drink. Shortly after his arrival, his sister-in-law, Mrs. Alexander Gordon, urged him to have his hair cut and to buy a fashionable coat, but he could not then bear to sacrifice his ambrosial locks, which continued to wave over his shoulders until the following March. If we can accept Sir Walter Besant's characterization of the period, the "long-haired Achæan" was no stranger to the streets of London as late as 1837: "brave is the exhibition of flowing locks; they flow over the ears and over the coat-collars; you can smell the bear's grease across the street; and if these amaranthine locks were to be raised you would see the shiny coating of bear's grease upon the velvet collar below."

Audubon had not been in England three weeks before he resumed his drawing and painting habits, at first in order to repay his friends for their kindness, and later as a means of support; at times he would devote every spare moment to this work, and he was then able to paint fourteen hours at a stretch without fatigue. On October 2 he recorded that he had made in less than twenty minutes a diminutive sketch of the Turkey Cock from his large twenty-three hour picture. This was for Mrs. William Rathbone, Senior, who later presented it to him in the form of a handsome gold-mounted seal, inscribed with his favorite motto, "America, my country."[326] The facility which Audubon displayed in producing his pictures of animal life—American wild turkeys, trapped otters, fighting cats, English game pieces, and the like, in a style both novel and individual, added much to his immediate popularity in England, as it later did to his purse. His painting devices are thus referred to in a journal entry for January, 1827:

> No one, I think, paints in my method; I, who have never studied but by piece-meal, form my pictures according to my ways of study. For instance, I am now working on a Fox; I take one neatly killed, put him up with wires and when satisfied with the truth of the position, I take my palette and work as rapidly as possible; the same with my birds; if practicable I finish the bird at one sitting,—often, it is true, of fourteen hours,—so that I think they are correct, both in detail and composition.

---

directed. I trust that future naturalists, regardful of the feelings which have guided me in naming this species, will continue to it the name of the *Rathbone Warbler*."

[325] Named after John Stevens Henslow, Professor of Botany in the University of Cambridge, whom Audubon had met in 1828, when Charles Darwin was still his pupil.

[326] This seal, the design of which has since been adapted for a bookplate, was long in use, and though at one time lost, is still in possession of the family. A copy of the large original, which was to serve as his first plate, was presented to the Royal Institution of Liverpool as an acknowledgment of its hospitality, for it had refused remuneration in any other form.

When he was painting pheasants and needed a white one as "a keystone of light" to his picture, a nobleman sent word that he would be given "leave to see the pictures" in his hall, but this Audubon characteristically refused, being determined to pay no such visits without invitation.

On the 10th of September, 1826, Audubon left Liverpool, in a hopeful mood, for Manchester, with the intention of visiting the chief cities of England and Scotland. He was fortified with a bundle of letters to a long list of distinguished people, including Baron von Humboldt, General Lafayette, Sir Walter Scott, Sir Humphry Davy and Sir Thomas Lawrence. His first step proved a disappointment, and when he finally left the City of Spindles six weeks later, he found himself poorer than when he had entered it. At Manchester, however, he added to his list of interested friends and possible patrons, and acting upon their suggestion, opened a subscription book for the publication of his long meditated work, to be called *The Birds of America*. The Rathbones, as well as other friends whose advice he esteemed, tried to dissuade him from the plan of publishing his drawings in their full size, which was that of life, on account of the great expense involved and the enormous bulk such a work would assume; but he could not bring himself to give up the idea, in which he received the support of the London bookseller, Mr. Bohn, who, after seeing Audubon's drawings reversed his opinion, saying that they must be brought out in their full size, and that they would certainly pay.

After coming to England Audubon often thought of the shifting scenes and strange contrasts his life had brought. One day he felt the pinch of poverty, but on the next fared sumptuously at the tables of the rich; now a rambler in the wilds of America, glad to accept the hospitality of the humblest prairie squatter, now the guest of some metropolitan aristocrat. "The squatter," he said, when writing in England, "is rough, true, and hospitable; my friends here polished, true, and generous. Both give freely, and he who during the tough storms of life can be in such spots may well say that he has tasted happiness."

While at Manchester Audubon was driven to the town of Bakewell, "the spot," he wrote in deference to his wife, "which has been honored with thy ancestor's name." Shortly after, on October 23, he started by stage for Edinburgh, and the distance of 212 miles was covered in three days; the fare was £5 5s. 5d., which he regarded as exorbitant, but he complained not so much of the charge as of the beggarly manner of the drivers, who never hesitated to open the door of their coach and ask for a shilling at the slightest provocation.

At Edinburgh Audubon was welcomed so warmly that he began to feel that ultimate success was at last within his reach. Professor Robert Jameson of the University did much to make his work known, and invited him to coöperate in an enterprise upon which he was then engaged;[327] this was pronounced by Dr. Knox of the Medical School to be a "job book," but whatever its merits may have been, Audubon decided after due reflection to stand on his own feet.

Not long after reaching the Scottish capital, Audubon made the acquaintance of Mr. W. Home Lizars, styled "a Mr. Lizard" by a snapshot biographer of a later day, a well known, expert engraver and painter, who engaged in various publishing enterprises. When Audubon had held up a few of his drawings for his inspection, Lizars rose, exclaiming: "My God! I never saw anything like this before." The picture of the Mockingbirds attacked by a rattlesnake particularly struck his fancy, but when he came to the drawing of the Great-footed

---

[327] See Vol. I.

Hawks, "with bloody rags at their beaks' ends, and cruel delight in their daring eyes," Lizars declared that he would both engrave and publish it. "Mr. Audubon," said he, "the people here don't know who you are at all, but depend upon it, they *shall* know." Lizars eventually agreed to engrave and bring out the first specimen number of *The Birds of America*, and about the 10th of November made a beginning with the first plate. On November 28, 1826, he handed Audubon a first proof of the Wild Turkey Cock, a subject chosen to justify the great size of the work, which was to be in double elephant folio, and which in point of size is perhaps to this day the largest extended publication in existence.[328] This and the second plate, which represented the Yellow-billed Cuckoo[329] in the act of seizing a tiger swallowtail butterfly on a branch of the paw-paw tree, were finished by December 10; the first number of five plates was ready some weeks later. Lizars engraved at Edinburgh the first ten of Audubon's plates, but most of these were subsequently retouched, colored and reissued by his successor in London, as will presently appear.

Figure 50. Plate I. *Wild Turkey* Meleagris Gallopavo. Linn. *Male.* American Cane. Miegia Macrosperma. Drawn from Nature by J. J. Audubon F.R.S.F.L.S. Engraved by W. H. Lizars Edinr. Retouched by R. Havell Junr.

---

[328] The plates as issued, untrimmed, measured 39½ by 29½ inches.

[329] Incidentally it may be noticed that the "tiger swallowtail" in this plate was possibly added for effect, for few of our birds, which habitually hunt moths, ever prey upon butterflies. I have seen the cabbage butterfly and a few of the smaller kinds brought to the nests of the Chebec and Wood Pewee but never a "monarch" or "papilio"; yet some affirm that the Kingbird will attack the "monarch."

When Audubon's pictures were exhibited at the Royal Institution of Edinburgh, their success was immediate, and like the appearance of a new Waverley novel, they became the talk of the town; the American woodsman had provided a new thrill for the leaders of fashion, as well as for the literati and the scientific men. The "noblest Roman of them all," Sir Walter Scott, refused to attend, but after having met the naturalist he wrote this in his journal: "I wish I had gone to see his drawings; but I had heard so much about them that I resolved not to see them—'a crazy way of mine, your honor.'"

Philarète-Chasles, a well-known French critic of the period, has left the following record[330] of the effect which this exhibition made on his impressionable mind:

> We have admired in the rooms of the Royal Society of Edinburgh the public exhibition of [Audubon's] original watercolor drawings. A magic power transported us into the forests which for so many years this man of genius has trod. Learned and ignorant alike were astonished at the spectacle, which we will not attempt to reproduce.
>
> Imagine a landscape wholly American, trees, flowers, grass, even the tints of the sky and the waters, quickened with a life that is real, peculiar, trans-Atlantic. On twigs, branches, bits of shore, copied by the brush with the strictest fidelity, sport the feathered races of the New World, in the size of life, each in its particular attitude, its individuality and peculiarities. Their plumages sparkle with nature's own tints; you see them in motion or at rest, in their plays and their combats, in their anger fits and their caresses, singing, running, asleep, just awakened, beating the air, skimming the waves, or rending one another in their battles. It is a real and palpable vision of the New World, with its atmosphere, its imposing vegetation, and its tribes which know not the yoke of man. The sun shines athwart the clearing in the woods; the swan floats suspended between a cloudless sky and a glittering wave; strange and majestic figures keep pace with the sun, which gleams from the mica sown broadcast on the shores of the Atlantic; and this realization of an entire hemisphere, this picture of a nature so lusty and strong, is due to the brush of a single man; such an unheard of triumph of patience and genius!—the resultant rather of a thousand triumphs won in the face of innumerable obstacles!"

Another French writer[331] remarked that Audubon produced the same sensation among the savants of England that Franklin had made at the close of the eighteenth century among the politicians of the Old World; his works, he added, should be translated into his native tongue, and produced in a form which would enable them to reach the library of every naturalist in France.

One after another the scientific, literary, and arts societies of the modern Athens elected Audubon to honorary membership; Combe, the phrenologist and author of *The Constitution of Man*, examined the naturalist's head and modeled it in plaster, for of course it proved to be a perfect exemplification of his system; Syme, the artist, did his portrait for Lizars to engrave. Meanwhile the press was giving such flattering accounts of the man and his work that Audubon confessed that he was quite ashamed to walk the street. At the annual banquet of the Royal Institution, held at the Waterloo Hotel and presided over by Lord Elgin, Audubon was

---

[330] Translated from *Études sur la Littérature et les Mœurs des Anglo-Americains au XIXe siècle*, "Audubon," pp. 66-106 (Paris, 1851). Philarète-Chasles, who wrote chiefly on American, English and European authors and books, has seventy volumes credited to him in the National Library at Paris.

[331] P. A. Cap, in *L'Illustration* for 1851. Cap's hint was taken by Eugène Bazin, who translated copious selections from the *Ornithological Biography*, which were published in two volumes in Paris in 1857.

toasted, and it required all his resolution to rise and, for the first time in his life, address a large assembly; this, however, he managed to do in the following words: "Gentlemen; my command of words in which to reply to your kindness is almost as limited as that of the birds hanging on the walls of your Institution. I am truly obliged for your favors. Permit me to say; may God bless you all, and may this society prosper." On the 10th of December he wrote: "My situation in Edinburgh borders on the miraculous," and he felt that his reception in that city was a good augury for the future. But the life that he was compelled to lead was extremely fatiguing, and he often longed to return to his family and to his favorite magnolia woods in Louisiana. "I go to dine," he wrote, "at six, seven, or even eight o'clock in the evening, and it is often one or two when the party breaks up; then painting all day, with my correspondence, which increases daily, makes my head feel like an immense hornet's nest, and my body wearied beyond all calculation; yet it has to be done; those who have my best interests at heart tell me I must *not refuse* a single invitation." But notwithstanding the tax which society always levies upon the lion's strength, he wrote almost daily in his journal or diary,[332] and its pages, from which we have been quoting, became a mirror of all that he saw, heard, or did. Audubon was generous with his time, as with everything else, and would never hesitate to lay aside his own work for the sake of a friend who was eager to acquire his method of drawing. But when his entertainment commenced with an invitation to breakfast, he began to be alarmed at the large share of his working hours which had to be surrendered to his friends. "I seem, in a measure," he said, "to have gone back to my early days of society and fine dressing, silk stockings and pumps, and all the finery with which I made a popinjay of myself in my youth.... It is Mr. Audubon here, and Mr. Audubon there, and I can only hope they will not make a conceited fool of Mr. Audubon at last."

In response to urgent appeals he began at this time to contribute to the scientific journals of the Scottish capital, a step which only served to remind him that the rose was more prolific in thorns than flowers. Dr. Brewster, however, in his *Journal of Science*, and John Wilson in *Blackwoods*, sang pæans in his praise, and there is no doubt that "Christopher North," so like and yet so unlike the American woodsman, did much to smooth his path in his own country as well as in Europe. Though keenly feeling the need of literary advice in those early contributions, Audubon was quite shocked at the alterations which Dr. Brewster had made in one of these articles, for though the editor had "greatly improved the style," he had quite "destroyed the matter."

On December 21, 1826, Audubon wrote to Thomas Sully that he would send him a copy of the first number of his *Birds*, with the request that he forward it in his name "to that Institution which thought me unworthy to be a member.... There is no malice in my heart," he continued, "and I wish no return or acknowledgment from them. I am now *determined* never to be a member of that Philadelphia Society." Let it be noted, however, that Audubon was elected to membership in the American Philosophical Society, when their recognition could no longer be withheld and when mutual animosities had died down. Three days later he recorded that all of his drawings had been taken from the walls of the Royal Institution, where they had been on exhibition a month, and that he was intending to present to the Society his

[332] See Maria R. Audubon, *Audubon and his Journals*.

large canvas of the Wild Turkeys, for which Galley, the picture dealer, had offered him a hundred guineas on the previous day.[333]

Among Audubon's early patrons were Lord and Lady Morton, and more than once he was invited to visit them in their beautiful country seat of "Dalmahoy," where a large, square, half-Gothic building, crowned with turrets and adorned with all the signs of heraldry, overlooked a beautiful landscape to Edinburgh, marked by its famous castle, seen in miniature on the horizon, eight miles away. Being somewhat apprehensive of meeting the former Chamberlain to the late Queen Charlotte, Audubon had imagined the Earl to be "a man of great physical strength and size"; instead, however, he saw

> a small, slender man, tottering on his feet, weaker than a newly hatched partridge; he welcomed me with tears in his eyes, held one of my hands, and attempted speaking, which was difficult to him, the Countess meanwhile rubbing his other hand. I saw at a glance the situation, and begged he would be seated ... and I took a seat on a sofa that I thought would swallow me up, so much down swelled around me. It was a vast room, at least sixty feet long, and wide in proportion, let me say thirty feet, all hung with immense paintings on a rich purple ground; all was purple about me. The large tables were covered with books, instruments, drawing apparatus, a telescope, with hundreds of ornaments.

After luncheon Audubon's "Book of Nature" was produced, and his drawings spread out and admired. Next day the Countess, who was "a woman of superior intellect and conversation," was given "a most unnecessary lesson" in drawing, for, said the naturalist, "she drew much better than I did; but I taught her to rub with cork, and prepare for water-color." Before he left the Countess wrote her name in his subscription book, and arranged that he should return and resume his instruction.

One of Audubon's early friends at Edinburgh was Captain Basil Hall,[334] traveler and writer, who was then about to start on a journey through the United States; he told the naturalist that he was a midshipman on board the *Leander* "when Pierce was killed off New York," at the time of Audubon's return with Rozier to America in 1806, when Captain Sammis, upon seeing the British frigate, "wore around Long Island Sound, and reached New York by Hell Gate." It was at Captain Hall's home that Audubon met Francis Jeffrey. The indomitable critic and reviewer was described as "a small (not to say tiny) man," who entered the room "with a woman under one arm, and a hat under the other." "His looks were shrewd," said the naturalist, his eyes "almost cunning" and though he talked much, he appeared unsympathetic. Their meeting was productive of no friendly feelings on either side.

Three months after reaching Edinburgh, the long awaited opportunity of meeting the greatest literary figure of the day came to Audubon unexpectedly, for he did not wish to be introduced in a crowd. Under date of January 22, 1827, he wrote that Captain Hall came to his

---

[333] Audubon's copy of this oil painting remained in the possession of his family until a few years ago, when it was sold for a much greater amount. It now adorns the beautiful ornithological museum of Mr. John E. Thayer, at South Lancaster, Massachusetts; it represents a cock and hen turkey in life size, adapted from the subjects of his two most famous plates, and is in an admirable state of preservation. Mr. Thayer's collection also embraces Audubon's large canvas of the Black Cocks, from the Edward Harris estate, a charming study of the Hen Turkey, with landscape setting, and, also in oils, several smaller panels of Flickers and Passenger Pigeons, which, if not the work of the naturalist, are copies after his originals, and possibly made by Joseph B. Kidd. (See Vol. I; and for a notice of Mr. Thayer's other Audubonian drawings, Vol. II , and Appendix II.)

[334] Basil Hall (1788-1844), noted for his travels in China, Korea, and on the coasts of Chili, Peru and Mexico, visited the United States in 1827-28; his *Travels in North America* appeared in 1829.

rooms and said: "Put on your coat, and come with me to Sir Walter Scott: he wishes to see you now." "In a moment," said Audubon, "I was ready.... My heart trembled; I longed for the meeting, yet wished it over." When they were ushered into Sir Walter's study, the great Scot came forward, and warmly pressing the hand of his visitor, said he was glad to have the honor of meeting him. Audubon's record of the meeting continues:

> His long, loose, silvery locks struck me; he looked like Franklin at his best. He also reminded me of Benjamin West; he had the great benevolence of William Roscoe about him, and a kindness most prepossessing. I could not forbear looking at him; my eyes feasted on his countenance. I watched his movements as I would those of a celestial being; his long, heavy, white eyebrows struck me forcibly. His little room was tidy, though it partook a good deal of the character of a laboratory. He was wrapped in a quilted morning-gown of light purple silk; he had been at work writing on the "Life of Napoleon." He writes close lines, rather curved as they go from left to right, and puts an immense deal on very little paper.... I talked little, but, believe me, I listened and observed.

Two days later Audubon paid Scott a second visit, this time with his portfolio, but little was recorded of this interview other than that it was more agreeable than the first, and that he greatly admired the accomplished Miss Scott, to whom he later sent as a gift the first number of his plates. Audubon's drawings were exhibited at a meeting of the Royal Society over which Sir Walter presided, and Scott was also in attendance at the Royal Institution when Audubon's large painting of the Black Cocks was shown. "We talked much" on this occasion, said the naturalist, "and I would have gladly joined him in a glass of wine, but my foolish habits prevented me." This restriction on wine was soon removed, as was that on whisky, whether of the Scotch or Kentucky brand, and during his later life in America Audubon was never a teetotaler by any means. While at the Exhibition Sir Walter pointed to Landseer's picture of the dying stag, saying, "many such scenes, Mr. Audubon, have I witnessed in my younger days." Audubon was doubtless too polite to express an opinion of that popular artist, though of that very picture he had written in his journal three days before that there was no nature in it, and that he considered it a farce; "the stag," he said, "had his tongue out, and his mouth shut! The principal dog, a greyhound, held the deer by one ear, just as if a loving friend; the young hunter had laced the deer by one horn very prettily, and in the attitude of a ballet-dancer was about to cast the noose over the head of the animal."

Scott and Audubon were kindred spirits in their love of sport, of wild and untameable nature, as well as of man in his Homeric relation to it. Shortly after their first interview the great Scotsman wrote this handsome tribute in his journal:

> January 22 [1827].—A visit from Basil Hall with Mr. Audubon, the ornithologist, who has followed that pursuit by many a long wandering in the American forests. He is an American by naturalization, a Frenchman by birth; but less of a Frenchman than I have ever seen—no dash, or glimmer, or shine about him, but great simplicity of manners and behaviour; slight in person, and plainly dressed; wears long hair, which time has not yet tinged; his countenance acute, handsome and interesting, but still simplicity is the predominant characteristic.

Of the later visit of which we just spoke we find this account:

> January 24.—Visit from Mr. Audubon, who brings some of his birds. The drawings are of the first order—the attitudes of the birds of the most animated character, and the situations appropriate; one of a snake attacking a bird's nest, while the birds (the parents) peck at the reptile's eyes—they usually, in the long-run, destroy him, says the naturalist. The feathers of these gay little sylphs, most of them from the Southern States, are most brilliant, and are represented with what, were it [not] connected with so much spirit in the attitude, I would call a laborious degree of execution. This extreme correctness is of the utmost consequence to the naturalist, [but] as I think (having no knowledge of *vertu*), rather gives a stiffness to the drawings. This sojourner in the desert has been in the woods for months together. He preferred associating with the Indians to the company of the Back Settlers; very justly, I daresay, for a civilized man of the lower order—that is, the dregs of civilization—when thrust back on the savage state becomes worse than a savage....
>
> The Indians, he says, are dying fast; they seem to pine and die whenever the white population approaches them. The Shawanese, who amounted, Mr. Audubon says, to some thousands within his memory, are almost extinct, and so are various other tribes. Mr. Audubon could never hear any tradition about the mammoth, though he made anxious inquiries. He gives no countenance to the idea that the red Indians were ever a more civilized people than at this day, or that a more civilized people had preceded them in North America. He refers the bricks, etc., occasionally found, and appealed to in support of this opinion, to the earlier settlers,—or, where kettles and other utensils may have been found, to the early trade between the Indians and the Spaniards.

Audubon was anxious to receive a written recommendation from the great "Wizard of the North" touching the merits of his work, the publication of which had just begun, but Sir Walter Scott sensibly demurred, on the ground that his knowledge of natural history was insufficient to qualify him to pass expert judgment. "But," he added, "I can easily and truly say, that what I have had the pleasure of seeing, touching your talents and manners, corresponds with all I have heard in your favor; and I am a sincere believer in the extent of your scientific attainments."

While Audubon was playing the *rôle* of society's pet lion at Edinburgh in the winter of 1827, he was painting to meet the expense of engraving his first plates, and writing at odd times of the day or night. On February 20 he recorded that his paper on the "Habits of the Wild Pigeon of America" was begun on the previous Wednesday, and finished at half past three in the morning; so completely, said he, was he transported to the woods of America and to the pigeons, that his ears "were as if really filled with the noise of their wings"; yet he added that were it not for the facts it contained, he would not give a cent for it, "nor anybody else, I dare say." Four days later, at the Wernerian Society, he read his paper on the rattlesnake, but the torrent of abuse which soon rewarded his efforts in this direction finally led him to reserve all literary efforts for a future and more propitious time.[335]

A large painting begun in January of this year, called "Pheasants attacked by a Fox," was probably a variant of the "Pheasants attacked by a Dog", the original of which is now in the American Museum of Natural History, New York City. This canvas, which was exhibited by

[335] See Chapter 28.

the Scottish Society of Artists in February, 1827, measured nine by six feet, and was the largest piece he had ever attempted. "Sometimes I like the picture," he said, and "then a heat rises in my face and I think it a miserable daub." "As to the birds," he added, "so far as *they* are concerned I am quite satisfied, but the ground, the foliage, the sky, the distance, are dreadful."[336]

In the spring of 1827 Audubon enjoyed the novel sensation of going to church in a sedan chair, and of hearing Sidney Smith preach. "He pleased me at times," he said, "by painting my foibles with care, and again I felt the color come to my cheeks as he portrayed my sins." Later there was an opportunity to meet the famous preacher with his fair daughter, and to show them his drawings of American birds.

The following letter[337] was sent at this time to his wife in America:

Audubon to his Wife

Edinburgh March 12th, 1827.

My Dearest Friend

I am now proud that I can announce thee the result of the last meeting of the Royal Society of Edinburgh. I was unanimously elected a Foreign Member of that Institution on the 5t.h. Instant and am at last an F. R. S..—Wilt thou not think it wonderful; to me it is like a dream, and quite as much so when I see the particular attentions paid me by all ranks of the best Society. On the 6t.h. I received the official Letter from the Secretary with the seal of the Society and the arms of Scotland—this along with my other diplomas and Letters, I assure thee enable me to be respected and well received in any portion of the Civilized World. Sir Walter Scott has also been so kind as to give me a Letter that I may exhibit wherever I may go==I have Two Letters from him very kind==all this I think will afford thee great Pleasure.

I am now preparing to leave Edinburgh and will do so in a few days, I am now anxious to visit London as soon as I possibly can, and yet want to spend a few days at New Castle, York, Liverpool, Dublin, then back again to England, go by Cambridge and Oxford.—If I meet the success that I expect in that Tour it is very probable that soon after my reaching London, I will write for thee to Come, and when I do so, my Lucy may come without the least Hesitation for I will then be ready to receive her!

Since my last of the 22d of February, I have received thine of the 31t of December, 3d of January and 8th of Do. this last mostly John's, I am particularly glad that thou hast left the Beech Woods, yet thou might as well have given me at once thy good reasons for doing so. I hope that at this Instant that I am writing, thou art snug and comfortably settled afresh.

The Trees and Segments have not yet arrived, but I hope to hear soon that they have—I have not a word about the Seeds reaching yet. do my Love always say by what vessel any thing comes, as John as concluded to take Lessons of Music I have no wish to sell my Gun but wish to give it him as his ow[n] in Fee Simple, as soon as he deserves it from thy own Hands. May God bless him!—if all continues well with me Victor and him

[336] Maria R. Audubon, *op. cit.*, vol. i, p. 204.

[337] Which I owe to the kindness of his granddaughter, Miss Maria R. Audubon; it is superscribed "Mrs. Audubon, St. Francisville, Bayou Sarah, Louisville, p Wm Penn;" it reached New Orleans on June 13, and is endorsed as answered on June 23.

may rise to eminence and therefore try Johny's Spunk. do beg or make him draw all kinds of Limbs of Trees or Flowers for me and whenever he kills a bird of any kind tell him to measure the Guts particularly and make a regular list of the names of the Birds, length and thickness of those Guts and their contents==[338]

I wrote a long letter to each Victor and N. Berthoud on the 27 February, but not a word from either of them as yet reached me. I was quite shocked to see thy last letter of the 8th of January without the print of thy new Seals, I am quite frightened at thy watch not having reachd thee, yet I hope every new Letter will bring me better tidings. I now collecting Letters from all my Friends here and will have God knows enough of them. I only hope I may soon be in a regular way of making a comfortable living for ourselves all:

All the papers and books I send thee mention my name. My work is lookd upon as unrivalled in any Country, I will soon know how it will pay.—I can only add that I will write to thee from all the places I visit==Let Victor have a copy of this==Collect all kinds of Curiosities whatever==try to send or bring with thee but send first if Possible Live Birds of hardy kinds such as Blue Jays by THEMSELVES. Red Birds Do. red wingd Starling Do, Partridges &c &c.—present my humble respects to Mr & Mrs Johnsons and remembrances to good Friend bourgeat—try to send me an account of the growing of Cotton from A to Z, written by an able Planter—I wish thee to make regular memorandums thyself respecting all about Habits & Localities &c &c.==thou wilst scarce believe that this day there [are] in many places 16 feet of snow. the weather has been tremendous—yet with all this no Invitation is ever laid aside and the other evening I went to Diner in a Hackny Coach drawn by 4 Horses, and to church on Sunday last in a Sedan chair to hear the famous Sidney Smith, curious diferences of manners here I assure thee.

I have seen and know personally all the great men of Scotland and many of England.

What a curious interesting book a Biographer—well acquainted with my Life could write, it is still more wonderfull and extraordinary than that of my Father!

Fear not my connecting myself in any way with Charles M. he is a mere worm on the hearth, and since he has abandoned his Grand Flora is out of my books—it has perhaps been an error in our Lives that thou didst not come with me. So much indeed do I now think so that I have advised Capn Hall to take his Lady and child with him. be sure to pave the way for them to Judge Mathews and N. Berthoud to whom I have given him letters to.—I send thee his Travels, read his interview with Napoleon; I write my Journal every day, it seems that that portion of it forwardd thee long ago as never reachd thee as thou dost not mention it. I am sorry for all these little misfortunes and can hardly a/c for them. I have not heard from H. Clay but will refresh his memory, I hope at the same time to receive a Letter from the President==I hope this day the last beautiful broach I sent thee as a new Years gift is shining on thy bosom, as I have witnessed the brightness of thy own sweet Eyes, oh my Lucy what would I give now in my possession for a kiss on thy Lips and——God for ever bless thee thine Husband and Friend for ever—

John J. Audubon

---

[338] John Woodhouse Audubon at this time was in his fifteenth year, and this injunction regarding the internal anatomy of birds, to which ornithologists had hitherto paid but little attention, was given three years before his father made the acquaintance of MacGillivray. (See Chapter 30.)

F.R.S.E.	Fellow	Royal Society Edinburgh—
F.A.S.—	Fellow	Royal Society antiquarians—
M.W.S.N.H.—	Member	Wernerian Society of Natural History
M.S.A.—	Member	Society of Arts of Scotland—
M.P.L.S.—	Member	Philosophical & Literary Society Liverpool
M.L.N.Y.—	Member	Lyceum of New York.

My Dear John—

I am very thankfull to you for your Letters continue to write from time to time, draw, and study music closely, there is time for all things—I give you my Gun with all my Heart best wishes, but earn it at your Dear Mamma's will—God bless You—

Your Father and Friend—

John J. Audubon

At Edinburgh Audubon met a young landscape painter, Joseph B. Kidd, and the two worked together for some time, Kidd receiving instruction in animal painting and Audubon hints on the treatment of his landscapes, which had always been a source of trouble to him. Kidd was Audubon's Edinburgh agent for a time, and later entered upon the ambitious project of reproducing all of his birds in oils, as will be noticed later.[339]

On March 17, 1827, when the second number of his *Birds* was in preparation, Audubon boldly issued his "Prospectus," contrary to the advice of some of his friends, who could see only egregious folly in such an undertaking and regarded it as foredoomed to failure. As everybody knows, it is easier to say things than to do them, but all these friendly critics sang a different tune later on, when they had seen more of the indomitable will and self-reliance of the man, who was to carry steadily forward to a successful issue a work which was in press nearly twelve years and which cost over $100,000 to produce. In Audubon's original prospectus of *The Birds of America* the specifications as to the form, size, and cost of the work, which had been determined for some months, underwent little change in subsequent editions of this printed statement.[340]

Audubon left Edinburgh for London on April 5, 1827, with locks shorn but energy unabated. He followed a roundabout course, visiting Belford, "Mitford Castle," Newcastle-upon-Tyne, York, Leeds, Liverpool, and Shrewsbury, at every point extending his acquaintance, showing his drawings to many, and adding appreciably to his growing list of subscribers. Several days were spent in hunting and drawing birds with the Selbys, at their

339 See Chapter 25.

340 The work, as originally announced, was to appear in parts of 5 plates each, at 2 guineas a part, and in order to distribute the expense to purchasers it was expected to issue but 5 parts a year. The plates, to be engraved on copper, were of double elephant folio size, and printed on paper of the finest quality, all the birds and flowers to be life-size, and to be carefully colored by hand, after the originals; any subscriber was at liberty to take a part or the whole. It was stated in the prospectus of 1829, when 10 parts had been published: "There are 400 Drawings, and it is proposed that they shall comprise Three Volumes, each containing 133 Plates, to which an Index will be given at the end of each, to be bound up with the volume. It would be advisable for the subscriber to procure a Portfolio, to keep the Numbers till a volume is completed." To avoid the expense entailed by copyright regulations in England, indices and all other letterpress were eventually omitted; the number of parts was extended to 87, or 435 plates, and the number of volumes to 4, a necessity imposed by the discovery of many new birds, even after the omission of the figures of the eggs, which Audubon had reserved for the close, and the undue crowding of many of his final plates. The "Prospectus" issued with the first volume of the text in 1831 contained a list of the first 100 plates, together with extracts of reviews by Cuvier and Swainson, and a list of subscribers to the number of 180.

beautiful country place called "Twizel House," at Belford, in Northumberland, where he was soon made to feel as much at home as with his older Liverpool friends, the Rathbones, at "Green Bank."

P. J. Selby, after whom Audubon named a Flycatcher which appeared in his second number, was an amateur artist and ornithologist, and at that time was engaged upon an extensive publication to which Audubon was invited to contribute, a single volume of plates and text having then been published.[341]

At Newcastle, where Audubon spent a week, he saw much of its grand old man, Thomas Bewick, "the first wood cutter in the world," and conceived a deep regard for him, which he afterwards expressed in one of his "Episodes." As they parted, this great son of nature held him closely by the hand, and for the third time repeated, "God preserve you!" "I looked at him in such a manner," said Audubon, "that I am sure he understood I could not speak."

As he proceeded southward, his subscription list augmented apace, Manchester alone giving him eighteen new names, and he began to feel more sanguine of success, if, he added, "I continue to be honest, industrious, and consistent."

---

[341] *Illustrations of British Ornithology*, by Prideaux John Selby. The British Museum copy of this work is in two large folio volumes (measuring about 25½ by 20½ inches), and was issued originally in numbers which appeared at irregular intervals. Vol. I, plates i-iv (of bills, heads, and feet), i-c (of land birds); most of the plates are by Selby, and many were etched by him and autographed, 1819-1821; plates xiv, xvi, and xx are by Captain R. Mitford, whose home, "Mitford Castle," near Morpeth, Northumberland, was visited by Audubon in April, 1827; published at Edinburgh by Archibald Constable & Co., and by Hurst, Robinson & Co., London, 1825(?)-1827. Volume II, plates i-ciii; printed for the Proprietor & published by W. H. Lizars, Longman, Rees, Orme, Brown, Green & Longman, London; and W. Curry, Junr. & Co., Dublin, MDCCCXXXIV. Quaritch, in offering a copy in 1887, at £55, stated that there were 383 figures, in 221 colored plates, and that the published price was £105. Newton (*Dictionary of Birds*, p. 27) says that the first series of these "Illustrations" was published in coöperation with Sir William Jardine, in 3 volumes of 150 plates, in 1827-1835, after which a second series was started by them, and completed in a single volume of 53 plates, issued in 1843. This was the "job book" mentioned earlier in this chapter, but neither Jardine's nor Jameson's name is mentioned in the volumes which I have examined.

In a letter to Audubon, dated "Sept. 13h 1830 Twizel [1?] House," and postmarked "Belford," Selby said: "I expect to bring my own work to a conclusion during the course of this winter having only the plates of another Number to finish. I am happy to add that the Work is doing well & is more than paying itself. The second Vol: of letter press will appear with the last No."

Two volumes of text were published in 1825 and 1833 respectively; the first, after readjustment to fit the "quinarian doctrine," to which Selby was a temporary convert, was issued in a second edition at London, in 1841; the second volume bore the imprint of Lizars, who soon after began to work for Audubon.

Selby's plates were for the most part rather crudely drawn, etched and colored, and could be commended only as the work of amateurs who strove for accuracy.

Chapter 23

AUDUBON IN LONDON

Audubon reached London on May 21, 1827, and put up at the "Bull and Mouth" tavern, but soon moved into more permanent lodgings at number 55 Great Russell Street, near the British Museum. Though for a long time eager to see the capital, no sooner had he reached it than he was anxious to be away and more homesick than ever for his family and his beloved America. London then seemed to him "like the mouth of an immense monster, guarded by millions of sharp-edged teeth," from which he could escape only by miracle.

He had brought with him a formidable array of letters addressed to the *élite* of the capital,[342] and he bore besides nearly a trunkful for the Continent, as well as general letters from Henry Clay, Andrew Jackson and others in America for our consular and diplomatic representatives in Europe. His epistolary basis for the acquisition of useful acquaintances could hardly have been better, and further testimonials were gathered at every stage of his progress to the city of his hopes, but Audubon's best letter of credit, which could be read by all the world, was an open, winning countenance. After he had wandered over London for the greater part of three days without finding a single individual at home, he was tempted to consign his valuable documents to the post, an error which he did not repeat, as it deprived him of the acquaintance of fully one-half of the people to whom they were addressed. One of these London letters which follows, written by Captain Basil Hall to John Murray, the noted publisher and founder of the *Quarterly Review*, is particularly interesting in showing that Audubon was far from pleased with the progress of his work in Edinburgh, and that he was then contemplating a change which was later effected.

Basil Hall to John Murray

Edinb 23rd Feby. 1827

My Dear Sir

This will be delivered to you by my friend Mr John Audubon, an American Gentleman who has been residing here this winter, & I beg in the most particular manner to introduce him to your acquaintance and to ask for him the advantage of your good offices.

---

[342] Among the sixty or more persons to whom Audubon carried written credentials at this time were the following: the Duke of Northumberland, Robert Peel, Sir Humphry Davy, Sir J. D. Aukland, Albert Gallatin, the American Minister, Sir Thomas Lawrence, David Wilkie, Dr. Buckland, Dr. Holland, Dr. Roget, Dr. Wollaston, William Swainson, Sir William Herschel, and his son, afterwards Sir John Herschel, John George Children, R. W. Hay, N. A. Vigors, Captain Cook, John Murray and Robert Bakewell.

Mr Audubon has spent [a] great part of his life in making a collection of drawings of the Birds of North America, & in studying their Habits, with the intention of publishing a Complete Ornithology of America. For such a work his materials, both in the shape of drawings and of written notes, are immense and he is now going to London in order to set this gigantic work in motion.

Mr Audubon, however, is not very well versed in the details of such matters, & therefore I beg of you to have the goodness to aid him with your advice on the occasion—to introduce engravers printers & so forth to him, and generally speaking to put him in the way of bringing out his work in an advantageous manner to himself.

I trust all this will give you no more trouble than you will be willing to take at my earnest solicitation.

I remain Ever, My Dear Sir,
Most Sincerely Yrs
Basil Hall.
John Murray Esqr

Audubon carried also a long letter from "Mr. Hay,"[343] dated at "16 Athol Crescent, Edinburgh, 15 March, 1827," and addressed to the care of his brother, Robert William Hay, of Downing Street, West, in which this curious statement occurs: "Mr. A. is son of the late French Admiral Audubon, but has himself lived from the cradle in the United States, having been born in one of the French colonies."

The document which was to prove of greatest service to him, however, was addressed to John George Children,[344] then in charge of the Department of Zoölogy in the British Museum and secretary of the Royal Society. Children assumed the management of Audubon's work when he returned to America in 1829 and again in 1831; to him and Lord Stanley, in 1830, the naturalist probably owed his nomination to membership in the Royal Society.

Soon after reaching London Audubon paid his respects to Sir Thomas Lawrence, for whom he had two letters, and made an appointment for showing his work to this famous artist. He was also gratified to receive the subscription of Lord Stanley and of Charles Lucien Bonaparte, who was then in London.

Audubon had not been in London a month before word was received from Lizars that all his colorers had struck work and that everything was at a stand. Accordingly, he began to search London for skilled workmen, and on June 18 wrote: "I went five times to see Mr. Havell, the colorer, but he was out of town. I am full of anxiety and greatly depressed. Oh! how sick I am of London!" Three days later another discouraging letter came from Lizars, who shortly after threw up his contract and left his patron in a sad predicament—with an enormously expensive work, still-born, on his hands, without adequate funds, and, in short,

[343] Probably the same that is referred to in his journals as "Mr. Hays, the antiquarian."

[344] J. G. Children (1777-1852) was early interested in chemistry, and at Tunbridge built a good laboratory, in which Humphry Davy conducted many of his early experiments, and while there was seriously injured in October, 1812. In 1824 Children discovered a method of extracting silver without the use of mercury. When Mr. Children, Senior, became insolvent through the failure of his bank, his son obtained a position at the British Museum; in 1816 he was librarian in the Department of Antiquities, but in 1823 he was transferred to a post in zoölogy which was eagerly sought by William Swainson; he was secretary of the Royal Society in 1826-27, and again in 1835-37. He resigned his position at the Museum in 1840, when Swainson was again an unsuccessful candidate, and was succeeded by J. E. Gray. Children was not a productive zoölogist, but has been described as a lovable soul, who was never soured by illness or other misfortunes, and who was as zealous in his friendships as in science. See "A. A." (Anna Atkins), *Memoir of J. G. Children, Esq.*

with all his cherished plans suspended in mid-air. Audubon no doubt realized that if his grand undertaking were to succeed at all, it must experience a new birth in London, where an expert engraver of the requisite enterprise and zeal must be found without delay. He closed his journal on the second day of July with the remark that he was too dull and mournful to write a line, and it was not opened again for nearly three months.

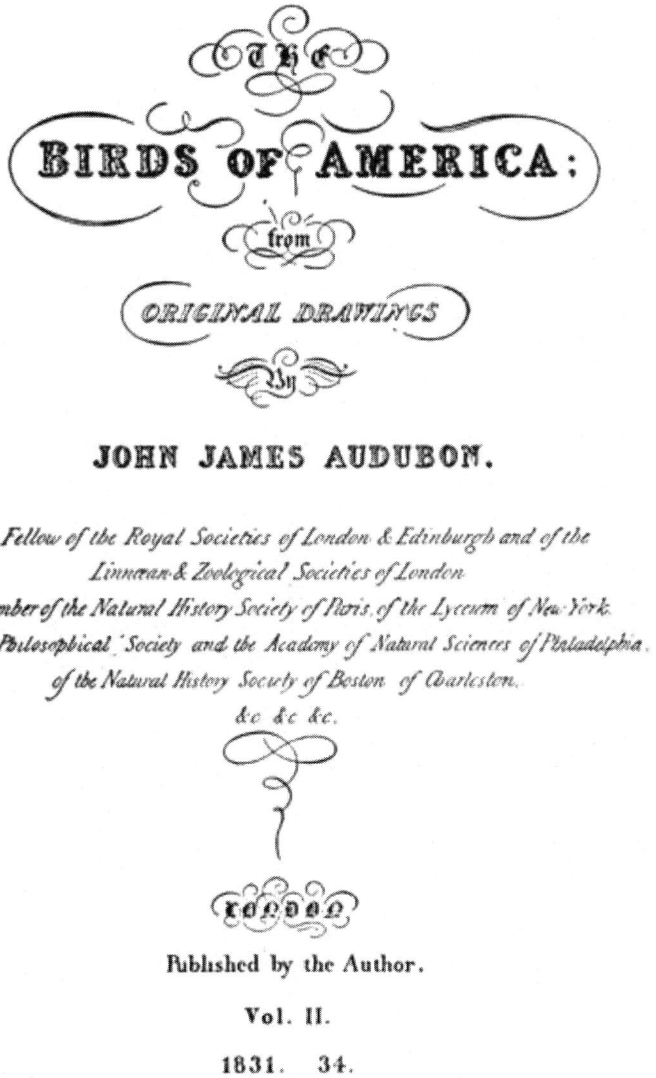

THE

BIRDS OF AMERICA;

from

ORIGINAL DRAWINGS

By

JOHN JAMES AUDUBON.

Fellow of the Royal Societies of London & Edinburgh and of the
Linnæan & Zoological Societies of London
Member of the Natural History Society of Paris, of the Lyceum of New York,
of the Philosophical Society and the Academy of Natural Sciences of Philadelphia,
of the Natural History Society of Boston, of Charleston,
&c &c &c.

LONDON.

Published by the Author.

Vol. II.

1831. 34.

Figure 51. Title Page of the Original Edition of "The Birds of America," Volume II, 1831-1834.

This gap in Audubon's record can now be filled in reference to some important particulars, for in the interval he made his greatest discovery in England, in Robert Havell, Junior, then a young and unknown artist of thirty-four, who through eleven years of the closest association with his new patron was to become one of the greatest engravers in aquatint the world has ever seen. Until recently the intimate story of Audubon's relation to the

Havells has been much obscured.[345] The reference in the journal record of June 19, just given, was undoubtedly to Robert Havell, Senior, who for many years was associated with his father, Daniel Havell, the first of five generations of artists of that name, in the engraving and publishing business, but who at this time was established independently at 79 Newman Street, London; he also conducted a shop called the "Zoölogical Gallery," at which were sold engravings, books, artists' materials, naturalists' supplies, and specimens of natural history of every sort. His three sons, Robert, George, and Henry Augustus, all became artists, but the eldest, who bore his father's name, was educated for a learned profession. Contrary to his father's injunctions and advice, Robert, who was bent on becoming an artist, abruptly left his home in 1825, determined to shift for himself. He began with an extensive sketching tour on the River Wye, in Monmouthshire, and produced numerous paintings which, as his biographer remarks, display all the charm found in the work of his distinguished cousin, William Havell. These won immediate recognition in London, where he received commissions from various publishers, including the house of Messrs. Colnaghi & Company.

Robert Havell, Senior, then in his fifty-eighth year, though deeply interested in Audubon's adventurous plans, felt himself too old to embark on so extended a work, which it was then believed would require from fourteen to sixteen years for completion; he volunteered, however, to do his best to find a substitute. With this in view, he applied to Mr. Colnaghi, the publisher, and was immediately shown the unsigned proof of a beautiful landscape, exquisitely drawn and engraved by one of the youthful retainers of his establishment. The elder Havell, after scrutinizing it carefully, exclaimed, "That's just the man for me!" "Then," replied the publisher, "send for your own son!" Through this singular coincidence, father and son became reconciled and a partnership between them was soon announced.

As a test of young Havell's skill, to follow the story of his biographer, Audubon gave him his drawing of the Prothonotary Warbler, which had already been engraved and issued by Lizars as Plate iii of *The Birds of America* earlier in that year. Havell finished the engraving in two weeks, when a proof was struck and the naturalist summoned. Audubon examined the print with the utmost keenness and deliberation; then he seized the sheet, and holding it up, danced about the room, calling out in his French accent: "Ze jig is up, ze jig is up!" The Havells, who at first thought this might signify disapproval, were quickly disabused when Audubon approached young Robert and, throwing his arms about his neck, assured him that his long-sought engraver had been found at last. Having given this story, I wish it were possible to confirm it, but a close examination of this plate proves either that the story is a fiction, or that some other drawing was used as a test of Havell's skill.[346]

---

[345] In the account which follows, as well as in numerous instances in Chapter 32, I am most indebted to George Alfred Williams, who in "Robert Havell, Junior, Engraver of Audubon's *The Birds of America*," (*Print-Collectors Quarterly*, vol. vi, no. 3, pp. 225-259, Boston, 1916), has given the only satisfactory account of the Havell family and the best analysis of the work of the great engraver.

[346] Mr. Charles E. Goodspeed, who recently sent me two of the original plates of the Prothonotary Warbler, one bearing the legend "Engraved by W. H. Lizars Edinr," and the other, "Engraved, Printed & Coloured, by R. Havell Junr," called attention to the identity of the two engravings. That these two impressions are absolutely identical in aquatint and line is proved by applying a magnifying glass to any part of their surfaces, and by counting and comparing the lines or dots within any selected area whatsoever; in short, they differ only in their legends, and in the coloring which was applied by different hands. That such methods should have been adopted for excluding Lizars' name is certainly surprising. In the first or Edinburgh impression of Lizars' original plate, the artist's legend reads: "Drawn by J. J. Audubon M. W. S.," and names of bird and plant appear at the bottom of the plate in three lines: "PROTHONOTARY WARBLER. *Dacnis protonotarius.* Plant

The part which this interesting family played in Audubon's success will be unfolded later.[347] Suffice it now to say that Messrs. Robert Havell & Son, in London, undertook afresh the production of *The Birds of America* in the summer of 1827. The partnership was divided or dissolved in 1828, when Robert, junior, who from the first did all of the engraving, took entire charge of that part of the business, and moved his engraving establishment around the corner to 77 Oxford Street; there it remained until broken up in 1838. Robert Havell, Senior, continued in charge of the printing and coloring until 1830, when he seems to have permanently retired, two years before his death in 1832, events which, as will be seen, are indirectly registered in the legends of some of Audubon's plates.[348]

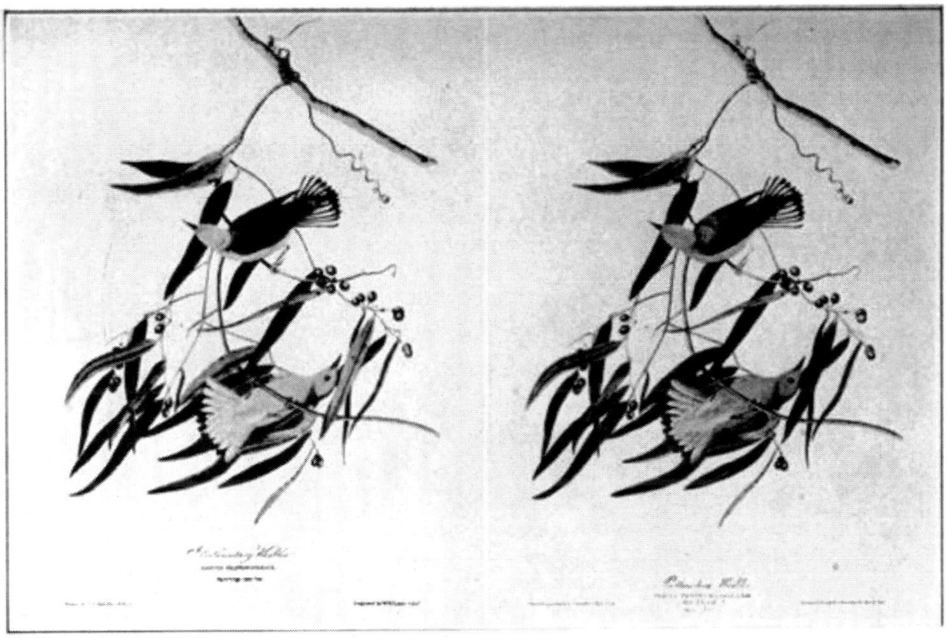

Figure 52. The Prothonotary Warbler Plates, "The Birds of America," Plate Xi, Bearing the Legends of the Engravers W. H. Lizars (Left) and Robert Havell, Jr. (Right), but Identical in Every Other Detail of Engraving, Any Apparent Difference Being due to the Coloring, Which Was Added by Hand.

Under the younger Havell's guiding hand, Audubon found that his illustrations could be produced in better style, more expeditiously, and at far less cost than in Edinburgh. When Lizars was later shown the third number which the Havells had produced, he called his assistants and observed how completely the London workmen had beaten them; he even offered to resume work on the engraving and at Havell's price, but Audubon was averse to further experimenting. "If he can fall," said he, "twenty-seven pounds in the engraving of each number, and do them in a superior style to his previous work, how enormous must his profits have been; a good lesson to me in the time to come, though I must remember Havell is

Vulgo Cane Vine." In the London edition the corresponding designations are: "Drawn from nature by J. J. Audubon F, R, S. F, L, S.," and PROTHONOTARY WARBLER. *Sylvia Protonotarius.* Lath, Male. 1. Female, 2. Cane Vine.," in four lines.

[347] See Chapter 32.

[348] See *ibid.*

more reasonable owing to what has passed between us in our business arrangements, and the fact that he owes so much to me."

This characteristic note was sent from Liverpool, December 6, 1827, to his agent, Daniel Lizars, father to W. H. Lizars, at Edinburgh:

> I will not ask if you have any new name for me, as I *might* be disappointed were I to expect an affirmative answer.
>
> If you see Sir Wm. Jardine tell him that Charles Bonaparte has left the U. S. forever, and has gone to reside in Florence, Italy.
>
> I have wrote to Mr. Havell to send you a No. 5, which I wish you to send to Professor Wilson, or indeed a whole set, to enable him to write the notice he has promised for me the 1st. of next month.

ENGRAVINGS.

IN submitting this List of Publications, R. HAVELL begs to state, he has on Sale a very extensive and well selected Assortment of ENGRAVINGS and WORKS OF ART, arranged in Portfolios, with the Prices affixed, comprising subjects after Wilkie, Turner, Martin, Lawrance, Newton, Burnett, and others. Lithographs, Studies of Animals, Figures and Heads, &c. &c.

Birds of America, by J. J. Audubon, vol. 1 and 2.
Ornithological Biography, by J. J. Audubon, vol. 1 and 2.
Antiquities of Mexico, by Lord Visconnt Kingsborough.
Hexandrian Plants, by Mrs. E. Bury.
African Animals and Scenery, by Daniels.
Birds of Paradise, by F. Le Vaillaint.
Floral Illustrations of the Seasons, by Mrs. E. Roscoe.
Views in the Ionian Islands, by J. Cartwright.
Court of Persia.
Albanian and Greek Costumes.
Aeronautical View of London.
Ditto ditto of Paris.
Panoramic View of King George's Sound.
Panoramic View of Corfu.
Foreign Costumes.
All Works on Natural History.

R. H. begs to observe that all Works entrusted to his care for Publication are Engraved, Printed and Coloured, under his entire inspection, on his Establishment, by which means they are not made public until ready for delivery.

Superfine Water Colours,

In Cakes or Boxes.

R. HAVELL begs to recommend his SUPERFINE WATER COLOURS, as being prepared with the greatest care, and solicits a Trial.

		£.	s.	d.
In Mahogany Boxes,	18 cakes, lock, palettes, marble slab, drawer, &c.	1	10	0
Ditto	ditto, 12 cakes	1	1	0
Ditto	ditto, 12 cakes, without drawer	0	18	0
Ditto	ditto, 12 cakes, lock and drawer	0	15	0
Ditto	ditto, 12 cakes, without ditto	0	13	0
Dittc	ditto, 18 cakes, sliding top	0	15	0
Ditto	ditto, 12 cakes, ditto	0	10	6
Ditto	ditto, 6 cakes, ditto	0	6	0

Cumberland Black Lead Drawing Pencils.

CAMEL HAIR AND SABLE BRUSHES.

Whatman's Superfine smooth rolled Drawing Paper, sold in sheets, quires or reams.

BRISTOL BOARDS OF ALL COLOURS AND THICKNESS.

Genuine Indian Ink.

English and French CRAYON PAPER of all Colours. Ditto ditto CHALKS.

Figure 53. Reverse of Panels of Robert Havell's Advertising Folder Reproduced on Facing Insert.

Figure 54. Outside Engraved Panels of an Advertising Folder Issued by Robert Havell, About 1834; the Printing on the Reverse Is Reproduced on the Facing Page.
From the Only Copy Known to Exist, in Possession of Mr. Ruthven Deane. It Is a Strip of Heavy Paper, 18 by 3⅝ Inches in Size, Printed on Both Sides, and Folded Twice, the Folded Size Being 4½ by 3⅝ Inches. One Side Bears the Four Panels, Engraved by Robert Havell, Reproduced on This and the Following Page; and the Reverse.

Audubon sent another letter to this agent, from London, January 21, 1828, when he was still waiting for an answer to his last: "When I write to any one I expect an answer, but when I write to a man I esteem, and to whom I entrust a portion of my business, I feel miserable until I hear from him.... I am extremely anxious to close my business for 1827, and cannot do so without receiving your a/c, and the money due by my subscribers."

The summer of 1827 was probably Audubon's most critical period in England. His work was then in the air and ruin of all his hopes seemed inevitable, but with palette and brush he again extricated himself from financial difficulties. At this time, he said, "I painted all day, and sold my work during the dusky hours of the evening as I walked through the Strand and other streets where the Jews reigned; popping in and out of Jew-shops or any others, and never refusing the offer made me for the pictures I carried fresh from the easel." He sold seven copies of the "Entrapped Otter" in London, Manchester, and Liverpool, and from seven to ten copies of some of his other favorite subjects; once when he inadvertently called at a shop where he had just disposed of a picture, the dealer promptly bought the duplicate and at the same price that he had paid for the first.

Figure 55. Inside Engraved Panels of the Advertising Folder Issued by Robert Havell, About 1834. The lower panel shows the interior of the "Zoological Gallery," 77 Oxford Street. Audubon's Plate of the Cock Turkey Is Being Examined at one of the Tables.

In the autumn of this year, when it was found that his agents were neglecting their business, Audubon determined to make a sortie to collect his dues and further augment his subscription list. He left London on September 16, and visited in succession Manchester, Leeds, York, Newcastle-on-Tyne, Alnwick Castle and Belford, to see the Selbys, finally reaching Edinburgh on the 22nd of October.

R. HAVELL begs to inform Zoological Collectors that, having an extensive correspondence, he is enabled to supply Natural Productions from all parts of the Globe.

Birds and Beasts Stuffed and Preserved

In the highest perfection, at his Establishment, and the greatest care taken to place the specimens in their natural attitudes and pursuits.

A GOOD ASSORTMENT OF INSECTS, BOTH BRITISH AND FOREIGN.

A great variety of Coloured and Black Eyes.

TAXIDURMIE, or the Art of Collecting & Preparing Objects of Natural History.

Maple, Gold and Black Cases,

FITTED-UP IN A SUPERIOR STYLE, WITH COLOURED SKY.

GLASS SHADES, OVAL, ROUND AND SQUARE, OF ALL SIZES.

The present Collection consists of the most rare LAND and WATER BIRDS from NORTH AMERICA, &c.

PICTURES AND PRINTS

Framed and Glazed, in Gold, Maple, and all kinds of Ornamental and Fancy Woods, Straining, Varnishing, &c.

DRAWINGS AND PRINTS LENT OUT TO COPY

ON THE FOLLOWING TERMS:

	£.	s.	d.
Yearly Subscription	3	3	0
Half-yearly ditto	1	11	6
Quarterly ditto	0	16	0
Weekly ditto	0	2	6

A GREAT VARIETY OF RICE DRAWINGS, AND RICE PAPER OF ALL SIZES.

Plain and Ornamental Albums.
Ditto ditto Scrap Books.
Ditto ditto Blotting Cases.
Ditto ditto ditto, with lock and key.

MINIATURE FRAMES AND GLASSES.

Sketch Books of all sizes.

ORDERS FOR EXPORTATION executed with promptitude, and on the lowest terms, at 77, OXFORD STREET.

Figure 56. Reverse of Panels of Robert Havell's Advertising Folder Reproduced on Facing Insert.

Audubon had set his mark at obtaining 200 subscribers by May, 1828, but he fell far short of realizing it. On August 9 he wrote: "This day seventy sets have been distributed; yet the number of my subscribers has not increased; on the contrary, I have lost some." At York he found that a number of his *Birds*, which had been forwarded from Edinburgh before he had taken his departure, "was miserably poor, scarcely colored at all"; and a copy of his first number which was later examined at the Radcliffe Library in Oxford was so unsatisfactory that he rolled it up and took it away, with the reflection that Lizars, whom he had paid "so amply and so punctually," could have made him a better return. The colorists gave no end of trouble, but he never hesitated to reject their work when it did not meet his requirements, and the defective plates were invariably sent back to Havell's shop to be washed, hot-pressed, and done over again. To such watchful care must be ascribed, in large measure, the high degree of perfection which his big work eventually attained. When it is remembered that upwards of one hundred thousand of his large plates had to be colored laboriously by hand, and that at one time fifty persons were engaged at the Havell establishment, we can understand the difficulties involved in maintaining a uniform standard of excellence in a work that was issued piecemeal and spread over a long period of time.

In August, 1827, Audubon wrote to Mrs. Thomas Sully of Philadelphia to announce the removal of his business to London. By this change he expected to save "upwards of an hundred pounds per annum, a large sum," as he remarked, "for a man like me." His third number had then been issued, and he expressed the hope that all would go smoothly after "this first year of hard trials and times," and that he would be able to send for his wife and one of his sons in the coming autumn or winter. He was then painting "a flock of Wild Turkeys for the king, who had honored him with his particular patronage and protection." When writing to his young son, John W. Audubon, on the 10th of the same month, he charged him to devote two hours daily to the preparation of bird skins, and to send him not only the skins but live birds and mussel shells, for which he would be duly paid. Said the father:

> I would give you 500 dollars per annum, were you able to make for me such drawings as I will want. I wish you would draw one bird only, on a twig, and send it [to me] to look at, as soon as you can after receiving this letter.... I should like to have a large box filled with branches of the trees, covered with mosses &c., such as Mama knows I want; now recollect, *all sorts* of Birds, males and females, ugly or handsome.

Audubon had come to London with the idea of having his work published under the patronage of King George IV; in order to gain a personal interview with the Sovereign he had brought a letter to Robert Peel, who was then the Home Secretary, but a change in the Cabinet had upset his plans and the letter was returned. He then applied to the American Ambassador, Mr. Albert Gallatin, who upon their first meeting addressed him in French and showed "the ease and charm of manner of a perfect gentleman"; but when the question of an audience with the King was broached, Gallatin laughed at the idea as preposterous. "The king," he declared, "sees nobody; he has the gout, is peevish, and spends his time playing whist at a shilling a rubber. I had to wait six weeks before I was presented to him in my position of ambassador, and then I merely saw him six or seven minutes." When Audubon then suggested that the Duke of Northumberland might interest himself in his behalf, Gallatin, who disliked the English heartily, replied: "I have called hundreds of times on like men in England, and have been assured that his grace, or lordship, or [her] ladyship was not at home, until I have grown wiser, and stay at home myself, and merely attend to my political business, and God knows when I will have done with that."

As the American Ambassador had predicted, King George evinced no ardent desire to meet the American woodsman, though he consented to take the work under his patronage and to become a subscriber on the usual terms; this plan, however, fell through, for the King, who was reported to have taken his copy, failed to pay for it. With Queen Adelaide, on the other hand, the naturalist was more successful, and in his "Prospectus" of 1831 she was announced as his special patron, with her name heading his list. Negotiations to interest the Queen were going on when the following note was sent to Audubon by Sir J. W. Waller, who occupied some position in the king's household and was spoken of as "oculist to his majesty":

UNDER THE SPECIAL PATRONAGE

OF

Her Most Excellent Majesty,

QUEEN ADELAIDE;

THE

BIRDS OF AMERICA,

ENGRAVED FROM

DRAWINGS

MADE IN

THE UNITED STATES AND THEIR TERRITORIES.

BY JOHN JAMES AUDUBON,

F.R.SS. L & E.

FELLOW OF THE LINNEAN AND ZOOLOGICAL SOCIETIES OF LONDON; MEMBER OF THE LYCEUM OF NEW YORK, THE NATURAL HISTORY SOCIETY OF PARIS, THE WERNERIAN NATURAL HISTORY SOCIETY OF EDINBURGH; HONORARY MEMBER OF THE SCOTTISH ACADEMY OF PAINTING, SCULPTURE AND ARCHITECTURE, &c.

PUBLISHED BY THE AUTHOR;

AND TO BE SEEN AT

MR R. HAVELL'S JUN. THE ENGRAVER,

77. OXFORD STREET, LONDON.

MDCCCXXXI.

Figure 57. Title Page of Audubon's Prospectus of "The Birds of America" for 1831.

Sir J. W. Waller to Audubon

Saturday 9 o clock [1830].

I have scarce an Instant as I am going to Town to breakfast with the Dk. of Gloucester, but yr. Letter is urgent & therefore I can only desire Mr. A. to send his Number immediately to the Stable Yard, directed to her Majesty, & the first moment I can see her, I will speak on the subject, but at this Moment I will not promise to mention it to the King for reasons I cannot put on paper.

Yrs. ever,
J. W. Waller

At Edinburgh Audubon was alarmed to find that subscribers were rapidly deserting him, six having cancelled their names without the formality of giving reasons. He hoped to supply their places at Glasgow, then a rich city of one hundred and fifty thousand people, but after a visit there of four days in November, 1827, he was obliged to return to Edinburgh with but one new name on his list.

On October 22 he expressed the resolve for the coming year "to *positively keep* a cash account" with himself and others, "a thing" he had "never yet done." The wisdom of that decision was apparent upon settling his accounts for 1827 with both Lizars and Havell, as appears from this note, written in his journal on January 17, 1828: "It is difficult work for a man like me to see that he is neither cheating nor cheated. All is paid for 1827, and I am well ahead in funds. Had I made such regular settlements all my life I should never have been as poor a man as I have been; but on the other hand I should never have published the "Birds of America." Again, for February 7 we find this record: "Havell brought me the sets he owed me for 1827, and I paid him in full. Either through him or Mr. Lizars I have met with a loss of nearly £100, for I am charged with fifty numbers more than can be accounted for by my agents or myself. This seems strange always to me, that people cannot be honest, but I must bring myself to believe many are not, from my own experiences."

Shortly after reaching London, as we have seen, Audubon had made the acquaintance of Sir Thomas Lawrence, then at the head of the Royal Academy and favorite painter of the Court and fashionable society. The friendship of this influential artist at a critical moment proved most fortunate, for Sir Thomas called repeatedly at his lodgings, and at each visit brought patrons who went away with some of his pictures but not without leaving a handsome toll of sovereigns in his lap; the "Entrapped Otter" again did duty by bringing him twenty-five pounds, while others returned from seven to thirty-five pounds. At a later time the artist visited the "Zoölogical Gallery," as the Havell establishment in Newman Street was then known, and saw Audubon's large paintings called "The Eagle and the Lamb," and "English Pheasants Surprised by a Spanish Dog" or "*Sauve qui peut.*" Audubon, who on this occasion missed seeing his distinguished visitor, had written in his journal three days before (December 23, 1828) that the paintings were what he called "finished," but that, as usual, he could not bear to look at either. Sir Thomas praised the "Eagle," admired an "Otter," which was later exhibited in London, but gave no opinion on the "Pheasants." Afterwards, however, when Audubon proposed to present this canvas to King George, the artist assured him that this picture was worth 300 guineas and that it was too good to be given away; if offered to the King, no doubt, said he, "it would be accepted and placed in his collections, but you would receive no benefit from the gift." According to a later record, this canvas was sold to Mr. John Heppenstall of Sheffield; whether it was ever delivered, or not, I do not know, but either the original or a copy, here reproduced, now forms the central figure in the large Audubon collection in the American Museum of Natural History in New York, and is an excellent illustration of the elaborate and ambitious character of Audubon's larger compositions. These fortunate windfalls came none too soon, for to follow the journal:

> Mr. Havell had already called to say that on Saturday I must pay him sixty pounds. I was then not only not worth a penny, but had actually borrowed five pounds a few days before to purchase materials for my pictures. But these pictures which Sir Thomas sold for me enabled me to pay my borrowed money, and to appear full-handed when Mr. Havell called. Thus I passed the Rubicon.

This was before the reform of the penal laws in England, when it seems to have been hard for a man to escape hanging, not to speak of being sent to prison for debt, the chief terror of life in certain circles. There were 223 capital offenses, and in 1829 in the city of London alone 7,114 persons were sent to the debtors' prison.[349]

Without the sale of his pictures in the summer of 1827, Audubon felt that he must certainly have become a bankrupt, yet he was periodically displeased with the results of his efforts in oil colors, and resolved to "spoil no more canvas" but to draw "in my usual old untaught way, which is what God meant me to do"; "I can draw," he continues, "but I shall never paint well." In the fall of 1828, however, he was again working in oils, and produced four large pieces, one of which was called "The Eagle and the Lamb," and two others which were doubtless variations of his "Pheasant" and "Otter" pictures. "It is charity," said the artist, "to speak the truth to a man who knows the poverty of his talents, and wishes to improve; it is villainous to mislead him, by praising him to his face, and laughing at his work as they go down the stairs of his house." Sir Thomas Lawrence had praised some of these pictures and had promised to select one for exhibition at Somerset House. As regards "The Eagle and the Lamb," which Audubon hoped would go to Windsor Castle, William Swainson would give no opinion; the same canvas, or else a replica, was in possession of the Audubon family in 1898.[350]

On December 14, 1827, Audubon wrote that, acting upon the advice of Mr. Maury, the American consul at London, he had presented a copy of his *Birds* to John Quincy Adams, the President of the United States, and another, through Henry Clay, to the American Congress; in order that the latter should be as perfect as possible, Havell was asked to do the coloring himself, but these proposed gifts do not appear to have been executed.[351]

New Year's, 1828, found the naturalist in Manchester, where but a few days before he had received the fifth and last number of his plates for 1827 and expressed himself well pleased with it. While returning to London by coach, he consented to take a hand at cards to accommodate his fellow passengers, but declined to play for money; "I never play," he confessed, "unless obliged to by circumstances; I feel no pleasure in the game, and long for other occupation." "I missed my snuff," he added, and whenever his hands went into his pockets in search of the box, he "discovered the strength of habit thus acting without thought"; but he remembered a resolution he had formed to give up the habit and stuck to it for a time at least; doubtless, like his later friend, John Bachman, he reformed more than once, for in a letter to Victor Audubon, of November 5, 1846, Bachman added this postscript: "To Audubon: The snuff—the snuff, it is here! I have just taken a pinch, and the ladies have blown *you* up—sky-high, for teaching me such a bad practice; I say, however, that you beat me all to pieces in that art."

---

[349] See Sir Walter Besant, *London in the Nineteenth Century* (London, 1909).

[350] See Maria R. Audubon, *Audubon and his Journals*, vol. i, p. 342, where the "Eagle and the Lamb" is reproduced.

[351] See Vol. I

Figure 58. "English Pheasants Surprised by a Spanish Dog." After Audubon's Original Painting, About Six by Nine Feet in Size, Now in Possession of the American Museum of Natural History, New York. Published by Courtesy of the American Museum of Natural History.

The first winter in London dragged heavily for the naturalist, who exclaimed in January, 1828: "How long am I to be confined in this immense jail"; when Daniel Lizars reported from Edinburgh the loss of four of his subscribers, he writes, "I am dull as a beetle. Why do I dislike London? Is it because the constant evidence of the contrast between the rich and the poor is a constant torment to me, or is it because of its size and crowd? I know not, but I long for sights and sounds of a different nature," such, we might add, as the flocks of wild duck which were occasionally seen from Regent's Park as they passed over the city and made him more homesick than ever. Audubon hated the city quite as cordially as Charles Lamb ever affected to detest the country, and when leaving it, afoot or by stage, it seemed as if he could never be rid of it. "What a place is London," he would say, but naïvely add: "many persons live there solely because they like it."

On February 4, 1828, Audubon was elected to membership in the Linnæan Society, and in November he presented it with a copy of his work, which was then well under way. This was noticed in a letter to Swainson, written on November 7, when no acknowledgment of the gift had then been received; and he mentioned also the sale of his picture of "Blue Jays" for ten guineas. At a meeting of the Linnæan Society not long after his election, copies of Selby's *Illustrations of British Ornithology* and of his own work were placed side by side for inspection, and "very unfair comparisons were drawn between the two"; had Selby, Audubon reflected, been given "the same opportunities that my curious life has granted me, his work would have been far superior to mine"; "I supported him," he added, "to the best of my power."

Revision of his older drawings demanded much of Audubon's attention during these years. On February 10, 1828, he began the Whiteheaded Eagle (No. 7, Plate xxxi), the original of which had been procured on the Mississippi, where the bird was represented as dining on a wild goose; now, he said, "I shall make it breakfast on a catfish, the drawing of which is also with me, with the marks of the talons of another eagle, which I disturbed on the banks of the same river, driving him from his prey." On the 16th of that month he was engaged with this drawing from seven in the morning until half after four, stopping only to take the glass of milk which his landlady would bring to him. This plate was engraved in the following April, and on May 1, 1828, a first proof was sent to the Marquis of Landsdowne, president of the Zoölogical Society, as a mark of appreciation by its author, who had become a member of that body in the preceding winter.

A striking characteristic of Audubon's work was its diversity, produced not only by attractive embellishments of many kinds, but by the moving force and action with which he ever sought to vitalize his subjects. It is therefore not surprising that he was nettled by an incident like this:

> February 28. To-day I called by appointment on the Earl of Kinnoul, a small man, with a face like the caricature of an owl; he said he had sent for me to tell me all my birds *were alike*, and he considered my work a swindle. He may really think this; his knowledge is probably small; but it is not the custom to send for a gentleman to abuse him in one's house. I heard his words, bowed, and without speaking, left the rudest man I have met in this land.

Audubon had not yet visited the great university towns of England, the support of which he knew would be a valuable asset, and on March 3, 1828, he set out by stage for Cambridge. His driver, he remarked, "held confidances with every grog-shop between London and Cambridge, and his purple face gave powerful evidences that malt liquor [was] more enticing to him than water." His reception at Cambridge was hearty; he was entertained by Professors Sedgwick, Whewell, and Henslow, dined repeatedly "in Hall" with the dons, and received the subscription of the librarian of the University. It is interesting to recall that young Charles Darwin, "the man who walks with Henslow," as some of the dons called him, was then an undergraduate at King's College, and that thirty-one years were to pass before modern biology was born in 1859, the year of the appearance of the epoch-making *Origin of Species*.

By the 15th of March Audubon was again in London, and on the 24th he started for Oxford. Dr. Williams, as he noted in his journal, subscribed for his *Birds* in favor of the Radcliffe Library, as did also Dr. Kidd for the Anatomical School; but, though hospitably treated by all, not one of the twenty-four colleges of that great University emulated their example, and the naturalist went away disappointed.

Upon his return to London in early April, Audubon received a call from John C. Loudon, editor of the *Magazine of Natural History*, and was invited to contribute to that journal. "I declined," he said, "for I will never write anything to call down upon me a second volley of abuse. I can only write *facts*, and when I write these, the Philadelphians call me a liar." He was then chafing under the criticism which his rattlesnake stories had produced.[352] On April 6 the persistent Mr. Loudon called again and offered Audubon eight guineas for an article, only to be again refused. Still unwilling to admit defeat, the editor proposed to engage William Swainson to prepare an extended review of the naturalist's work, and in this he succeeded so

[352] See Chapter 28.

well that Audubon immediately relented and sent him a paper.[353] Swainson offered to write the review for a copy of the work at its cost price, and Audubon replied in the following letter:[354]

Audubon to William Swainson

London, April 9th 1828.

My Dear Sir,

Mr. Loudon called on me yesterday and showed me a letter from you to him, in which many very flattering expressions respecting myself and my works you are so kind as to offer to review the latter so as to have your opinion in writting in time for the first no. of the magazine that will appear next month.—you also desire that I should send you a sett of the works as far as publishing which you wish to keep provided I will let you have it at the price it costs me. I assure you my Dear Sir, that was I to take you at your word it would be a sore bargain for you as the a/m would be very nearly double that for which it is sold to my subscribers.—therefore you will permit me to alter your offer and to say that if it suits you to pay 35 shillings per number I will be contented; I would be still more so was I rich enough to present it to you.—

It is the only set on hand at present except one which I must have to exhibit.—

The answer respecting the Shrieke [Shrike] has I hope met with your wishes.—

Ever since I became acquainted with our mutual friend Dr. Fraill [Traill] I have had a great desire to see and speak to you & I regret that I never have had an opportunity. My time is so completely taken up that it is with difficulty that I can enjoy a day's rest—Should you come to town pray call on me when I may have the pleasure of shaking your hand and to assure you verbally that I am truly and sincerely yours best

John J. Audubon

95 Great Russell St.
Bedford Sq.

353 The seventh which he had contributed to the scientific press of Europe, entitled "Notes on the Bird of Washington (Falco Washingtoniana), or Great Sea Eagle," now believed to have been mistaken by him for an immature stage of the true "bird of freedom," the White-headed Eagle. It was dated "London, April, 1828," and was published in Loudon's *Magazine* for July of that year.

354 From the originals in possession of the Linnæan Society of London. Swainson's scientific correspondence was taken with him to New Zealand, where it remained fifty years, until returned by his daughter, who sent it to Sir Joseph Hooker; it was finally purchased by a number of Fellows of the Society, and presented to its historical collections. It consists of 934 letters written by 236 correspondents, from 1806 to 1840. Of the 24 letters written by Audubon, and dated 9 April, 1829, to 11 January, 1838, none has been previously published. Dr. Albert Günther, who has given a summary of their contents (*Proceedings of the Linnæan Society*, 112th Session, 1900) found them rather disappointing, since they dealt mainly with personal and domestic matters, and were written in a style characterized as "fantastic and unnatural." Through the kindness of my esteemed friend, George E. Bullen, Esq., of the Hertfordshire County Museum, St. Albans, and through the courtesy of the Council of the Linnæan Society and its secretary, Dr. Daydon Jackson, I am able to reproduce transcripts of the most interesting of these letters, which readers in America will, I believe, find interesting because of their personal details. I am indebted also for their good offices to John Hopkinson, F.L.S., and to William Rowan, Esq.

From the context of the nine letters which are here reproduced without change, it is evident that Audubon paid little attention to grammar, syntax, or orthography, but if the reader will compare the letters written before and after 1830, or before and after his first serious discipline in English composition, he will find marked improvement in all these respects.

Thus began an intimate friendship between William Swainson and John James Audubon which lasted until 1830, and their intercourse did not wholly cease before 1838. In his use of English at this time Audubon was not far behind Swainson, whose mother tongue it was. Swainson, according to Dr. Günther, was "extremely careless in orthography and loose in his style of writing: he persistently misspelt not only technical terms, but also the names of foreign authors, and even of some of his familiar friends and correspondents; he knew no other language but his own, and the application of Latin and Greek for the purpose of systematic nomenclature was a constant source of error."

At this time Swainson was living in semi-retirement at a farmstead of considerable size, called "Highfield Hall,"[355] near Tyttenhanger Green, a small settlement, off the highroad, two miles southeast of the historic town of St. Albans, in Hertfordshire; though his letters were always dated from "The Green" at Tyttenhanger, his associations were with the more considerable village of London Colney, but a mile to the south, on the road to Barnet. Audubon had brought a letter of introduction from Dr. Traill, a valiant champion of Swainson at Edinburgh, but was unable to go to the country to deliver it. Swainson, however, attended promptly to the review, and on April 11, 1828, sent it to Mr. Loudon, who published it in the May number of his *Magazine*.[356]

Swainson's review was extremely laudatory, and Audubon reproduced extracts from it in later editions of his "Prospectus." To quote a characteristic paragraph, he said that the naturalist's ornithological papers printed in one of the Scotch journals, are as valuable to the scientific world, as they are delightful to the general reader. They give us a rich foretaste of what we may hope and expect from such a man. There is a freshness and an originality about these essays, which can only be compared to the animated biographies of Wilson.... To represent the passions and the feelings of birds, might, until now, have been well deemed chimerical. Rarely, indeed, do we see their outward forms represented with anything like nature. In my estimation, not more than three painters ever lived who could draw a bird. Of these the lamented Barrabaud [Barraband], of whom France may be justly proud, was the chief. He has long passed away; but his mantle has at length been recovered in the forests of America.

Audubon spent four days with Swainson and his family at Tyttenhanger, from May 28 to June 1, 1828, when they talked birds and made drawings; Audubon also showed Swainson "how to put up birds in *his* style, which delighted him." The friendship between these men, though very intimate while it lasted, received a sudden check two years later, when Audubon was about to publish the letterpress to his plates, as will be related farther on.[357]

---

[355] Swainson's house has been kindly identified by my friend, Mr. George E. Bullen, to whom I am indebted also for an interesting photograph, taken from an old print. Mrs. Swainson, who died February 12, 1835, was buried in the parish church, with which she was closely identified, at London Colney, and a tablet to her memory is still to be seen there. Swainson probably preferred the historic associations of Tyttenhanger, a name originally applied to the manor and manor house of the Abbot of St. Albans, a famous abbey property acquired before the Conquest, with a history extending over six hundred years, but he did not live there. The oldest resident now on the spot, a man over ninety, told Mr. Bullen that as a boy he often collected butterflies, moths and other specimens of natural history which he took to "Highfield Hall," and was always paid by one of the Swainson children. Since Swainson's time the original house, which was approached by a long walk, has become almost unrecognizable, having received an addition to one side; the grass land which then surrounded it has been converted into beautiful lawns.

[356] See Bibliography, No. 95.

[357] See Chapter 29.

Though his hands were already more than full at this time, Audubon seems to have played with the idea of publishing a work on the birds of Great Britain, but on May 1 he wrote to Swainson that the plan did not meet with favor, and later he relinquished all claims in such a project to his assistant, William MacGillivray.[358]

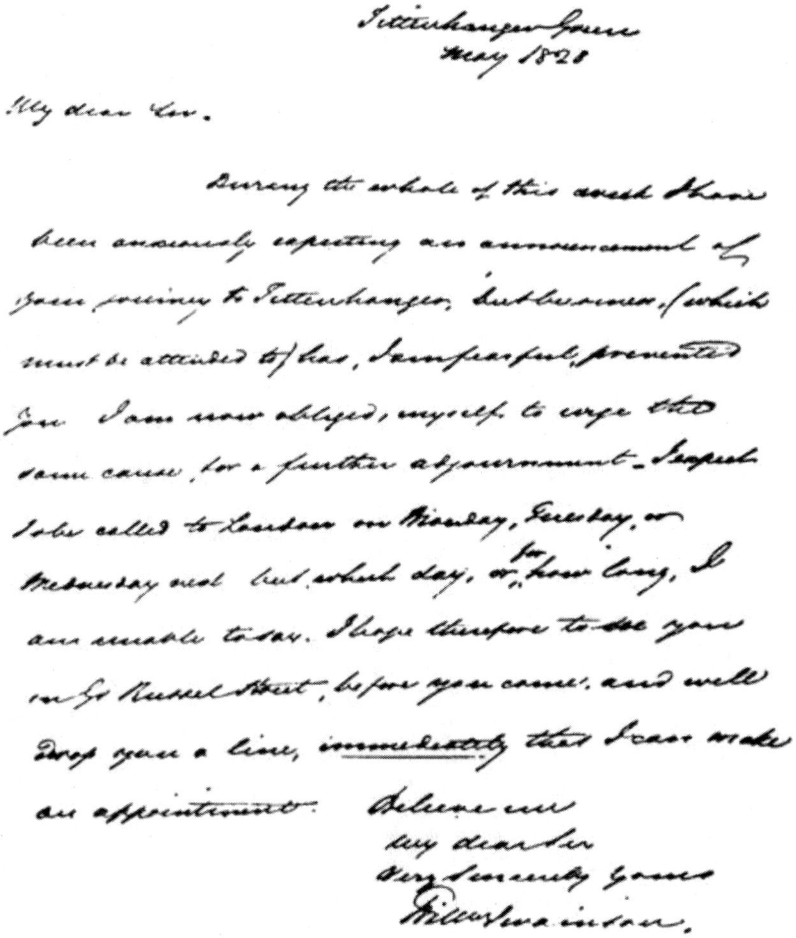

Tittenhanger Green
May 1828

My dear Sir.

During the whole of this week I have been anxiously expecting an announcement of your journey to Tittenhanger, but business, (which must be attended to) has, I am fearful, prevented you. I am now obliged, myself, to urge the same cause, for a further adjournment. I expect to be called to London on Monday, Tuesday, or Wednesday next but which day, or for how long, I am unable to say. I hope therefore to see you in Gt Russell Street, before you come; and will drop you a line, immediately that I can make an appointment.

Believe me
My dear Sir
Very sincerely yours
Willm Swainson.

Figure 59. Letter of William Swainson to Audubon, May, 1828. From the Deane MSS.

In the spring of 1828 Audubon began to think of returning to the United States, to renew or revise his drawings and extend his researches. "I am sure," he said, "that now I could make better compositions, and select better plants than when I drew mainly for amusement." In order to raise the necessary funds, he resorted again to picture painting, his never failing resource, and worked in oil colors daily from morning light until dusk, unless called to Havell's to decide some question of necessary detail. The following letters to Swainson shed further light on this work and on the progress of *The Birds of America*, the eighth number of which was published early in July:

[358] See Vol. II.

Audubon to William Swainson

London, July 1st 1828.

My dear Sir.—

I have been expecting to have the pleasure of seeing you for upwards of a week, having mentioned in your last note that you intended spending a couple of days in London before the end of June.—When are you coming?—the beautifull lamb came quite safe and is now on the canvas (in efigy) for ages to come—I bought a superb Golden Eagle from Mr. Cross that also has helped to fill it —— [Here apparently some words have been deleted, and it is impossible to read them.] I long to shew them to you.—I have finished the picture of the Turkeys, and painted a white headed eagle—in fact I have worked from 4 every morning until dark—but the best news I have to tell is; that I have received 4 letters from my wife, one dated 2nd of May, all well—but not quite settled about coming before the end of summer. I have changed quarters and am now at 79 Newman Street Oxford Street, in Mr. Havell's house where I have taken 3 rooms and feel more comfortable although I have not the little piece of ground to walk on.—I imagine the country to be now quite beautiful and had I time to spare would walk out to see you Mrs S & the dear little folks at Tittenhanger Green.—I received a visit on Saturday last of the whole of Lord Milton's family who after complimenting the author of the "Birds of America" very kindly subscribed for two copies of the work.—I have mended my pen—I should have sent the Blackwood magazine to you, but I so much expected to see you here that it is yet on my table, and will keep it untill you come.—All my exertions to procure live grouses have been abortive here—I have written to Scotland to a friend and perhaps will have some soon.—The 8th number is now printing and colouring and will be out this month—the 9th is began.—If you are hungry or thirsty when you come to town please make for my [here a word is omitted], and I will try to manage matters in this way.—May I ask what you are doing?—I saw Dr Fraill's [Traill's] son a few days ago—he inquired after your son and family.—I expect a copy of Loudon's magazine this evening. I feel anxious to see what sort of a cut the Doves make, as well as the birds of Washington.—

With sincerest regards & esteem to yourself and Lady—
I am yours most truly
John J. Audubon.

79 Newman Street,
Oxford Street.

Audubon to William Swainson

London Thursday July 1828.

My dear Mr Swainson,

Although your last note said that you knew not when I should have the pleasure of seeing you in town, I have hoped every morning to see you that day.—When will you come?—There is a talk of my picture of the Eagle and the Lamb going to her Majesty, Sir Walter Waller has been written to on the subject and everything is in train to lead poor I like a lamb to Windsor Castle!—I am told the picture is a grand one but you, my dear Sir, have not said so! When you come I will show you 13 grouses pretty fairly grouped on one canvas, with seven pheasants with a Fox on another, etc. etc. I have worked hard this month from 4 p.m. untill 7 a.m. [sic] every day—I regretted that your brother did not come to see me—I have a great desire to see you but I cannot at present leave town.—My 8th No. is just out.—The 9th & 10th are engraving.—I have sent word to my son to land [?] & bring some skins for you & perhaps you may have a rare assortment bye and bye.—I hope your Lady and dear Children are all quite well Pray remember me kindly to them.——I wish to name a bird after you in the 1st No. of 1829 & wish you to choose a name. Believe yours ever and truly obliged

J. J. Audubon

79 Newman Street,
Oxford Street.

By the 9th of August eight pictures had been begun, but none was finished, and the number of his subscribers had fallen to seventy. At about this time Captain Basil Hall[359] returned from his journey through the United States, and brought direct news from Victor Audubon, who was then at Louisville, from Dr. Richard Harlan and Thomas Sully, to all of whom the naturalist's letters had been delivered the previous year. Towards the end of the month Audubon received the following note from the secretary of the Zoölogical Society, N. A. Vigors, who was also anxious to obtain from him an article for his *Journal*:

N. A. Vigors to Audubon

Bruter Ct
Aug. 23, 1828.

My dear Sir:—

I hope you do not forget your promise of giving us a paper for the Zoölogical Journal. We should be much gratified by having your name with us: and, if possible, should wish to have whatever you may favour us with within the next ten days. I have been but a few hours in town, and shall leave town again tomorrow for a few days, or I should have called upon you to speak personally upon the subject. I believe I have already mentioned, that we are in the habit of remunerating those of our correspondents who wish for payment for their labours, at a rate not exceeding £10.10.0 per sheet.

A letter from you in answer will reach me, if sent to Bruter Ct; before Wednesday on which day a parcel will be forwarded to me from thence.

Believe me my dear Sir,
Yours faithfully,
N: A: Vigors.

[Addressed]
J. J. Audubon Esq.,
69 Great Russell St.;
Bloomsbury.

[Readdressed]
Newman Street, Oxford Street

[359] See Vol I.

Audubon refused this request, saying that "no money can pay for abuse," and this time he did not retract.

Without immediate prospect of seeing his family, for neither Mrs. Audubon nor her sons were enthusiastic over the proposal that they should go to England, the naturalist was momentarily depressed; he turned to Swainson for advice, at the same time suggesting that they visit Paris together. Audubon wrote in his journal for August 16, 1828, that he had invited Swainson to accompany him to France, whither his friend had expressed a desire to go when the subject had been broached at Tyttenhanger; on the 25th of that month he added: "I do not expect much benefit by this trip, but I shall be glad to see what may be done." The letter just referred to follows:

Audubon to William Swainson

London, Wednesday Augt. 13, 1828.

My dear Mr. Swainson,

I reached my lodging in great comfort by the side of your amiable Docr Davie two hours and a half after we shook hands—I wish I might say as much of my Journey through Life.—I have had sad news from my dear wife this morning, she has positively abandoned her coming to England for some indefinite time, indeed she says that she looks anxiously for the day when tired myself of this country I will return to mine and live although a humbler (Public) Life, a much happier one—her letter has not raised my already despondent spirits in somethings and at the very instant I am writing to you it may perhaps be well that no instrument is at hand with which a woeful sin might be committed—I have laid aside brushes, thoughts of painting and all except the ties of friendship—I am miserable just now and you must excuse so unpleasant a letter—Would you go to Paris with me? I could go with you any day that you would be please to mention, I will remain there as long and no longer than may suit your callings—I will go with you to Rome or anywhere, where something may be done for either of our advantage and to drive off my very great uncomfortableness of thoughts—My two sons are also very much against coming to England, a land they say where neither freedom or simplicity of habits exist and altogether uncongenial to their mode of life.—What am I to do? As a man of the World and a man possessed of strong unprejudiced understanding I wish that you would advise me.—But now on your account I will change the subject—I called on Newman two days ago & to the following enquiries he gave me yesterday the following answers

What the price of

½ doz best Pure Lake	dowards [?]	answer	12/—
½ doz best Carmin	"	"	20/—
½ doz best UltraMarine	"	"	84/—
½ doz best Vermillion	"	"	6/—
½ doz best Terra di Verona	"	"	4/—

As I thought the above prices enormous I have declined advising chalks for you & will await your advent.—

Should you not feel inclined to go to France at present which by the bye is the very best season on account of seeing the vintage etc. etc.—please write to me so or come to town which would be still more agreeable & talk the matter over as I think I would persuade you to absent yourself for a month or so—I hope your kind lady continues quite well & your Dear Little ones—

Believe me yours most sincerely
John J. Audubon.

Please write by return of Post—
79 Newman Street
Oxford Street.

On this journey to Paris Audubon was accompanied by Mr. and Mrs. Swainson and an American artist, named Parker, who had been at work on a portrait of the naturalist in oils. For Audubon it was mainly a canvassing tour; Parker hoped to obtain orders for portraits, and Swainson, new ornithological material at the great museum in the Jardin des Plantes, for a work upon which he was then engaged.[360]

The party set out on the 1st of September, traveling by way of Dover and Boulogne, and reached Paris on Thursday, September 4. They alighted at the Messagerie Royale, Rue des Victoires, and, after looking up lodgings, went at once to the Jardin des Plantes to pay their respects to Cuvier. The Museum of Natural History was closed, but they knocked and asked for the Baron. "He was in," said Audubon, in the journal of his Paris experience,

> but, we were told, too busy to be seen. Being determined to look at the great man, we waited, knocked again, and with a certain degree of *firmness* sent in our names. The messenger returned, bowed, and led the way upstairs, where in a minute Monsieur the Baron, like an excellent good man, came to us; he had heard of my friend Swainson and greeted him as he deserves to be greeted; he was polite and kind to me, though my name had never made its way to his ears. I looked at him, and here follows the result: age about sixty-five; size corpulent, five feet five, English measure; head large; face wrinkled and brownish; eyes gray, brilliant and sparkling; nose aquiline, large and red; mouth large, with good lips; teeth few, blunted by age, excepting one on the lower jaw, measuring nearly three-quarters of an inch square.[361]

They were immediately invited to dine on the following Saturday at six o'clock, and later saw Cuvier at his home, at his Museum, and at the Academy of Sciences, over which he presided.

Geoffroy Saint-Hilaire pleased Audubon greatly and proved to him by his conversation that he understood perfectly the difference between the French and the English. The Duke of Orleans, who then occupied the Palais Royal, seemed to him the finest physical type of man he had ever met. "He had my book brought up," said the naturalist, "and helped me untie the strings and arrange the table, and began by saying that he felt great pleasure in subscribing to the work of an American, for he had been most kindly received in the United States and should never forget it." When the plate of the Baltimore Orioles was held up to view, the Duke exclaimed: "This surpasses all I have seen, and I am not astonished now at the

[360] *Fauna Boreali-Americana; or the Zoology of the northern parts of British America*; Part Second, "The Birds;" by William Swainson and John Richardson (London, 1831).

[361] Maria R. Audubon, *op. cit.*, vol. i, p. 306.

eulogiums of M. Redouté." He conversed in both English and French, had much to say of American cities and rivers, and added: "You are a great nation, a wonderful nation." The Duke wrote his name in Audubon's subscription book, promised to try to enlist a number of the crowned heads of Europe in his behalf, and gave him besides a number of orders for pictures of animals.

Audubon had already made friends with the veteran painter of flowers, Pierre Joseph Redouté, and when it was proposed that they should exchange works, the "Raphael of Flowers" consented, gave Audubon at once nine numbers of his *Belles Fleurs*, and promised to send "*Les Roses*."

Figure 60. Audubon. After a Portrait in Oils, Hitherto Unpublished, Painted About 1826 W. H. Holmes for Walter Horton Bentley, of Manchester, and in 1913 in Possession of His Grandson, John Conway Bentley, Formerly of Glasgow. In the Original Audubon Is Represented in a Green Coat, a Crimson Cloak with Deep Fur Edging Thrown over One Shoulder, and with Portfolio in Hand. From a Photograph in Possession of Mr. Ruthven Deane.

During this visit of eight weeks Parker painted portraits of both Cuvier and Redouté; Swainson worked steadily at the Museum, where Isidore Geoffrey Saint-Hilaire gave him the use of his private study; while Audubon, for the most part, was driving from post to pillar in his not altogether successful efforts to extend his subscription list. As already intimated, his greatest success in Paris was in winning the friendship and endorsement of Cuvier, who reported upon his work at a meeting of the Royal Academy of Sciences held on September 22.[362] Audubon has related how on this occasion he had an appointment to meet the Baron in the library of the Institute at precisely half past one o'clock; he waited; the hall filled, and the clock ticked on, but the great savant did not appear. Finally, said Audubon, after an hour had passed, "all at once I heard his voice, and saw him advancing, very warm and apparently fatigued. He met me with many apologies, and said, 'Come with me'; and as we walked along, he explaining all the time why he had been late, while his hand drove a pencil with great

[362] See Vol I.

rapidity, and he told me that he was actually now writing the report on my work!"[363] Cuvier's published report, which was extremely laudatory, showed little signs of haste. After speaking of Audubon's talents and accomplishments he said:

> The execution of these plates, so remarkable for their size, seems to us equally successful in the drawing, the engraving, and coloring, and though it may be difficult to represent relief in a colored print with as much effect as in painting proper, this is no disadvantage in works on natural history; naturalists prefer the true color of objects to those accidental shades which result from the diverse inflections of light; necessary though these be for completing the truth of a picture, they are foreign as well as prejudicial to scientific accuracy.[364]

By November Audubon was once more in London, busy at painting to fill his orders and his purse. On the 11th of the month, we find Swainson, whose own exchequer was empty, writing to Audubon for a loan; this letter, and one soon to follow, illustrate some of the characteristics to which we have referred:

William Swainson to Audubon

Tuesday 11 Nov. 1828.

I had written the enclosed, my dear Mr. Audubon, before your letter of Monday reached me. It has come this instant, Dreams, you know, must always be interpreted contrawise, we might have lifted up our arms, as you saw in your dream but, if you had not awoke, it was no doubt to have shaken hands! But that my regard for you may be evinced, I will bring myself to lay under an obligation, which I would only ask for one of my own family. I was that moment thinking to which I should write, to ask the loan of 80 £ for a few months, and now I will ask it of you. If you was aware of the peculiar feelings which we Englishmen have on such occasions, perhaps you would smile, but so it is that we never ask any one, from whom we have the least idea of a refusal. Now, did I not believe you to be a sincere friend, do you imagine I should have told you I was in want of Money much less have asked you to lend me some. The fact is, I have suffered a severe loss during my being in Paris, what little I had on hand has been spent there and in making preparations for the publication of my Zool. Illustrations. Two or three months however, hard work will bring me round again & repay you.

Let me see your letter to the President of the Zool. Soc. before it goes, and you shall see mine.

I shall be most thankful for the Grouse. I send 2 drawings to Havell to be engraved spur him on for I want to have everything ready before the new year.

Yours most sincerely,
W. Swainson.
John J. Audubon,

Esq. 79 Newman St.

---

[363] Maria R. Audubon, *op. cit.*, vol. i, p. 323.
[364] See Bibliography, No. 93a.

In December the Swainsons invited Audubon to dine with them at Christmas; in his letter Swainson said:

Why are you so sad? I would lay ten shillings that old Havell has been disappointing you as he has done me. He is in matters of business a complete daudle—an old woman, and I have done with him. His son I think better of he has a good idea of punctuality in business.... In one of your walks I hope you have thought about the French Wine that we talked so much about and have ascertained the particulars from your friend, so that we may order a cask. I hope you have not mistaken the price,—for if not, nothing that can be drank in this country is one half as cheap.

In the following letter Swainson refers to the second series of his *Zoölogical Illustrations*,[365] the sale of which was irritating him, and to N. A. Vigors, with whom he had entered upon a notorious controversy in 1828:

William Swainson to Audubon

18 January, 1829.

My Dear Mr. Audubon,

I write this in utter uncertainty whether it will find you in London. My first number has now been out three weeks—it has been seen and universally admired, and how many copies do you think the Publisher has sold? now pray guess as the Americans say. 100—no. twentyfive, no. fifteen, no. ten? yes. positively ten copies and *no more*, has been sold. I blush almost to confess this mortification to even, *you*, but so it is. Now, my dear Sir, what am I to think of the "generally diffused taste," as the phrase is, for Natural History.

This allthough vexing to *me*, may be a consolation to *you*, who are able to exhibit on what I call your *Red Book* the names of a good portion of 150 subscribers to a 200 guinea Book. Think yourself my friend exceedingly well off.

The amount of sale must be kept silent, it would be a nice nut to crack for V [igors]. & his friends.

I shall be able to do without the water birds, if you have not found any.

I have had a most extraordinary letter from Waterton, which will highly amuse you. The man is mad—stark, staring mad.

Yours very faith'ly

W. Swainson.

Can you tell me any safe expeditions made of sending and receiving letters and Parcels from Philadelphia.

J. J. Audubon Esq.

79 Newman St.
Oxford St.

Early in 1829 Bonaparte wrote from Rome, where he had then settled, and the following letter shows that he had then heard of Audubon's visit to France, and was keenly interested in his success:

---

[365] The three volumes of this series bear date of 1832-33, but the preface is inscribed "Tittenhanger Green St. Albans, 24th July, 1829."

Charles L. Bonaparte to Audubon

Rome January 10th 1829.

Dear Sir,

I received in due time your favours of November 3d. & December 21 st. & now come to thank you for them, wishing you or rather expressing to you at the occasion of the renewal of the year, the warm wishes I constantly have for your health, happiness & especially for the success of your work. From the contents of your letter I clearly perceive that one at least of my letters to you must have miscarried. Nothing could be more interesting to me than the narrative of your journey to France, though I had heard from other quarters the good & well deserved reception you met with. Your letter of August 20 th. never came at hand, & it must have been the same with at least one of mine to you. What you mention about Temminck quite astonishes me! ... I thought he would have undertaken even a journey to see you & your drawings!!! Please let me know when you write whether the Ornithological Illustrations of Jardine, Vigors & Co are stopped or still going on.—The animals I spoke to you of were reported as delivered to you by Mr Gray of the British Museum who had received them for me from the U. States. Is it not so? ... Corvus Cornix with us is very fond of the sea shore & feeds occasionally on fish, but I never observed it had the singular habits of C. ossifragus at least as described by Wilson.

I am surprized at Messrs J B's conduct; I have always found them extremely kind and well disposed towards me; & although we have settled our accounts I had no reason to believe they would refuse our box. However we can do without their interference quite as well, & I hope you have already forwarded the box to Leghorn recommending it to the care of my agent in that port. Messrs F. & A. Filuchs.(?) I shall keep a good lookout for it being extremely anxious to see your new number. I should never have done if I was [to] repeat [to] you all the praise given to your work by our Italian artists & men of science!... I shall merely state that on my part I prefer the plate of Goldfinches to any other, birds and plants, being life itself; & that I am most anxious to see Astur Stanleyi which I strongly suspect to be my Falco Cooperii.... By this time, however you may have been able to ascertain the fact ... please let me know how the thing stands. It is only by your letter that I hear of my work (2 d) being in London: I have not yet seen a copy myself nor did I know positively that it had been published. You must surely have received one from myself at all events, for I directed Messrs Gay & Lea to let you have one of the very first out. Let me know whether you have it & your opinion about it.—I think you are right in going to Russia, especially as in giving them the American Birds you will probably give us the Russians, some of which are hardly known. Try to get for me Pyrrhula longicauda, P. rosea & Scalopax—thalina, the latter especially. I shall not loose sight of the portrait, but it will be still more difficult to get the signature. I will however endeavor from some of my relations. You were right in supposing me "dans les bras de la paix & le bonheur d'un heureux père de famille" but greatly mistaken to think I was taking "le plaisir des sciences." Settling and other cursed worldly affairs have so much taken up my time, that I have not looked a specimen or a book since I am in Rome my small library itself & my Cabinet have not even been arranged & I tremble to find all my birds destroyed when the happy day will come to look into them. In the meantime an addition has been made six weeks ago to my small family. I have another son who has received the names of Lucien Louis Joseph Napoleon & better than that who is the porthrait of health itself. I am sure you will divide my happiness & excuse my delay in answering you principally on that account. I am in debt with half the scientific world & this has been the first letter I scratched since I am in Rome!... I hope to be more regular & less in a hurry in future ... though God knows!... I will not however close this letter without mentionaing the pleasure I had the other day in getting you a new subscriber & that among the English themselves.! The Earl of Shrewsbury & his good Lady highly admired your work the other day at my house & were so pleased with it that they said they would write immediately to add their name to the list. The Earl of Shrewsbury is as you know the first Earl of Great Britain a catholic & what is more to you a man of great taste. His not having heard of your work shows that you have not made enough noise about it: & I am sure his name will be followed by a great many others to which Mr. Chapittar (Lord Shrewsb. friend) has promised me to show the work & deliver the prospectuses. Did you hear of the death of poor Mr Barnes killed by a stag (?). It is a great loss for the Queen. I remain, Dear Sir, begging you the London news

Your most obliged friend
Charles L. Bonaparte.

[Addressed] Mr. J. J. Audubon
79 Newman Street
Oxford St.
London
Inghilterra.
[Endorsed] Answered Feby. 8th. 1829.
J. J. A.

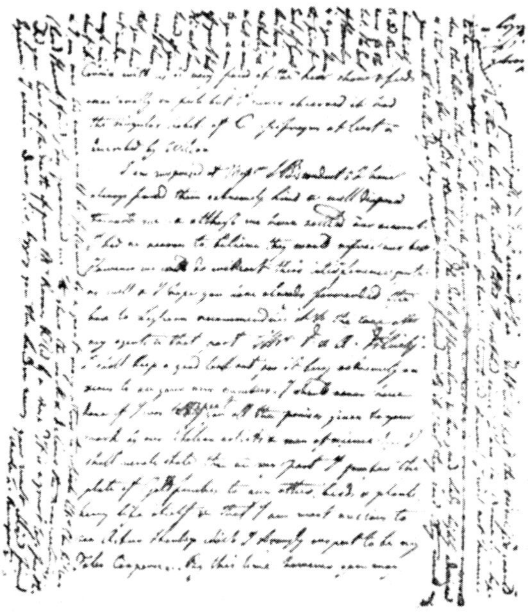

Figure 61. Part of Letter of Charles Lucien Bonaparte to Audubon, January 10, 1829. From the Howland MSS.

Audubon continued to work on his paintings during the winter of 1828-9, hoping to put his affairs in such order that he might be able to start for America in the following year.

Chapter 24

First Visit to America in Search of New Birds

Audubon laid his plans to visit America in 1829 with unusual care, and was fortunate in being able to entrust his publication to the competent hands of John George Children, of the British Museum. This was to be actually his third voyage to the United States, but it was the first which he made from English soil, and after he had become known as an ornithologist and animal painter. He wished to renew at least fifty of his earlier drawings and to obtain new materials of every description. Although he was naturally anxious to see his wife, from whom he had been absent for nearly three years, and his boys, the elder of whom had been left at Shippingport five years before, he felt constrained to devote to his work every moment that could be spared.

When writing to his wife of his difficulties and prospects at this period, he assured her that he would act cautiously, with all due diligence and sobriety, and continued:

> Thou art quite comfortable in Louisiana, I know; therefore wait there with a little patience. I hope the end of this year will see me under headway sufficient to have thee with me in comfort here, and I need not tell thee I long for thee every hour I am absent from thee. If I fail, America will still be my country, and thou, I will still feel, my friend. I will return to both and forget forever the troubles and expenses I have had; when walking together, arm in arm, we can see our sons before us, and listen to the mellow sounding thrush, so plentiful in our woods of magnolia.[366]

A little later in 1829 he also wrote: "I have finished the two first years of publication, the two most difficult to be encountered." At that time he fully expected that fourteen years would be required for the completion of his task, owing to the many difficulties experienced, especially in securing competent workmen, as well as the necessity of distributing the expense for the benefit of his subscribers.

When Havell had been provided with all the drawings needed for the remainder of the year 1829 and the first issue of 1830, Audubon sailed from Portsmouth on the 1st of April, 1829, in the packet ship *Columbia*, which reached New York on the opening day of May. "I chose the ship," he said, "on account of her name, and paid thirty pounds for my passage."

He paused in New York to exhibit his drawings at the Lyceum of Natural History, of which he had become a member in 1824, but soon hurried to Philadelphia, and finally settled

[366] Mary F. Bradford, *Audubon*.

down for work at Camden, in New Jersey, later known to fame as the home of "the good gray poet." There, at a boarding house kept by a Mr. Armstrong, he remained three weeks, from about May 23 to June 13, hunting and painting every day. From Camden he went to Great Egg Harbor, then a famous resort of both land and water birds in great variety, and for three weeks more he lived and worked in a fisherman's cabin by the sea. It is interesting to recall that Alexander Wilson, in company with George Ord, had spent a month at this point in the spring of 1813.

The following letter[367] from Swainson was probably the one to which Audubon replied from New Jersey on September 14:

William Swainson to Audubon

My dear Mr. Audubon

I welcomed the news of your arrival in America yesterday, and as I am making up a packet for Liverpool today, I seize the opportunity of wishing you joy and happiness in the new world. I am surprised and disappointed as not receiving one line from Ward it is at the best negligent, and somewhat ungrateful. Hope you have begun your studies among the birds on a better plan than formerly, that is, in preserving the skins of every one on which there is the least doubt whether the bird is young or old, particularly the former. If you are to give scientific descriptions and definitions of the species this precaution is absolutely necessary. What your Americans do with their money I know not, Mr. Lea tells me he cannot procure one purchaser for my new Illustrations: here it is now going on very well.

You asked me what you can do for me in America. I will tell you. Send me a cart load of shells from the Ohio, or from any of the Rivers near New Orleans. The very smallest, as well as the very largest—all sizes. I have been long expecting those which your son promised you for me near twelve months ago! but I have heard nothing of them! you may spend a few dollars for me and send people to fish the shells at the dry season, when the waters are low, that is the best time.

Things go on here much as usual, but I have not been in London since Xmas. The first volume, containing the Quadrupeds, of Dr. Richardson's work, is out. I am now busy in preparing the second, which contains the Birds. Let me particularly direct your attention to the manners of the Cedar Bird, Ampelis Americana. I suspect it feeds much on Insects in default of fruit, but what is desirable, is to know the way in which it captures Insects, whether as a flycatcher i.e., by seizing them on the wing, or like the Gold crest—by picking them up among the branches or leaves. I am now in close correspondance with Charles Bonaparte, & a most valuble correspondant he is.

Mrs. Swainson is just recovering from her confinement after giving me another little son I am happy today they are both going on well.

Wilson I believe mentions two birds very like the Red eyed Flycatcher, this is a point deserving your attention, but the manners of these birds are much more important. I feel convinced there are several species of my Genus Ammodramus shore finch, in the So. States, they all have narrow pointed tails, like the seasidefinch of Wilson. I further suspect there is more than one species confounded with the Towee Buntling.

[367] Published originally by Ruthven Deane, *The Auk*, vol. xxii, 1905.

I hope soon again to hear more fully from you, and of your ornithological acquisitions. The dear little ones are quite well.

Yours very sincerely,
Wm. Swainson

The Green 26 June 1829.
Mr. John J. Audubon
care of
Mess. Thomas E. Walker & Co.
Merchants.
New York [Philadelphia]

On the 4th of July Audubon returned to Philadelphia and prepared for a longer sojourn in the Great Pine Forest, or Great Pine Swamp, as it was sometimes called, in Northumberland County, Pennsylvania. In this letter to his son we shall find an account of his plans and accomplishments:

Audubon to his son, Victor

Philadelphia, July 5th., 1829.

My dear Victor:—

I have been in America two months this day, and not a word from you have I had in answer to my several letters, dated New York, and at this place. I am also without answer from your Mama, but do not feel so surprised as I know that about 2 months is the time necessary to have a return from Louisiana.

I have come to take your Mama over to England, if her wish inclines her to do so, and have wrote fully to her, giving her all the particulars respecting my situation that I thought could possibly be trusted to a letter.

I have also come to America to redraw some of my earliest productions, and am now closely engaged at this. I remained near this city for 3 weeks, and since have spent 3 more at Great Egg Harbour, from which place I returned yesterday. I have already 13 drawings by me. I have letters from London, up to 30 th. April, when all my business was going on well with an increase of 4 subscribers. I have no news to transmit; on the contrary, I was in hopes that ere this I should have had at least one long letter from you. I beg you will write me when you last heard from your Mama. Dircct your letter to the care of Messrs Thos. E. Walker, & Co, merchants here, who know all my movements, and will see anything forwarded to wherever I may choose to go to.

I hope your uncle Berthoud & family are all well; present them my best regards, and to all others who may feel interested in my welfare, and believe me

your affectionate father,
John J. Audubon.

I have bought a good gold time-keeper, intended for you, and a copy bound, of my work, and wish to know how it can be forwarded. God bless you.

After outfitting himself in Philadelphia, Audubon proceeded to Mauch Chunk; his provisions for this journey to the forest consisted of a "wooden box containing a small stock of linin, drawing-paper, my journal, colors and pencils, together with twenty pounds of shot, several flints, a due quantum of cash, my gun 'Tear Jacket,' and a heart as true to nature as ever." From Mauch Chunk he traveled fifteen miles into the heart of the wooded hills, and was received into the family of Jedediah Irish, lumberman and philosopher, whose praise was celebrated in a later "Episode."[368] "What pleasure," said the naturalist, "I had in listening to him, as he read his favourite poems of Burns, while my pencil was occupied in smoothing and softening the drawing of the bird before me. Was this not enough to recall to my mind the early impressions that had been made upon it by the description of the golden age, which I here found realized?"

During his stay in the forest Audubon paid particular attention to the smaller land birds, such as finches, warblers and flycatchers, and many of the original drawings which were made in the summer of 1829 still bear his penciled designations of time and place.[369] About ten weeks[370] were spent in the woods, from late July until the 10th of October, when the naturalist returned to Philadelphia and settled again for a time in Camden. At this period he was enjoying the best of health and spirits, and he worked during the entire season under the highest pressure of which he was capable. At Camden, October 11, 1829, he wrote:

> I am at work, and have done much, but I wish I had eight pairs of hands, and another body to shoot the specimens, still I am delighted at what I have accumulated in drawings this season. Forty-two drawings in four months, eleven large, eleven middle size, and twenty-two small, comprising ninety-five birds, from Eagles downwards, with plants, nests, flowers, and sixty kinds of eggs.[371] I live alone, see scarcely any one, besides those belonging to the house where I lodge. I rise long before day, and work till night-fall, when I take a walk, and to bed.

---

[368] See "The Great Pine Swamp," and "Great Egg Harbour," *Ornithological Biography*, vol. i, p. 52, and vol. iii, p. 606.

[369] Though the year is not usually indicated on the originals, the following drawings probably belong to this period:

- Black Poll Warbler, New Jersey, May.
- Wood Pewee Flycatcher, New Jersey, May.
- Small Green-crested Flycatcher, New Jersey, May.
- Golden-crowned Thrush, New Jersey, May.
- Warbling Flycatcher, *Vireo gilvus*, New Jersey, May 23.
- Yellow-breasted Chat, New Jersey, June 7.
- Sea Side Finch, Great Egg Harbour, June 14.
- Marsh Wren, New Jersey, June 22.
- Bay-winged Bunting, Great Egg Harbour, June 26.
- Canada Flycatcher, Great Pine Swamp, August 1.
- Pine Swamp Warbler, Great Pine Swamp, August 11.
- Black and Yellow Warbler, Great Pine Swamp, August 12.
- Hemlock Warbler, Great Pine Swamp, August 12.
- Autumnal Warbler, Great Pine Swamp, August 20.
- Connecticut Warbler, New Jersey, September 22.

Mottled Owl, New Jersey, October.

[370] Though Audubon said that he spent only six weeks in the forest, the indications upon his drawings imply a longer period.

[371] At this time Audubon intended to figure, in full size and natural colors, the eggs of the "Birds of America," for which the concluding numbers of his plates had been reserved, but when the time came, these numbers had to be given over to new acquisitions, so the eggs were eventually crowded out.

At about the middle of October Audubon set out to join his family in the South. Crossing the mountains by mail-coach to Pittsburgh, where he met his former partner in business, Thomas Pears, he descended once more his favorite river, the Ohio. It was no longer necessary to rough it on a flatboat or to sleep on a steamer's deck; it was to be "poor Audubon" no longer. To be sure, he was not rich, but he had made his way and his mark, and the attention which he now began to receive when traveling in his adopted land must have gratified his heart. He paused at Louisville to visit his two boys, the elder of whom, Victor, was then a clerk in the office of his uncle, William G. Bakewell, while John was with another uncle, Nicholas A. Berthoud. Hastening on he reached Bayou Sara on November 17, where he finally joined his wife, who was living at the home of William Garrett Johnson, in West Feliciana Parish, near Wakefield. Some account of this journey is given in the following letter,[372] written on the eighteenth to Dr. Richard Harlan; in the postscript Audubon gives the first reference to a new hawk which he proposed to name after his friend, and which has given no little trouble to ornithologists ever since:[373]

Audubon to Dr. Richard Harlan

[Superscribed] Richd Harlan Esqr. M. D. &c &c &c
Philadelphia Pensa

St Francisville Louisiana November 18th 1829—

My Dear Friend.—

You will see by the data of this the rapidity with which I have crossed *two thirds* of the United States. I had the happiness of pressing my beloved wife to my breast Yesterday morning; saw my two sons at Louisville and all is well.—from Philadelphia to Pittsburgh I found the Roads, the Coaches, horses Drivers and Inns all much improved and yet needing a great deal to make the traveller *quite comfortable*—The slownesse of the stages is yet a great bore to a man in a hurry—I remained part of a day at Pittsburgh where of course I paid my respects to the Museum! I was glad to see the germ of one—it is conducted by a very young man named Lambdin—I made an arrangement with him [place of seal—paper gone] &c. &c. &c. at Cincinnati I also visited the Museum [paper gone] it scarsely improves since my last view of it, except indeed by wax figures and such other shows as are best suitable to make money and the least so to improve the mind.—I could not see D [illegible] my time was very limited.—The Ohio was in good order for Navigation and I reached Louisville distant from you about 1,000 Miles in one week.==as you spoke of travelling westwardly I give you here an a/c of the Fare.—to Pittsburgh all included 21$.—to Louisville 12$.—and 25$ more to Bayou Sarah where I Landed. 30$ is the price from Louisville to N. Orleans.==our Steam Boats are commodious and go well—but my Dear Friend the most extraordinary change has taken place in appearance as I have proceeded.—The foliage had nearly left the Trees in Pensylvania, the Swallows had long since disapeared severe frost indeed had rendered Nature gloomy and uninteresting—Judge of the contrast: I am now surrounded by *Green*

[372] At one time in possession of Dr. R. W. Shufeldt, who received it from Mrs. Audubon; given verbatim by Elliott Coues, *Bulletin of the Nuttall Ornithological Club*, vol. v, 1880.

[373] Harlan's Hawk, or the Black Warrior, is now regarded as a southern variety of the Red-tailed Hawk, and is designated under the trinomen, *Buteo borealis harlani*.

Trees and Swallows gambole around the house as in Pensylvania during June & July==The mock bird is heard to sing and during a Walk with my Wife yesterday I collected some 20 or 30 Insects==that is not all, a friend of mine here *says* that he has discovered 2 or 3 *New Birds*!!!—new Birds are new birds our days, and I shall endeavour to shew you the Facts Simile when again I shall have the pleasure of shaking your hand—although so lately arrived, I have established the fact that Mrs A. and myself will be on our way towards "Old England" by the 15th of Jan.y. we will ascend the Mississipi and after resting ourselves at Louisville with our sons and other relatives about one month and then proceed with the Rapidity on the Wild Pigeon should God grant our wishes!—

have you seen or heard anything of Ward?—have you the little sketch of Dear?==we had a passenger on Board the Huntress named *Potts* from your City who knows you well as lively young Gentleman; has a Brother (a Clergyman) established and married at Natchez.—

I will begin Drawing next week having much scratching with the Pen to perform this one, and I am also desirous to make [paper gone] Large Shipment of aborigines both animal and vegetal as soon as possible.—Turkeys, Aligators, Oppossums, Paroekett, and plants, as Bignonias &c &c &c. will be removed to the Zoological Gardens of London, from the Natural ones of this Magnificent Louisiana!—meantimes I will not forget *my Friends in Phila.* no I would rather forgive all, to all my Ennemies there.—assure Dr. Hammersley that Ivory Billed and Peleated Woodpeckers will be skinned, and who knows but I may find something more for him.—I will give free leave to Dr. Pickering to chuse amongst the Insects and who knows but I may find something new for him. remember me most kindly to both, nay not in the common manner of saying "Mr Audubon begs to be remembered" no not [at] all. This way Mr A remembers *you* and *you* and *I* will remember *you* and *you* and *I* always!!—

May I also beg to be remembered in humble words to a fine pair of Eyes; divided, not by the Allegany Mountains; but by a nose evidently imported from far *East*, to a placid forehead, to a mouth speaking happiness to——[dash nearly across page.]

Should you see Friend Sully remember me to him also—and should you see George Ord Esqr. *Fellow* of all the Societies *Imaginable* present him my most humble——[dash line more than across the page.]

Should you see that good woman where I boarded at Camd'den tell her that I am well and thankful to her for her attentions to me.—

I cannot hope the pleasure of an answer from you here but you may do so, and I say pray do so, directed to the care of N. Berthoud Esqr Louisville Kentucky.—by the bye my sons are taller than me, the eldest one so much altered that I did not know him at first sight, and yet I have *Eyes*—

God bless You, Your Friend

John J. Audubon.

[The following is written across the first page:]

I reopened my letter to say that I have Just *now* killed a Large *New Falcon yes positively a new Species of Hawk almost black about* 25 Inches Long and 4 feet broad tail square Eye yellowish White, Legs and Feet bare short & strong.——I will skin it!!!—remember me to Lehman——

What I have said about the Hawk to You must be *Lawful* to *Academicians* and you will please announce *Falco Harlanii* by John J. Audubon, F. L. S. L.

The following extracts are from a letter[374] written by Swainson, January 30, 1830, and sent to Havell in London to be readdressed:

William Swainson to Audubon

I know not in what part of the Wilds of America you may now be wandering, but I hope you are fully intent upon your great object, and that you are not only making drawings, and taking notes, *but preserving Skins*, of all your little favorites. Don't forget the *Shrikes*, of which I have strong suspicions there are 2 or 3 species mixed up with the name of Loggerhead. Should you be in the land of the *Scarlet Ibis*, do pray procure a dozen or two of the best skins, they are the most magnificent birds of No. America, and are said to be common towards New Orleans.

You will learn from the Newspapers how uncommonly severe is our winter the snow has now been upon the ground five weeks and it is still falling. I manage, however, to walk out every day, and thus have acquired better health than I have enjoyed for many years.

Previous to your embarking to England, which I hope you will do very early in the spring you must do me one favor. Bring me two Grey Squirrels alive, and a cage full of little birds, either the painted or non-Pareil finch the Blue finch, or the Virginian Nightingale, as they are called, 3 or 4 of each to guard against casualties by death on the voyage. I do not care one farthing whether they sing or not, so that I presume they may be got for a mere trifle. The Squirrels would delight the little people beyond measure, and would prove a never-failing source of amusement to them. I believe you have other kinds than the *grey*, so that any will do. If you cannot get them pray supply their place by two Parrots of America.

We continue pretty well at the Green. Seldom go to town, but I find people begin to discover the true character of V [igors]. and many that were formerly his friends now speak very differently of him. His father having died the property has come to him. He has now taken a fine house in the Regents park, and holds *conversaziones* (in humble imitation of those of the President of the Royal Society) every Sunday evening *during the season*!! all this is very grand, and he appears to have abandoned writing any more papers on ornithology, since I have begun to point out his errors.

Ward wrote to me since my last, he is a poor weak fellow, with a good natural disposition, but so little to be depended upon, that he is turned round by every feather, after inserting that he could not go on "in my service" as he called it, under *ten dollars* a week, he now says he should be most happy to receive *four*. He says not a word of his marriage, which proves his wish to deceive one. I have done with him.... I hope you have got me *lots of River shells*.

About the beginning of the year 1827 Mrs. Audubon gave up her "Beechwoods" school, and thereafter took a position as governess in the home of Mr. William Garrett Johnson, whose plantation, called "Beechgrove," was situated in the same parish. An anonymous writer thus referred to this house in 1851.[375]

---

[374] Published by Ruthven Deane, *The Auk*, vol. xxii, 1905.
[375] Thomas B. Thorpe, *Godey's Lady's Book*, vol. xlii, 1851.

In the hospitable mansion of W. G. J——, in the parish of West Feliciana, if one will look into the parlor, they will see over the piano a cabinet sized portrait, remarkable for a bright eye and intellectual look. The style of it is free, and there is an individuality about the whole that gives assurance of a strong likeness. Opposite hangs a proof impression of the bird of Washington, a tribute of a grateful heart to an old friend. The first is a portrait of Audubon painted by himself; the other is one of the first [of his] engravings that ever reached the United States.

There Audubon spent nearly two months at the close of 1829, and followed his usual occupations of hunting and drawing, while his wife prepared for their contemplated journey to Europe. He is said to have drawn at this time the "Black Vulture attacking a herd of Deer," several large hawks, squirrels, and heads of deer which were never finished.

Although Audubon's business affairs in England had been left in charge of his trustworthy friend, John G. Children, his engraver, Havell, had become alarmed at the loss of subscribers and the failure of certain of their agents, and particularly M. Pitois of Paris,[376] to render due returns. Havell, as it proved, was unduly disturbed, but his gloomy accounts tended to hasten the naturalist's departure, a circumstance that was later deplored. These matters are clearly reflected in the following letter written from the Johnson home in Louisiana when the Audubons were preparing to leave it; particularly interesting are the included statements through which it was hoped that a competent successor might be secured for the duties of the position which Mrs. Audubon had so ably filled:

Audubon to Robert Havell

Beech Grove, Louisiana
Decr 16th 1829

My Dear Mr. Havell.—

I received yesterday from New York your letter of the 29th. Sept. which must have reached Philadelphia 3 days after my departure for home==

I am sorry that Bartley should have made you suffer a moment by sending you the intelligence of the failure of the several subscribers you mention in your favor—it cannot be helped—there is none of your fault and *I* must repair these matters when I reach England again==

I am considerably more sorry and much vexed that Sowler should have failed in his *written* promise to accept your Dfts.—even in a case of the diminution of subscribers he could certainly have sent you a progressional amount—I am now *almost* sure that Pitois has failed or acted the Rogue==

We are making all preparations in our power to leave Louisiana on the 5 or 10th. of Jan.y and we will proceed as fast as Steam Boats, Coaches and the weather will admit of and we will sail for England from New York with all possible dispatch. I have made a shipment of Forest trees to England that I hope will turn to good account as they are to be

[376] While in Paris in 1828, Audubon wrote on October 26 that he had received a call from "a M. Pitois, who came to look at my book, with a view to becoming my agent here; Baron Cuvier recommended him strongly, and I have concluded a bargain with him. He thinks he can procure a good number of subscribers. His manners are plain, and I hope he will prove an honest man." See Maria R. Audubon, *Audubon and his Journals*, vol. i, p. 339.

presents to Public Institutions &c and that I think it necessary to be *remembered myself.*—

We are both well—our sons are at Louisville, Kentucky where we will see them about the 20th of next month.—I sent you in my letter a proposal for your sister and should you not have received it I send it you again here in Mrs A.'s. hand writing.—*I would advise your sister to come* if the *money* is an object.—I think that besides she will be comfortable with the family Johnson—if she thinks fit to wait until we see her, we can tell her all about it.==

I have received only one letter from friend Children during all this absence against my very many—I hope the insects I sent him by the Annibal have reached safely.—have no news to give you—Keep up a good heart—we will be in London as soon as possible.—I have not had a letter from Miss Hudson for a long time—I hope her mother & her are well—Remember me kindly to your Dear Wife and Little ones—Mrs Audubon joins me in all good wishes—If you see Parker my remembrances to him==I will carry with me some Drawings that I know will make the *graver* and the *Acid* Grin again.—

Believe me your friend—

John J. Audubon.

When you present my sincere regards to friend Swanson [Swainson] tell him that I have had only one letter from him and that I am now quite unable to say where Mr Ward is==I had a letter from Henry Havell[377] the other day merely acknowledging the money I have paid him—he was in New York, I hope quite well—

[Enclosure]

A friend of ours here named Wm. Garrett Johnson (a cotton planter) a gentleman who resides in a perfectly healthy and agreeable part of the country, desires that I should write to England to procure for him a Governess, one who can teach music, drawing and the usual branches of education to young Ladies. Mr. Johnson will pay the sum of one thousand dollars per annum, board, lodging &c, also and considered in all respects as a member of the family, to any lady who will undertake occupation (the sum is about 230£) the governess will have to instruct ten or twelve young persons of various ages, and may make the arrangement for five years if desirous of it. I have thought this would suit your sister precisely, and for my part knowing the family Johnson as I do I should think it an excellent thing for her. if not I will look for some one when I am in England, Sailing from England direct for New Orleans, steam Boats reach the place of Mr Johnson in two days.

Duplicate.

I, Wm. Garrett Johnson do authorize my friend J. J. Audubon to make the above proposition and do by these present obligate myself to comply with them punctually and particularly.

Wm. Garrett Johnson.

[Addressed]
Mr Robt Havell Jur
Engraver
79 Newman Street
Oxford Street
London
England

[377] Henry Augustus Havell, a younger brother of Robert Havell, Junior.

"On January 1, 1830," said the naturalist, "we started for New Orleans, taking with us the only three servants yet belonging to us, namely, Cecilia, and her two sons, Reuben and Lewis. We stayed a few days at our friend Mr. Braud's, with whom we left our servants, and on the seventh of January took passage on the splendid steamer *Philadelphia* for Louisville, paying sixty dollars fare."[378] After a long visit with their sons, on the seventh of March they ascended the Ohio to Cincinnati, and at Wheeling took the mail-coach to Washington. At the national capital Audubon met the President, Andrew Jackson, and was befriended by Edward Everett, at that time a leader in the House of Representatives. "Congress," said the naturalist, "was then in session, and I exhibited my drawings to the House of Representatives, and received their subscription as a body." He also recorded that he obtained three subscribers in Baltimore, and left for Philadelphia, where they remained a week. The following note, which Edward Everett gave Audubon for New York, is particularly interesting, since it expressly states that at that time the ornithologist had not received a single subscriber in the United States:

Edward Everett to Dr. Wainwright

Washington 18 March 1830

My dear Sir,

Allow me to introduce to your acquaintance, the bearer of this letter, Mr. Audubon of Louisiana. His drawings of American Birds, of which he will show you some, will I am sure command your approbation, as they have the applause of Europe.—I am sorry to say, that he has not yet procured a single subscriber, in the United States of America. Will not one of your Institutions in New York—or your wealthy and liberal individuals—take a copy? I pray you endeavor to procure him at least one subscriber, in New York.—

Yours with great regard
E. Everett.
Rev Dr Wainwright

Audubon had evidently reconsidered his expressed intention of presenting a copy[379] to Congress, and to Edward Everett belongs the credit of subscribing to *The Birds of America* in behalf of the Congressional Library. At about this time also he obtained another subscriber at Washington, in the person of Baron Krudener, the Russian envoy, but later experienced difficulty in collecting his dues.[380]

---

[378] See Lucy B. Audubon, ed., *Life of John James Audubon, the Naturalist*, p. 203. Since black slaves were the only domestics available in the South at that time, it is probable that the "servants" referred to were employed by Mrs. Audubon at her "Beechgrove" school.

[379] See Vol. I.

[380] See Vol. II.

Chapter 25

AUDUBON'S LETTERPRESS AND ITS RIVALS

On the 1st of April, 1830, Audubon and his wife sailed from New York in the packet ship *Pacific*, bound for Liverpool, where they landed after a voyage of twenty-five days. Upon returning to London the naturalist found that upon the 18th of the preceding March he had been elected to membership in the Royal Society, an honor for which he felt indebted to Lord Stanley and his friend Children, of the British Museum; after paying the entrance fee of £50, he took his seat in that body on the 6th of May. The painting of pictures was at once resumed to meet his heavy expenses, but towards the end of July he started with Mrs. Audubon on a canvassing tour, in the course of which his plans suddenly were changed so that London did not see him again for nearly a year.[381] On this journey they touched at Birmingham, Manchester, Leeds, York, Hull, Scarborough, Whitby, New Castle, and Belford, to visit the Selbys, and on the 13th of October reached Edinburgh, where they were soon comfortably settled in the naturalist's old lodging place, the house of Mrs. Dickey, Number 26, George Street.

Audubon was now ready to begin the text of his *Birds of America*, to be called *Ornithological Biography*, which is often referred to as his "Biography of Birds." This work, which was eventually extended to five large volumes of over three thousand pages, was published at Edinburgh from 1831 to 1839. He had made crude beginnings with this in view as early as 1821, and on October 16, 1830, he wrote: "I know that I am not a scholar ..." but, "with the assistance of my old journals and memorandum-books, which were written on the spot, I can at least put down plain truths, which may be useful, and perhaps interesting, so I shall set to at once. I cannot, however, give scientific descriptions, and here must have assistance." To supply this need, as we have seen already, he had earlier applied to William Swainson, but the negotiations with that naturalist were soon broken off, and led to a sharp and acrid discussion upon the authorship of the work itself.[382]

---

381 His correspondence with William Swainson from this point, and the history of his letterpress so far as that naturalist was concerned, will be unfolded later.

382 See Chapter 28.

Figure 62. Mrs. Dickie's "Boarding Residence," 26 George Street, Edinburgh, Where Audubon Occupied Apartments and Painted and Wrote in 1826-27 and 1830-31. A Large Public Building Now Occupies the Site. After a Photograph in Possession of Mr. Ruthven Deane.

Figure 63. The Academy of Natural Sciences of Philadelphia. After an Old Print; Reproduced from *Cassinia* for 1910.

By a rare stroke of genius or good fortune, Audubon chose for his assistant a young Scotch naturalist, William MacGillivray, who had been introduced to him by another naturalist, James Wilson, soon after he reached the Scottish capital. MacGillivray agreed "to revise and correct" his manuscript at the rate of two guineas per sheet of sixteen pages, and in the latter part of October, 1830, they set to work. We shall soon have occasion to speak more fully of his debt to this estimable Scotchman,[383] and will only add here that a better trained or more competent helper than MacGillivray could hardly have been found in Great Britain or elsewhere.

No sooner had Audubon begun to write than it was learned that "no less than three editions of 'Wilson's Ornithology' were about to be published, one by Jameson, one by Sir W. Jardine, and another by a Mr. Brown." The outlook could not be considered encouraging, but this intelligence only nerved him to greater effort, and he was determined to push his own publication with such unremitting vigor as to anticipate them all. "Since I have been in England," he wrote in his journal, "I have studied the character of Englishmen as carefully as I have studied the birds in America, and I know full well that in England novelty is always in demand, and that if a thing is well known it will not receive much support." Audubon worked continuously at his *Biography*, rising before the dawn and writing all day, while the able worker at his side carried his efforts far into the night, and in three months the first volume was ready for the printer; Mrs. Audubon meanwhile copied their entire manuscript to be sent to the United States in order to secure the American copyright. When this work was offered to the publishers at Edinburgh, however, not one of them, said the naturalist, would offer a shilling for it, but this did not deter him from publishing it at once and at his own expense.[384] On March 13, 1831, he wrote: "The printing will be completed in a few days, and I have sent copies of the sheets to Dr. Harlan, and Mr. McMurtie, at Philadelphia, and also one hundred pounds sterling to Messrs. T. Walker & Sons, to be paid to Dr. Harlan to secure the copyright, and have the book published there."

The following friendly letter from one of Wilson's editors belongs to this period:

Sir William Jardine to Audubon

Jardine Hall 3rd Decr. 1830—

My dear Sir,

I only learnt a few days since that you were to winter in Edinburgh, and perhaps since you are not Hurried for time in Trovelly [?] will come out to spend a day or two with me—If you can come out before the 10 th. when I shall have the pleasure of shewing you some Blackgame Shooting—The season expires on the Tenth of the Month partridges have bred so ill that there is scarsely any in the whole country, and pheasants have been so lately introduced that they are yet rather scarce—In a wet day you may have your easel & brushes I should wish much to hear your account of Wilson during the times you hunted with him—and also some account of the New Species you figure in the american Ornithology—

---

[383] See Chapter 30.

[384] The first volume of the *Ornithological Biography* in the European edition bears the imprint of "Adam Black, 55 North Bridge, Edinburgh;" in the four subsequent volumes this was changed to "Adam and Charles Black," while the entire work was printed by "Neill & Co., Printers, Old Fish Market, Edinburgh."

I am happy to learn you intend figuring the learned Men of America as accompanyment to your work particularly the ornithologists, do you know the painter of the portrait of Wilson—I have three portraits of him in the House, and also a profile taken by the machine I should like to have your opinion of them one of the portraits was painted from an original that went to America—

I shall expect to hear you are coming soon—Mr Lizars will tell you about coaches—&c

With best regards believe me
Sincerely yours
Wm Jardine

[Addressed] J. Audubon Esqr
Care of W. H. Lizars Esqr
3 James Square
Edinburgh.

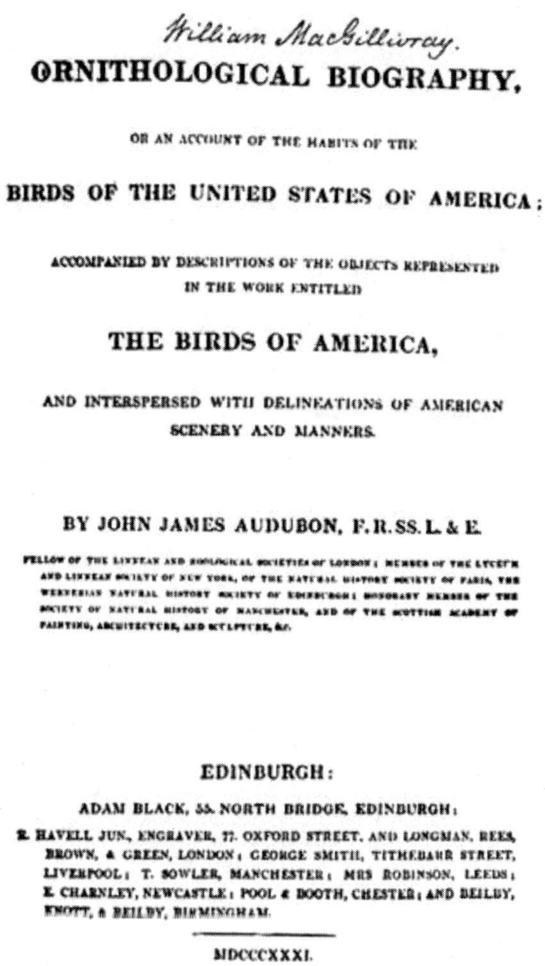

William MacGillivray.

ORNITHOLOGICAL BIOGRAPHY,

OR AN ACCOUNT OF THE HABITS OF THE

BIRDS OF THE UNITED STATES OF AMERICA;

ACCOMPANIED BY DESCRIPTIONS OF THE OBJECTS REPRESENTED IN THE WORK ENTITLED

THE BIRDS OF AMERICA,

AND INTERSPERSED WITH DELINEATIONS OF AMERICAN SCENERY AND MANNERS.

BY JOHN JAMES AUDUBON, F.R.SS.L. & E.

FELLOW OF THE LINNEAN AND ZOOLOGICAL SOCIETIES OF LONDON; MEMBER OF THE LYCEUM AND LINNEAN SOCIETY OF NEW YORK, OF THE NATURAL HISTORY SOCIETY OF PARIS, THE WERNERIAN NATURAL HISTORY SOCIETY OF EDINBURGH; HONORARY MEMBER OF THE SOCIETY OF NATURAL HISTORY OF MANCHESTER, AND OF THE SCOTTISH ACADEMY OF PAINTING, ARCHITECTURE, AND SCULPTURE, &c.

EDINBURGH:

ADAM BLACK, 55. NORTH BRIDGE, EDINBURGH;

R. HAVELL JUN., ENGRAVER, 77. OXFORD STREET, AND LONGMAN, REES, BROWN, & GREEN, LONDON; GEORGE SMITH, TITHEBARN STREET, LIVERPOOL; T. SOWLER, MANCHESTER; MRS ROBINSON, LEEDS; E. CHARNLEY, NEWCASTLE; POOL & BOOTH, CHESTER; AND BEILBY, KNOTT, & BEILBY, BIRMINGHAM.

MDCCCXXXI.

Figure 64. Title Page of Volume I of the "Ornithological Biography."
From a Copy Presented by Audubon to William MacGillivray and Bearing the Latter's Signature.

Audubon was not outstripped by his Edinburgh rivals, who to all appearances had planned to cover the field of American ornithology so thoroughly as to render his work a drug on the market, if not to make it superfluous. Whether this were really true or not, there is no doubt that Audubon's activity furnished the stimulus to the sudden appreciation of the work of his predecessor that was manifested in Edinburgh at this very moment of time. It will be interesting to see just what these rival enterprises were. Professor Jameson, who had been of great service to Audubon at the beginning of his undertaking, prepared a pocket edition of Wilson's and Bonaparte's *Ornithology*, with miniature plates which were issued separately, and the two works, which were intended to go together, were published in 1831.[385] Sir William Jardine brought out an edition of Wilson's and Bonaparte's work, in three large volumes, with plates engraved by W. H. Lizars after the originals and carefully colored by hand.[386] This was thoroughly legitimate enterprise, but the climax was reached when Captain Thomas Brown began to publish an "Audubonized edition" of Wilson's and Bonaparte's plates, or an attempt to present their plates of American birds in the Audubonian manner, to the extent at least of showing the characteristic flowers, trees, and insects of the American continent, a plan to which some of Audubon's earlier critics in Philadelphia had offered strenuous objection. Brown's large atlas of plates[387] was issued in parts, from 1831 to 1835,

---

[385] *American Ornithology, or the Natural History of the Birds of the United States.* By Alexander Wilson and Charles Lucien Bonaparte. Edited by Robert Jameson ... Regius Professor of Natural History in the University of Edinburgh. Appearing as vols. lxviii-lxxi of *Constable's Miscellany*, 4 vols., 18mo., Edinburgh and London, 1831. This was the fourth (?) edition of Wilson's work, and the first (?) to appear in Europe; with portrait of Wilson and vignettes on titles engraved by Lizars, memoir of Wilson by W. M. Hetherington, and extracts from Audubon, Richardson, and Swainson.

The plates of this edition were issued in numbers, under title of *Illustrations of American Ornithology*; reduced from the work of Wilson; 18mo., Edinburgh and London (1831). In a notice of the first number which appeared in the *Caledonian Mercury* (Edinburgh) for Oct. 29, 1831, it was stated that the plates were issued in small size to be bound up with Jameson's edition of the text, and that they were intended "for a different class of purchasers from those likely to take the folio edition, then being brought out by the publishers of Constable's Miscellany. The plates were engraved in line and executed in a very superior style, both plain and colored."

[386] *American Ornithology; or Natural History of the Birds of the United States*, by Alexander Wilson, with a Continuation by Charles Lucien Bonaparte, Prince of Musignano. The Illustrative Notes and Life of Wilson by Sir William Jardine, 3 vols., 8vo., London and Edinburgh, 1832.

The second (?) European edition of Wilson and Bonaparte, with 97 hand-colored plates engraved by Lizars. The *Caledonian Mercury* in noticing the work, October 29, 1831, said: "It must be highly gratifying to the friends and connections of poor Sandy Wilson to see such honor, at last, paid to his memory in his native land."

[387] *Illustrations of the American Ornithology of Alexander Wilson and Charles Lucien Bonaparte, Prince of Musignano.* With the addition of numerous Recently Discovered Species, and Representations of the Whole Sylva of North America. By Captain Thomas Brown [etc., etc.]. Folio, with engraved title, engraved dedication, index, and 124 engraved and hand-colored plates. Edinburgh, Frazer & Co., 54 North Bridge, William Curry, Jun'r & Co., Dublin & Smith, Elder & Co., 65 Cornhill, London, MDCCCXXXV.

It is stated by the editor of this extraordinary work that he had added 161 birds, and that 87 have been considerably enlarged. There are 167 representations of American trees and shrubs, said to have been copied for the most part from Michaux' *Silva*. The striking *Hibiscus grandiflorus* (plate xli) was taken without acknowledgment from Audubon's drawing of the Blue-winged Warbler (*The Birds of America*, plate xx). For the most part the figures of birds are redrawn from Wilson and Bonaparte and given new positions and backgrounds. A few of the plates, as that of the California Vulture (no. 1), bear the legend, "Drawn by Captn. Tho. Brown;" all are uneven, and many extremely poor in execution, the fourteen by W. H. Lizars being the best. J. B. Kidd, for a time associated with Audubon is credited with four plates; other engravers employed on the work were James Turvey, who executed the elaborate title, Samuel Milne, James Mayson, R. Scott, J. & J. Johnstone, E. Mitchell, William Davie, S. A. Miller, John Miller, Audw. Kilgour, Wm. Warwick, and W. McGregor. Plate xiv, the Snowy Owl, *Strix nyctea*, engraved by the editor, has the interest of a caricature. Some plates show as many as fourteen birds in a medley of brilliant foliage, flowers and fruits. The violence of

and was intended as a further companion to Jameson's text for all who could afford that expensive form of illustration. By a curious coincidence Audubon's *Ornithological Biography* (vol. i), Jameson's edition of Wilson and Bonaparte (vol. i), and Brown's *Illustrations* (pt. i), were all noticed on the same page of the *London Literary Gazette* for April 9, 1831. "This day is published," so reads the advertisement of Audubon's work, "price 25s. in royal octavo, cloth, Ornithological Biography...." If the desire of these various editors were to cripple the work of the American naturalist, their efforts were certainly vain, for he was able to make his way against all competitors. Brown's work was a failure, so few copies having been distributed that it is doubtful if more than one ever came to this country, and only one is known to be in possession of any large library in England.

Audubon's initial volume of the *Biography* was well received and drew forth immediate and unstinted praise from many sources. He was anxious that MacGillivray should contribute some account of it to the London *Quarterly Review*, then under the editorial management of John Gibson Lockhart, but his suggestion was coldly received and drew forth the following declaration of independence from his able, if as yet undistinguished, coadjutor:[388]

> With respect to the review, I can only say that if Mr. Lockhart is so doubtful as to my powers, he may doubt as long as he lists. I shall not submit any essay of mine to his judgment. If you had informed me that he or the conductor of my other review would print a notice of your works, I should have agreed to write one with pleasure, but under existing circumstances I shall not, it being repugnant to my feelings and contrary to my

---

the coloring is often such as to destroy the effect of the best plates, and gaudy butterflies flit through the pages as if they were the common food of every species, not excluding the American grouse

Captain Brown's *Illustrations* were said by a writer in the *Edinburgh Literary Journal* for April 9, 1831, "to form a companion to the letterpress in *Constable's Miscellany*; price, colored, 15 shillings; plain, 10s. 6d. A few in elephant folio (same size as Selby's *British Ornithology*); colored, 1 guinea. To be completed in 10 parts, each containing 5 colored plates; 22 inches long by 17 inches broad, being considered more than double the size of the original work." The first number of this work was reviewed in the *London Literary Gazette* for October 8, 1831, when it was said that in it were represented 25 birds, 13 forest trees, and 12 insects; the completed work would comprehend "all the forest trees of America, with their fruits, together with the principal insects of the country," as well as all the birds that had been discovered up to the time of issue.

Brown's piratical work must have had a very limited circulation, since it is now so rare that not even the British Museum possesses a copy, and, so far as known, it is not found in any public library of the United States. I was told at Wheldon's, the London shop devoted to works on natural history, that but two copies had ever been handled, and that they commanded a high price. The work was originally sold at £26. The only copy known to me is in the library of the Zoological Society in London, from which the present citation is made; on one of its fly-leaves is written this note: "I have seen the wrapper of No. 1 of this work. It is dated 1831. There is no information as to its contents. C. Davis Sanborn. 22.5.05." This copy was referred to by Dr. Theodore Gill; see *The Osprey*, vol. v, pp. 31 and 109 (Washington, 1900 and 1901). Dr. Walter Faxon has traced two other copies, one formerly in possession of Professor Alfred Newton, and another, but very imperfect set, in a private library at Tarrytown, New York. According to Faxon, a single brown paper wrapper preserved in the Tarrytown copy bears a full printed title, which differs, however, from that which was subsequently engraved for the completed work; for fuller citation, see "A Rare Work on American Ornithology," *The Auk*, vol. xx (1903), pp. 236-241.

Mr. Ruthven Deane has written me that several years ago he secured in New York a fragment of this work, consisting of the paper wrappers of four Parts, Nos. 1-4, the last three of which contained five plates each; there were in addition 10 scattered plates, making 25 plates in all; the price of "21 Shillings" is printed on each of the wrappers, which also bear the date "1831," but no titles.

Another pirated work, *Illustrations of the Genera of Birds*, by the same author, was begun in 1845, but met with even less success, and was never completed; this was taken from *A List of the Genera of Birds*, published in 1840 by George Robert Gray, and according to Alfred Newton (*A Dictionary of Birds*, London, 1896, p. 30, note) was "discreditable to all concerned with it."

388 See Ruthven Deane, *The Auk*, vol. xviii (1901). The extract is from a letter dated "Edinburgh, 22 Warriston Crescent 7th May, 1831."

practice and principles to sue for favor with any man. I have already written three reviews of your books which have been printed, and when I am applied to for a fourth I shall write it too, with "an elegance of style, a power of expression, and knowledge of the subject" equal to those usually displayed by the editor of the Quarterly.

Some of the criticism, whether friendly or hostile, which this work eventually evoked will be considered in a later chapter.

Shortly after his arrival in London, Audubon received a call from Joseph Bartholomew Kidd, a young artist whom he had met at Edinburgh the previous March, and was attracted so much by his "youth, simplicity and cleverness" that he again invited him to paint in his rooms. On the 31st of March, 1831, an agreement was made with Kidd[389] to copy some of his drawings in oils and put in appropriate backgrounds. "It was our intention," said Audubon, "to send them to the exhibition for sale, and to divide the amount between us. He painted eight, and then I proposed, if he would paint the one hundred engravings which comprise my first volume of the Birds of America, I would pay him one hundred pounds." In 1832 Captain Thomas Brown gave this notice of the undertaking in the *Caledonian Mercury*:

> About a year ago Audubon conceived the grand idea of a Natural History Gallery of Paintings, and entered into an agreement with Mr. Kidd to copy all his drawings of the same size, and in oil, leaving to the taste of that excellent artist to add such backgrounds as might give them a more pictorial effect. In the execution of such of these as Mr. Kidd has finished, he has not only preserved all the vivacious character of the originals, but he has greatly heightened their beauty, by the general tone and appropriate feeling which he has preserved and carried throughout his pictures.

Kidd worked intermittently on some such scheme for about three years, and produced numerous pictures on canvas or mill-board. He was thus engaged in 1833 when he wrote to ask for an advance of from twelve to fourteen pounds on account of an accident that had befallen him on the 16th of May of that year. Kidd said in his letter that while he was attending a sale of Lord Eldin's pictures, the floor of the building suddenly gave way with a crash and precipitated the whole company, together with the furniture, into a room below; that he had sustained many bruises himself, not to speak of a dislocated arm, but what with blisters, cupping, nurses and remedies of all sorts, he was then slowly mending. Another of their projects was to publish Kidd's copies of Audubon's drawings as individual pieces, and a notice of this appeared in *Blackwood's Magazine* for 1831. John Wilson, in reviewing Audubon's work in the magazine for that year said: "it is expected that there will be completed by Audubon, Kidd, and others,—Four Hundred Subjects. Audubon purposes opening, on his return [from America], an Ornithological Gallery, of which may the proceeds prove a moderate fortune!" All such plans, however, seem to have been delayed or frustrated, and a misunderstanding with Kidd brought them suddenly to a close in 1833. Audubon's explicit directions under this head were given in a letter to his son Victor, written at Charleston on Christmas Day of that year.[390]

---

[389] Kidd, who was twenty-three at the time he began to work for Audubon, died in 1889, when he had attained his eighty-first year.

[390] See Chapter 27.

When his letterpress was finished, Audubon left Edinburgh with Mrs. Audubon on April 15, 1831. Newcastle, York, Leeds, and Manchester were again visited, and a pause of several days was made at Liverpool before proceeding to London, when, as the naturalist recorded, they "traveled on that extraordinary road, called the railway, at the rate of 24 miles an hour." In May[391] they visited Paris, Audubon no doubt wishing to collect the money due from his agent there, as well as to introduce his wife to the unrivaled attractions of the great city. Upon returning to London in July he had the pleasure of again meeting his *fidus Achates*, Edward Harris,[392] of Moorestown, New Jersey, and immediately began to put his affairs in order for a long period of absence.

While Audubon was in Paris, the following letter[393] was written by his staunch friend and supporter in Congress, Edward Everett, who, as has been seen, fully appreciated the national character of his great undertakings. The effort of this able advocate to give *The Birds of America* free passage to their native land, however, do not appear to have been successful until two years later, as a letter to be quoted in due course clearly indicates.

Edward Everett to Audubon

Charlestown, Mass., May 19th, 1831.

My dear Sir

I duly received your favor of the 1st. of Nov. accompanied with some copies of the Prospectus, and a few days since your letter of the 5th. March reached me. I owe you an apology for being so tardy in my reply to the former letter. It reached me at Washington, while I was confined with a severe illness, with which, since Oct. last, I have till lately been much afflicted. I was, most of the session, in such a state of health, as to be kept at my lodgings, and when in my place, in the House of Representatives, little able to attend to business. As soon as I went abroad, after the receipt of your letter, I consulted some of the most influential members of Congress, as to the probability of being able to pass a bill for the free introduction of your work. Last winters session was the short session, terminating by the Constitution on the 3rd of March. At this session, it is always very difficult to pass any bills, originating during the session. The time is regularly taken up by bills, prepared the previous winter. In addition to this circumstance, more than half of the last session was taken up, by an impeachment before the Senate. A procedure, which suspended during its continuance, the legislative business of the two Houses, and left no time for scarce anything, beyond the annual appropriation bills for the support of the government. Under these circumstances, the gentlemen, whom I consulted, were of opinion with me that it was impossible, for want of time, to pass a bill in your favor, and that it was therefore better not to attempt it, at the late session, but to reserve it for next winter, when it can be brought up seasonably, and with good hope of success. I shall take

---

391 An indication of the time of this visit is given by the following inscription written in the copy of the first volume of the *Ornithological Biography*, which was presented to Cuvier at this time:

To
Baron G. Cuvier,
with the highest respect of the author.—
Paris—17 th. May, 1831.

392 On Wednesday evening, July 27, 1831, Audubon sent the following note to Mr. Harris: "Come to meet me *tomorrow, precisely* at *twelve* o'clock, at our lodgings, 121 Great Portland street."

393 For the perusal of this letter the reader is indebted, as in so many other instances, to Mr. Ruthven Deane.

great pleasure to seize the first moment, at the opening of the next session, to bring the subject before Congress.

The portions of your work, which arrived at Washington before I left it, were publicly exhibited in the library, and attracted great attention and unqualified admiration. The same is true of the copy received by the Boston Athenaeum. The plates were specially exhibited in the great hall of the Athenaeum, to the entire satisfaction and delight of those who saw them.

The copy-right law authorizes any citizen of the U. States to take out a copy-right of his work, on depositing a printed copy of the title page in the office of the District Court. I infer from your letter of the 5th. of March, that you had sent copies of the printed sheets of your work to Drs. Harlan and M. Mertrie [McMurtie] of Philadelphia with a view of having the copy-right.

I have distributed a part of your prospectuses, and shall do the same with the rest, in the manner that may seem most likely to promote your interest. I regret to say, that I have not yet been able to add another, to the list of your subscribers.

You mention, in each of your letters, the little picture you were so kind, as to propose sending me. This alone leads me to say, that whenever it comes to hand, it will be most welcome: but that, engaged as you are in laboring in the cause of science and of America, you must not feel obliged to consume one hour of your precious time at the sacrifice of those higher objects.

I am happy to be able to say to you, that my health, though not wholly restored, is greatly improved, and that if you will continue to favor me with your commands, I will prove myself, hereafter, a more punctual correspondent.

I look forward with sincere pleasure, to the prospect of meeting you again, on this side of the Atlantic, and with my respectful compliments to Mrs. Audubon, I beg leave, dear sir, to tender you the assurance of my high respects, and with it my most friendly salutations.

Edward Everett.

P.S. Since the foregoing was written, I have received your favor of the 23d. of April. I beg leave particularly to thank you for your kindness in reference to the picture. I shall prize it, not merely on account of its scientific value and beauty as a work of art, (both of which I feel assured it will be found to possess) but as a token of your friendly regard. It will give me great pleasure to furnish you any letters in my power, for your adventurous south western tour. These I shall have the pleasure of handing you, when we meet this side the water.

You were elected in November last a fellow of the American Academy of Arts and Sciences, on the nomination which I had the honor to submit to that body. Owing to a change in the secretaryship a delay arose in preparing your diploma, which will however be forwarded in a few days.

Upon balancing his accounts with The Birds of America at about this time, Audubon thought it was truly remarkable that $40,000 should have passed through his hands for the completion of the first volume.

Who would believe that once in London I had only a sovereign left in my pocket, and did not know to whom to apply for another, when at the verge of failure; above all, that I extricated myself from all my difficulties, not by borrowing money, but by rising at four o'clock in the morning, working hard all day, and disposing of my works at a price which a common labourer would have thought little more than sufficient remuneration for his work? To give you an idea of my actual difficulties during the publication of my first volume, it will be sufficient to say, that in the four years required to bring that volume before the world, no less than fifty of my subscribers, representing the sum of fifty-six thousand dollars, abandoned me! And whenever a few withdrew I was forced to leave London, and go to the provinces, to obtain others to supply their places, in order to enable me to raise the money to meet the expenses of engraving, coloring, paper, printing ...; and that with all my constant exertions, fatigues, and vexations, I find myself now having but one hundred and thirty standing names on my list.

INDEX

A

B

F

G

H

I

J

K

L

M

Q

R

S

T

U

V

W

Y

Z